COLLIN COUNTY COMMUNITY COLLEGE

3 1702 00334 1017

Understanding Flowers and Flowering

Collin College Library
PRESTON RIDGE CAMPUS
Frisco, Texas 75035

D0813620

QK	Glover, Beverly J.
653	
G56	Understanding flowers
2007	and flowering.

$75.00

Understanding Flowers and Flowering

An Integrated Approach

Beverley J. Glover

OXFORD

UNIVERSITY PRESS

OXFORD
UNIVERSITY PRESS

Great Clarendon Street, Oxford OX2 6DP

Oxford University Press is a department of the University of Oxford.
It furthers the University's objective of excellence in research, scholarship,
and education by publishing worldwide in

Oxford New York

Auckland Cape Town Dar es Salaam Hong Kong Karachi
Kuala Lumpur Madrid Melbourne Mexico City Nairobi
New Delhi Shanghai Taipei Toronto

With offices in

Argentina Austria Brazil Chile Czech Republic France Greece
Guatemala Hungary Italy Japan Poland Portugal Singapore
South Korea Switzerland Thailand Turkey Ukraine Vietnam

Oxford is a registered trade mark of Oxford University Press
in the UK and in certain other countries

Published in the United States
by Oxford University Press Inc., New York

© Oxford University Press 2007

The moral rights of the author have been asserted
Database right Oxford University Press (maker)

First published 2007

All rights reserved. No part of this publication may be reproduced,
stored in a retrieval system, or transmitted, in any form or by any means,
without the prior permission in writing of Oxford University Press,
or as expressly permitted by law, or under terms agreed with the appropriate
reprographics rights organization. Enquiries concerning reproduction
outside the scope of the above should be sent to the Rights Department,
Oxford University Press, at the address above

You must not circulate this book in any other binding or cover
and you must impose the same condition on any acquirer

British Library Cataloguing in Publication Data

Data available

Library of Congress Cataloging in Publication Data

Data available

Typeset by Newgen Imaging Systems (P) Ltd., Chennai, India
Printed in Great Britain
on acid-free paper by
Antony Rowe, Chippenhem, Wiltshire

ISBN 978–0–19–856597–0   978–0–19–856596–3 (pbk.)

10 9 8 7 6 5 4 3 2 1

Preface

PRESTON RIDGE LIBRARY
COLLIN COLLEGE
FRISCO, TX 75035

Flowers are the features of plants that most endear them to human beings. We grow flowers in our gardens, we display them in our homes, and they feature prominently in our artistic history. Scientists, too, have been fascinated by flowers, and there is a long tradition of botanical study of floral structure and floral ecology. However, it is only recently that the tools of molecular genetics have been applied to flower induction, development, and morphology. The data arising from these studies, when combined with painstaking observation and analysis of the interactions of plants with their pollinators, are beginning to provide the first truly integrated understanding of both how and why flowers take the forms we so admire.

Traditionally, flowers are studied from one of a number of viewpoints. Molecular biologists may study the genetic control of flower induction or flower development, usually focusing on one, or a very small number of, model species. Evolutionary biologists may study how flowers evolved, the forms of the earliest flowers, or the development of flowers of ancient lineages alive today. Pollination biologists may study the interactions between plants and their pollinating animals, the natural selection exerted by those pollinators, and how these pressures affect plant population dynamics. However, it is becoming clear that these various disciplines each have enormous power to inform and shape the work of the others. An understanding of how flower colour is controlled biochemically and genetically, for example, can be of great benefit when studying how pollinator-imposed selection might have influenced the evolution of petal colour within a group of plants.

I hope that this book might serve as a starting point for those interested in taking such an integrated approach to the study of flowers. My aim is to provide students and researchers studying one aspect of floral biology with an overview of other important aspects of flowers, both to help them to set their own work in context and to encourage them to consider experiments which might lead to greater integration of the field. In particular, I hope that this book will encourage dialogue between floral biologists of all varieties, with a long-term view to ensuring a continuing increase in interdisciplinary studies.

The book is divided into three main sections. Section I is introductory, giving some necessary background to the evolution of flowers and to the history of scientific thought on flowers and flowering. Section II considers the molecular mechanisms that control flower induction, flower development, and floral mating type, providing coverage of the genetic material available for shaping by natural selection. This section is initially focused on a very few species of model plants, looking at the molecular similarities which unite all flowers. In the later chapters it considers the development and reproductive strategies of plants from a range of species. Section III extends this analysis much further, considering the explanations for the differences between flowers, rather than their similarities. This section moves between molecular explanations for flower morphology and the ecological consequences of that morphology, in an attempt to integrate what we know both of how and why different flowers take their different forms. Finally, the epilogue attempts to draw out some themes which persist throughout the book, suggesting possible future directions for the field.

Many people have contributed to the final shape of this book, and I am particularly grateful to members of my own research laboratory for enthusiastic support and helpful discussions at many points in the process. John Parker, Caroline Dean, David Hanke, Cathie Martin, Jeff Ollerton, Rob Raguso, and Nick Waser read various sections in detail, and I must thank them all for the time and care that they took and for their excellent suggestions and advice. Many people were kind enough to provide me with images for figures. While these are acknowledged in the relevant figure legends, I particularly thank Nick Waser, Enrico Coen, and the Cambridge University Botanic Garden for their great generosity in sharing images, Kit Wilkins for help with graphics packages, Matthew Dorling and Heather Whitney for photographic assistance, Mike Webb for biochemical pathways, and Rosie Bridge for the line drawings. Thanks also to Roy Barlow and Don Manning for their excellent cover design. At Oxford University Press, Ian Sherman has provided steadfast support for this project, dealing with various changes to the schedule with calm good humour while Helen Eaton and Christine Rode have kindly shepherded me through the production process. On a personal note I must thank Jocelyn, Duncan, and Katie Taylor for lending me the space and quiet to really begin writing, rather than just thinking about it. Finally, I thank my parents for their endless encouragement, and Stuart, David, Rosie, and Sam, who put up with me during this process with remarkable patience and love.

Contents

PRESTON RIDGE LIBRARY
COLLIN COLLEGE
FRISCO, TX 75035

INTRODUCTION

CHAPTER 1

The Evolution of Flowers

The oldest fossil flower currently known is 127 million years old. Flowers, and the plants that produce them (angiosperms or flowering plants), are relatively recent innovations in evolutionary terms. The first land plants, which did not possess flowers, arose around 470 mya, but fossil evidence indicates that only after another 340 million years did the angiosperms appear. However, following their appearance in the fossil record of the early Cretaceous period, the angiosperms spread geographically from their point of origin in the tropics and diversified dramatically to become the ecologically dominant plant group in the great majority of terrestrial habitats. This extraordinary radiation took a mere 40 million years, made more extraordinary for the number of species it generated. The 300–400,000 species of extant angiosperms represent the most species-rich plant group by far, and are exceeded in numbers in the speciose animal kingdom only by the arthropods. Furthermore, the differences in growth habit, morphology, and life history within the angiosperms are vast, leading Darwin to describe the speed and scale of this recent radiation as 'an abominable mystery'. It is not possible in a book of this size to provide a full analysis of the extensive literature on the origins and radiation of angiosperms and their flowers. The aim of this chapter is to provide an overview of the key issues surrounding the origin of the flowering plants.

1.1 The origin of flowering plants

Angiosperms are defined by a number of features, of which possession of a flower is only one. Some of these traits are shown in Fig. 1.1. They have fully protected and enclosed ovules with two layers of integuments, wood with vessels in it, and distinctive pollen with columnar structures providing support for the surface layer. In addition, only angiosperms undergo double fertilization, whereby two genetically identical sperm cells are released into the ovary with one fertilizing the egg and the other fusing with the central cell to form the endosperm. Traditionally, fossil evidence was all that was available to probe the origins of the angiosperms, but, more recently, molecular data obtained from extant species have been used to inform the debate.

1.1.1 Fossil evidence

Fossil evidence dates the origin of the angiosperms firmly to the early Cretaceous, with the oldest fossil flowers (127 mya), angiosperm fruits (121 mya), angiosperm pollen (130 mya), and angiosperm leaves (120 mya) all supporting this conclusion (Hughes 1994; Krassilov and Dobruskina 1995; Brenner 1996; Friis et al. 1999). The oldest fossil families suggest that the early angiosperms were primarily understorey plants and species with early successional roles, but that their subsequent radiation quickly led to the adoption of late successional ecological positions, presumably as their size and woodiness increased. The scarcity of angiosperm wood in the early Cretaceous fossil record, along with the apparently small size of early angiosperm seed and leaves, has led to the conclusion that these early plants were small, herbaceous species with a weedy lifestyle (Willis and McElwain 2002). The first angiosperms appear to have originated in tropical regions, only spreading to higher latitudes after 20–30 million years (Barrett and Willis 2001). The earliest angiosperm fossils (of pollen) have been found in modern-day Israel and Morocco, land that lay just north of the equator in the early Cretaceous

Figure 1.1 Some defining features of angiosperms. (a) Enclosed ovules, enfolded within the carpel. (b) Wood with true vessels as well as the narrower tracheids found in gymnosperms. (c) Columnar pollen exine, shown in cross section. (d) Double fertilization, with two sperm nuclei arriving in the pollen tube. One fertilizes the egg cell and the other fertilizes the central cell with its two nuclei, generating the triploid endosperm.

period. The subsequent spread of the angiosperms into higher latitudes was very rapid, accompanied by an increasing dominance of those areas occupied (as measured by the relative proportions of angiosperm and other pollen types retained in the fossil record) (Willis and McElwain 2002).

The identity of the last common ancestor of the angiosperms and earlier plant groups has long been a source of debate. The most closely related extant group is the gymnosperms, traditionally viewed as a cluster of three or four divisions. The first of these divisions, the conifers (Coniferophytes), is familiar, and its members dominate many high-latitude forests. The cycads (Cycadophytes) are currently less prominent, but fossil records indicate that they were once ecologically highly significant. The Ginkgophytes are currently represented by only one surviving species, *Ginkgo biloba*, and are thought to be closely related to both conifers and cycads. The remaining group, the Gnetophytes, was, until recently, considered the most likely precursor of the angiosperms, on the basis of both fossil and morphological evidence (Crane *et al.* 1995). There are currently three extant genera of Gnetophytes, which share a range of morphological similarities with some angiosperms. In particular, leaf venation in *Gnetum* closely resembles that of angiosperms, their xylem does contain vessels, and some Gnetophytes produce reproductive structures containing both male and female parts. However, recent molecular studies have refuted this hypothesis (see below), and have even questioned the status of the Gnetophytes as a division, suggesting that they may be part of the Coniferophytes (Qiu *et al.* 1999).

Fossil evidence has also been unable to resolve the identity of the earliest angiosperm, since evidence from different fossil deposits varies. The presence of small, unisexual flowers with few flower parts as well as large, bisexual flowers with many flower parts in deposits of similar ages has led to a range of suggestions as to the identity of the oldest angiosperm (Endress 1987). The larger fossil flowers were reminiscent of those of modern-day Magnolia or waterlily, and both the Magnoliales and the Nymphaeales have been proposed as the ancestral angiosperm. However, the smaller fossil flowers resemble those of the Piperales (containing the true peppers), another group which has been proposed as ancestral. As with the identity of the last common ancestor of angiosperms and gymnosperms, molecular data have recently thrown new light on this problem, proposing an entirely different ancestral angiosperm (see below).

1.1.2 Molecular evidence

Molecular evidence for the origins of angiosperms is based on the comparison of DNA sequences or fingerprints from extant species. However, use of a molecular clock, which calculates the age of divergence of two sequences according to the number of differences between them, can provide estimates of the date of evolutionary events. This technique assumes that mutation rate is constant in a particular type of DNA sequence, which may not always be true and is difficult to test. It is certainly the case that the mutation rate of sequences from different regions of the genome can vary significantly.

Three different molecular studies have all indicated an origin of the angiosperms considerably earlier than the 130 mya that the fossil record suggests. Analysis of chloroplast DNA of phylogenetically

MOLECULAR

Figure 1.2 Molecular and fossil evidence suggest different times for the origin of angiosperms. Molecular evidence (above) indicates an earlier origin of angiosperms, and of divergence of monocots and dicots, than does fossil evidence (below).

distant extant species indicates that the monocotyledonous angiosperms diverged from the dicotyledonous species 250–200 mya (Wolfe *et al.* 1989), therefore pushing the origin of angiosperms back to at least the Triassic period. A similar study of nuclear DNA of nine extant species dated the monocot/dicot divergence to 300 mya, indicating an angiosperm origin in the Carboniferous period (Martin *et al.* 1989). More recently, a comprehensive survey of five independent DNA sequences from chloroplast, mitochondrial, and nuclear DNA of 106 plant species placed the origin of the angiosperms in the late Carboniferous, 290 mya (Kenrick 1999; Qiu *et al.* 1999). To reconcile the discrepancies between these molecular data and the fossil record, we must consider the possible reasons why the fossil record might be incomplete, while recognizing that the molecular studies do not give identical results and are not infallible. If the origin of angiosperms was indeed earlier than the Cretaceous period, it is possible that the fossil record does not reflect this because early angiosperms were not woody, were relatively rare within their communities, and were predominantly found in dry environments not conducive to fossil formation. It is also important to distinguish between plants which are members of the angiosperm total group and those that are members of the angiosperm crown group. Angiosperm total group members are defined as those that show some features of modern-day angiosperms, and it may be these plants that the molecular data are estimating at around 300 million years old. Angiosperm crown group members show

all the features that define a modern angiosperm, and it is possible that these did not appear until much later, consistent with the appearance of flowers in the fossil record. Molecular data aside, the earliest compelling evidence for the existence of flowers is still the presence of fossilized Magnolia-like flowers in 127- to 120-million-year-old deposits from Portugal (Friis *et al.* 1999; Fig. 1.2).

The molecular phylogeny of Qiu *et al.* (1999) has also shed light on questions of relatedness between gymnosperm groups and the early angiosperms (Fig. 1.3). To the astonishment of many botanists, the extant Gnetophytes resolved in a single clade with the conifers, refuting the suggestion that their ancestors were the precursors of the angiosperms. Instead, the gymnosperms have been shown to be monophyletic and to have diverged from the total land plant lineage around 300 mya. They can therefore be thought of as the sister group of angiosperms, and do not necessarily provide much relevant information about the likely morphology of the ancestral angiosperm. The phylogeny also indicates that the large-flowered Nymphaeales are indeed ancestral to many other plant groups, but are not the sister group of all the other angiosperms. This position was awarded to *Amborella trichopoda*, the only extant species of the Amborellaceae and a native of New Caledonia (Fig. 1.4; see also Plate 1). *Amborella* is a small, weedy shrub, supporting the idea that the basal angiosperms were understorey or early successional species, but indicating that they were probably woody, not herbaceous, despite the relative scarcity of early fossilised angiosperm wood. Very

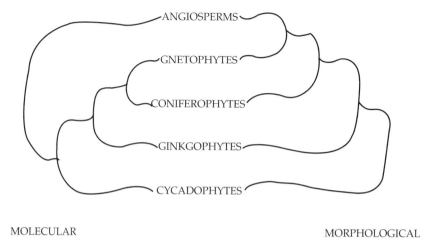

ANGIOSPERMS

GNETOPHYTES

CONIFEROPHYTES

GINKGOPHYTES

CYCADOPHYTES

MOLECULAR   MORPHOLOGICAL

Figure 1.3 Molecular and morphological evidence suggest different relationships between the angiosperms and the various groups of gymnosperms. Molecular data (left) place the angiosperms as sister group to all the gymnosperms, while traditional morphological analysis (right) placed the angiosperms and the gnetophytes as sister taxa.

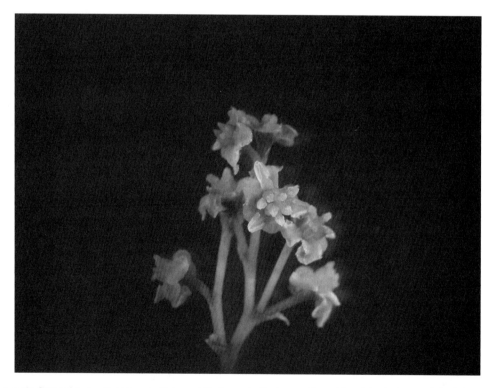

Figure 1.4 The flower of *Amborella trichopoda*. Photograph kindly supplied by Sangtae Kim and Pam Soltis (University of Florida).

recently, the position of *Amborella* as the most basal extant angiosperm was queried by a study that sequenced its entire chloroplast genome. Goremykin *et al.* (2003) looked at the conservation of 61 protein-coding genes common to the 13 land plant fully sequenced chloroplast genomes. Their phylogenetic analysis placed the origin of *Amborella* later than the origin of the monocts, clustered with other dicots. Whereas Goremykin *et al.* (2003) used far more genes in their analysis than did Qiu *et al.* (1999), those genes were derived from only one of the three plant genomes (chloroplast, as opposed to mitochondrial or nuclear). As such, their conclusions have not generally been felt to supercede those of Qiu *et al.* (1999), although their data do emphasize the uncertainties associated with attempting to reconstruct evolutionary pathways using extant species.

1.2 Early reproductive structures

The reproductive structures of most plant lineages prior to the angiosperms, including most of the gymnosperms, were unisexual. The evolution of seeds freed plants to reproduce in the absence of a film of external water (previously necessary to allow fragile free-swimming male gametes to fuse with a static egg). In seed plants the female gametophyte is surrounded by parental sporophyte tissue, usually derived from bracts. These enfolding bracts act to protect the ovule and may also be a protective coat when the seed is dispersed. In gymnosperms the ovule is not completely enclosed in sporophyte tissue, but is protected within a chamber, into which wind-blown pollen is drawn after being trapped by a drop of secreted liquid. In cycads, dioecy is the rule, with female plants producing clusters of ovules on the edges of modified leaves called megasporophylls. The whole female reproductive structure forms an ovulate cone. Male plants produce cones of differently specialized leaves, the scale-like microsporophylls, arranged in a spiral phyllotaxis to produce a cone. These microsporophylls possess a pollen sac on their abaxial surface, each of which produces numerous pollen grains. Similar systems operate in many other gymnosperms, although the conifers are monoecious, producing male and female cones on the same individual plant.

1.3 The first flowers

Despite the absence of a clear picture of the last common ancestor of angiosperms and gymnosperms, there are clearly some key innovations which arose in the angiosperm lineage and gave rise to flowers, such as the combining of male and female reproductive structures within a small space on the same shoot and the production of perianth organs (Theissen *et al.* 2002). Indeed, it is these innovations that give us the modern definition of a flower. Analysis of the development of these innovations allows us to build a general picture of the evolution of the first flowers.

The typical angiosperm flower is hermaphrodite. Although there are many examples of unisexual flowers (either on hermaphroditic plants (monoecy) or on unisexual plants (dioecy)), the majority of angiosperm flowers produce both male and female reproductive organs. This is in marked contrast to the reproductive structures of most other plant lineages, including the gymnosperms, where unisexual reproductive cones are the norm, either both forms on a single plant or each form on a separate plant, but the combination of male and female organs on a single shoot is rare. The development of a single structure containing both male and female reproductive organs was therefore a key event in the evolution of flowers. If the ancestor of angiosperms and gymnosperms produced its male and female reproductive structures in separate cones, each composed of a spiral of organs, then the development of a bisexual flower requires either the production of ectopic female organs at the top of the male cone or the production of ectopic male organs at the base of the female cone. All angiosperm flowers contain an outer whorl of male reproductive organs and an inner whorl of female reproductive organs, suggesting that this evolution of bisexual flowers occurred only once (Cronk 2001). A model in which bisexual flowers evolved by the production of ectopic ovules in the centre of a male cone (the mostly male hypothesis) is currently favoured on the basis of two sets of data (Frohlich 2002). First, ectopic sterile ovules are produced by male reproductive branches of the Gnetophyte *Welwitschia*, and appear to attract pollinating insects to the male branches by exuding the same droplets of liquid secreted by fertile female ovules (Endress 1996). This provides an adaptive explanation for the retention of ectopic ovules in a male cone. Secondly, analysis of sequence and expression patterns of genes regulating gymnosperm cone development indicates that an angiosperm gene required for flower production is most closely related to a gymnosperm gene involved in male (but not female) cone development (Frohlich and Parker 2000). The *Leaf* gene of *Welwitschia*, *Gnetum*, *Pinus*, and *Ginkgo* is very similar to the angiosperm *LEAFY* gene, which encodes a protein crucial in determining the floral nature of the meristem. The *Leaf* gene has been studied in *Pinus radiata*, where it is expressed in male cones only. This implies that the angiosperm flower is descended from the ancestor of the male gymnosperm cone, secondarily acquiring ectopic ovules.

The typical angiosperm flower produces two whorls of perianth organs—often, but not always, differentiated into protective green sepals and brightly coloured petals. The production of these perianth organs was another angiosperm innovation. However, it is not necessary to assume that the first angiosperm flower possessed a perianth. Indeed, some authors have argued to the contrary, proposing a perianth-less structure containing both male and female reproductive organs as the first flower (Theissen *et al.* 2002). Some fossil evidence does seem to support this hypothesis, including the morphology of the fossil *Archaefructus* (Sun *et al.* 1998). The development of the perianth may have occurred in a number of ways. One hypothesis proposes that the perianth developed by loss of reproductive function of the outermost male organs, a model that has molecular support from studies of Ranunculid flower development (Kramer and Irish 1999). Alternatively, bract-like structures may have been modified to produce sepals, with petals then evolving as a result of outward expansion of the spheres of action of some genetic components of stamen development and their subsequent combination with elements of the sepal developmental programme (Baum 1998).

However they developed, the first angiosperm flowers with perianth organs almost certainly did not have separate sepals and petals, but an undifferentiated and unspecialized perianth. Analysis of the fossil record and of extant basal angiosperms suggests that the anthers of these flowers made a lot of pollen, while the ovules may not yet have been entirely enclosed by carpel tissue (Dilcher 2000). Such carpel tissue as had arisen would have enfolded individual groups of ovules, with the multiple fused carpels of many modern flowers still to evolve.

It should be noted that similar flower-like structures also evolved in the now-extinct Bennettitales, a group of plants widespread between 248 and 140 mya (Willis and McElwain 2002). Some species of this group combined both male and female reproductive organs on a single axis, and appear to have enclosed those organs within a circle of bracts, superficially like angiosperm flowers. However, there is debate as to whether these bracts opened, like the perianth organs of flowers, or whether structural constraints would have prevented this. It is also unclear whether the bracts contained pigmentation, believed to be a significant factor in the success of angiosperm flowers.

1.4 Floral diversification

The first angiosperm flowers were borne by a lineage which subsequently underwent a dramatic radiation, including an astonishing diversification of floral form. It has been hypothesized that this process was initiated by the association of male and female organs on the same axis, which increased the effectiveness of animal pollination (previously less significant than wind pollination for most plant species). An alternative hypothesis is that the appearance of coloured pigments in the new flowers had a dramatic effect on their attractiveness to animals, again establishing biotic pollination as the norm. Studies of the colour vision of a range of arthropods have inferred that the Cambrian ancestors of today's insects possessed the same trichromatic colour perception system as that used by modern pollinating insects to discriminate between flowers (Chittka 1996). If insects were preadapted for viewing flower colour 500 mya, then the appearance of coloured pigments in the perianth organs of the first flowers may have led rapidly to a strong association between flowers and animal pollinators. Increased animal pollination is likely to have resulted in increased outcrossing, in part through the dual service provided by an animal in both removing pollen from the anthers and applying non-self pollen to the stigma of a single flower in a single visit. This increase in outcrossing will have increased the amount of recombination between plant genomes and therefore the speed of angiosperm evolution (Dilcher 2000). Outcrossing can be both advantageous and disadvantageous to plants, having consequences for the genetic make-up of populations and effective population size (Barrett 2002; Charlesworth 2006; also see Chapter 12). However, in many situations selection will strongly favour adaptations that maintain increased outcrossing, resulting in the appearance and maintenance of a suite of novel colours, shapes, structures, scents, and rewards. At the same time, animal pollination is more likely than abiotic pollination to result in reproductive isolation, particularly where specific pollinating animals are attracted only to a subset of floral attributes but the wind is random in its dispersal of pollen. Such reproductive isolation may have facilitated speciation and divergence of form, further increasing the speed of the angiosperm radiation. Finally, full closure of the carpel and the ability to recognize and reject self pollen through biological means may have evolved in concert to reinforce outcrossing (Dilcher 2000).

That rapid evolution coincided with the recruitment of animal pollinators is not in doubt, with the advantages of outcrossing and the ability to exchange genetic information between widely spaced individuals representing major benefits in terms of increased genetic diversity and thus 'evolvability'. What is less clear is the extent to which angiosperm and flower radiation can be directly attributed to a subsequent coevolution between angiosperms and pollinating insects. This attractive hypothesis has informed much of pollination ecology for decades, and can best be summarized by viewing flowers with a perianth as an angiosperm's way of manipulating an insect into carrying its pollen around for it. Once the flower has acquired the ability to manipulate one insect, it may radiate (genetic variability permitting) into as many different forms as there are insects with different preferences to manipulate.

There is some fossil evidence to support the hypothesis that angiosperm radiation was the result of angiosperm–insect coevolution. Fossils of early Cretaceous flowers show some features indicative of animal pollination, including a larger size than necessary for wind pollination (Crane *et al.* 1995). The first Lepidoptera appear in the fossil record at the same time as the first angiosperms, and some groups of the Hymenoptera also appear to have evolved at the same time. However, there is little correlation between the origin or speciation of other insect groups and the origin and speciation of the angiosperms (Willis and McElwain 2002). This suggests that angiosperm–insect coevolution may have played an important role in the great diversification of some groups and floral forms, but that a significant part of this radiation must instead be attributed to the high rates of outcrossing of the new plants and their consequent rapid acquisition of new habitats and new niches.

CHAPTER 2

Historical Interpretations of Flower Induction and Flower Development

The modern analysis of flowers and flowering rests on a wealth of literature concerned with the description and interpretation of plant form. Chief among these works is Goethe's foliar theory, which proposes that all aerial plant organs are analogous to a single organ, which he calls the leaf. The foliar theory has underpinned all work on flower development, including modern molecular genetic analyses, as well as providing a frame of reference for evolutionary studies. This chapter describes Goethe's theory with reference to the flower, and its use and expansion by twentieth-century botanists. An analysis of the differences and similarities between vegetative and floral organs is presented, which provides a framework for a section that interprets the foliar theory in an evolutionary context. The second half of the chapter moves on to consider historical interpretations of the transition from the vegetative state to the flowering state, describing the mechanisms proposed by early plant physiologists to explain this transition. Again, this overview is designed to provide context for further chapters on the current state of knowledge with regard to this floral transition.

2.1 The foliar theory of the flower

The scientific study of plant morphology, and, in particular, the study of comparative morphology, both of different organs within a plant and of the same organ from different plants, was effectively the invention of the great German philosopher Johann Wolfgang von Goethe. In 1790 Goethe published his seminal essay on the metamorphosis of plants (translated and discussed by Arber 1946), in which he proposed that all plant organs could be thought of as equivalent or analogous to a single type organ, which he called the leaf. However, it is important to establish that he was not proposing a developmental or evolutionary concept, with other plant organs descended from a leaf. Indeed, it may have helped to separate his thinking on equivalence from later evolutionary thinking on relatedness had he used a different name for his hypothetical type organ. Nageli (1884) suggested 'phyllome' as a more suitable name, a suggestion with which later morphological botanists such as Arber (1937) strongly concurred. Because Goethe's theory implied no developmental or evolutionary progression, it is effectively reciprocal. Thus it is equally easy to describe leaves as analogous to petals as to describe petals as analogous to leaves. In essence, Goethe's view was that reproduction (and the development of reproductive organs) ought to be treated as a function of the entire plant and that the flowering shoot could not be understood except in relation to the vegetative shoot of the same plant (Arber 1937). This holistic approach was a new development in a field that had relied previously on analysis of individual components of organisms. Its genius lay in the detail with which it compared different parts of the same plant, combined with the simplicity of its approach. This approach involved no speculation about the meaning behind the analogy of different plant organs. It provided, however, a data set, free of preconceptions, which was then adopted by later evolutionary and developmental biologists and has provided remarkable support for their theories.

Goethe's analysis of the equivalence of the different plant organs to a basic vegetative unit was summarized by Arber (1937), from which the following

**PRESTON RIDGE LIBRARY
COLLIN COLLEGE
FRISCO, TX 75035**

Figure 2.1 Analogous floral and reproductive organs in tobacco (*Nicotiana tabacum*). (a) The inflorescence shoot (left) is analogous to the vegetative shoot (right). (b) The bract (below) is analogous to the leaf (above). (c) The sepal (below) is analogous to the leaf (above). (d) The petal (below) is analogous to the leaf (above). (e) The stamen (below) is analogous to the leaf (above).

analysis of each of the reproductive structures is derived and expanded. Pairs of analogous structures are shown in Fig. 2.1 (see also Plate 2).

2.1.1 Inflorescence shoot

The inflorescence shoot consists of the shoot apical meristem (SAM), the stem which it lays down from its rib meristematic region, and the bracts, flowers, and secondary branches produced laterally from that stem. Its similarities to the vegetative stem from which it derives are clear. There is little difficulty in accepting such similarities since the two are merely different phases in the life cycle of the SAM, which itself undergoes no major reorganization of structure in the transition from the vegetative to the flowering stage, although many changes in patterns of gene expression within that structure have been observed. The SAM usually remains indeterminate in the inflorescence shoot, allowing indeterminate growth similar to that shown previously by the vegetative shoot. The inflorescence shoot's direction of growth and its function—the production and support of lateral organs—are also the same as those of the vegetative shoot.

Differences between the inflorescence shoot and the vegetative shoot are primarily related to the organs that they produce, and these will be considered individually below. However, within the shoot itself a number of changes may occur. In some species phyllotaxy in the inflorescence shoot is altered, with organs being produced in a pattern different from that of the vegetative shoot. Secondly, the distance between lateral organs may vary. In many species the inflorescence shoot is characterized by internodes shorter than those of the vegetative shoot. Finally, the details of cellular differentiation within the epidermis of the stem itself may be altered. For example, the vegetative stem of *Antirrhinum majus* (snapdragon) is entirely hairless, but the inflorescence stem is covered with short multicellular trichomes with glandular heads.

2.1.2 Floral shoot

The floral shoot consists of the floral meristem and the lateral organs derived from it. It should not be considered analogous to the main vegetative shoot (derived directly from the SAM) but to the axillary shoots which develop from the flanks of the SAM. In the same way, the floral meristems develop on the flanks of the inflorescence meristem. The floral shoot usually arises in the axil of a leaf-like organ, the bract, while the axillary vegetative shoot always arises in the axil of a leaf. So the architecture of the floral shoot, and its positioning with respect to the main axis of plant growth, is very similar to that of a vegetative axillary shoot. The function of the floral shoot is also analogous—the production and support of lateral organs.

However, the differences between the floral shoot and an axillary vegetative shoot are not solely restricted to the organs that they produce, very different though those may seem. There are again three significant differences between the floral shoot and a vegetative shoot. First, phyllotaxy is almost always altered in the floral shoot, with whorled phyllotaxy the norm in flower development (with notable exceptions), irrespective of the usual phyllotaxy adopted by the leaves of the plant. Secondly, and more significantly, there is an enormous difference in the extent to which the shoot elongates. The axis of a vegetative shoot will almost always elongate, so that the leaves are separated by well-defined internodes. In contrast the floral axis very rarely elongates at all after the production of the first sepal, producing organs so closely placed that fusions both within and between different whorls of floral organs are common. This difference was noted by Goethe, and has been described as the primary divergence between floral and vegetative shoots (Arber 1937). However, there are examples of primary vegetative shoots adopting a similar form—many species in families such as the Brassicaceae adopt a rosette form in the juvenile phase, with almost no internodal elongation at all. The loss of internodal expansion in the floral shoot can therefore be thought of as a reversion by the axis to a more juvenile growth form. Such retention of juvenile characteristics in mature organisms is termed neoteny, and has been hypothesized to be responsible for many apparent saltational leaps in the animal fossil record, including the divergence of humans from apes. A third, and equally significant, difference between the flowering shoot and a vegetative shoot is the determinacy of the meristem and consequently of the shoot. The early morphologists were concerned with the fate of the meristematic cells within a flower, believing that consumption of the meristem by the terminal organs (the carpels) would indicate a significant difference from the activity of the axillary shoot. However, careful analysis has shown that many species retain an apex to the floral axis, clearly distinct from the floral organs (Fig. 2.2; Arber 1937). Instead of consumption by the carpels, then, the floral meristem is simply determinate and ceases to produce more organs after the flower is complete. It is likely that this determinacy, unusual in a plant meristem, is secondarily imposed. Genetic analyses have revealed that floral meristem determinacy can be broken by mutation of a single gene (called *AGAMOUS* in *Arabidopsis thaliana*; Coen and Meyerowitz 1991) and the meristem returned to an indeterminate state (Fig. 2.3, see also Plate 3). The analogy between floral and vegetative meristems is therefore not broken by the determinacy of the floral meristem, a secondarily imposed character controlled by a single gene.

Figure 2.2 The shoot apical meristem still present between the carpels in a flower of *Ranunculus acris* (redrawn from Arber 1937).

2.1.3 Bracts

If the evidence suggests that the vegetative and reproductive shoots are analogous, the various appendages borne by these shoots are increasingly difficult to relate to one another. The bract is the easiest case, being positionally and morphologically often barely distinguishable from a leaf. Bracts, like leaves, are lateral organs produced by the SAM and which subtend a meristem. They are also in possession of dorsoventrality, are usually laminate in form, and are frequently green and photosynthetic.

Figure 2.3 The indeterminate nature of the *AGAMOUS* mutant flower (b), which produces flowers within flowers, confirms that determinacy has been imposed upon the wild type floral meristem (a) by the action of this single gene. Photographs kindly supplied by Ian Furner (University of Cambridge).

The differences between bracts and leaves are limited to details of size, shape, pigmentation, and cellular differentiation. In many species the bracts are smaller than the foliage leaves, perhaps reflecting the transition of the plant from an organism primarily concerned with photosynthetic production to an organism concerned with investing the products of that photosynthesis in reproduction. In some plants the bracts have become modified to attract pollinators to the flowers, and may therefore contain anthocyanin or other pigments, either as well as or at the expense of chlorophyll. Similarly, plants which use their bracts in this way, such as some species of *Cornus*, may develop light-focussing papillae on their bracts (Kay *et al.* 1979), while other species may show different trichome distributions on their bracts relative to their leaves.

2.1.4 Sepals

The first whorl of perianth organs, containing the sepals, is often morphologically very similar to a whorl of leaves. The sepals are produced as lateral appendages from the floral meristem, and are usually laminate in form and green in colour (except in those species, such as members of the Liliaceae, where both whorls of perianth organs are specialized for pollinator attraction and are sometimes known as tepals). Aside from this particular difference, related to adoption of a novel function, the sepals usually differ from leaves only in their size, often being much smaller.

2.1.5 Petals

The second whorl of perianth organs can, in species with a simple undifferentiated perianth, be morphologically very similar to leaves too. However, in many species, particularly in some of the higher eudicots with zygomorphic (or bilaterally symmetrical) flowers, the petals can be morphologically very distinct both from leaves and from one another. However great the differences in morphology between petals and leaves may seem, they are nonetheless analogous in terms of their position as lateral appendages from a meristem. Other similarities include dorsoventrality and a laminar form. Both petals and leaves can clearly be seen to distinguish their dorsal surface from their ventral surface. The leaves of most angiosperm species differentiate stomata at a considerably higher frequency on their ventral epidermis than on their dorsal epidermis. Internally, the palisade mesophyll, the columnar cells which provide the extended surface area for photosynthesis, are located on the dorsal side of the leaf. Similarly, many petals develop specialized epidermal cells on their dorsal surfaces, and may even be differentially pigmented or patterned on dorsal and ventral surfaces. The laminar form, produced by growth perpendicular to an initial axis, is also common to both leaves and petals.

If the similarities between petals and leaves are in major architectural and developmental characteristics, the differences are largely in the detail of function. Petals are not usually green and photosynthetic, but instead may contain pigments that make them stand out against vegetation. This can be interpreted as a secondary adaptation to their specific function as pollinator attractants. Similarly, petals usually have no palisade mesophyll (since they do not photosynthesize) and often develop elaborate papillate epidermal cells to enhance their apparent colour and texture. These differences, like changes in size and shape, do not reflect major differences between petals and leaves, but rather minor modifications of a similar ground plan, and therefore do not perturb Goethe's vision of the analogous nature of leaves and petals.

2.1.6 Stamens

The outer whorl of reproductive organs, the stamens, was also considered analogous to leaves by Goethe. Their similarities to leaves lie in the presence of chlorophyll and in their growth form, which consists of elongation in a single plane (although usually with little or no laminar growth). Their differences may seem more striking, but can again be attributed to minor adaptations to changing function. The absence of a laminar form may at first seem significant. However, analysis of mutants with perturbations in leaf development has shown that, in order for a leaf to develop a lamina, it must first have dorsoventrality (Waites and Hudson 1999). Put simply, a leaf cannot grow sideways unless it can detect which way is up, which way is down, and thus

which way is sideways. There is no need for a stamen to have dorsoventrality, as it is simply a filament supporting the pollen-containing locules. In the absence of dorsoventrality a laminar form cannot develop, so the lack of a lamina can be interpreted as a consequence of the loss of dorsoventrality. That leaves the locules themselves as the only significant difference between stamens and leaves.

A further argument in favour of the analogy of leaves and stamens is the inter-relationship between petals and stamens. A variety of authors, beginning with Goethe himself, observed significant similarities between stamens and petals. It had also been noted that the absence of stamens in species which produce some female-only flowers was often associated with a reduction in petal development. Modern molecular genetic analyses have confirmed the association between petals and stamens, with both requiring the activity of 'B-function' genes (see Chapter 9) for their development (Coen and Meyerowitz 1991). It follows, then, that if petals are analogous to leaves, and petals and stamens are variations on the same theme, then stamens must also be analogous to leaves.

2.1.7 Carpels

The female reproductive structures have traditionally presented the greatest difficulty to people intent on interpreting the flower in a foliar context. Goethe himself had some difficulties with the carpel, and it was not until microscopical techniques advanced considerably that it was possible to see how the development of each carpel was analogous to the development of a leaf. Each carpel is essentially composed of fused leaf-like structures, with ovules developing on their adaxial sides. Ovules aside, the only differences between carpels and leaves are the development of specialized stigmatic cells and the patterning of the vasculature. The stigmatic cells, and the stylar tissue beneath them, are clearly adaptations to the function of the carpel, and can be interpreted as secondary consequences of the novel function. The significance of the patterning of carpel vasculature has been the subject of much debate in the literature. The essential problem is that leaves have a dominant mid-vein with secondary vasculature branching outwards from it, while carpels have dominant lateral veins with secondary veins branching inwards from them. While it is not clear how the distribution of vasculature could have altered, it has generally come to be accepted that this pattern reflects the lateral positions of the ovules, and the need to supply them with nutrients, and is thus likely to be a secondary consequence of ovule development.

Overall, then, comparative morphology shows us that Goethe's foliar theory of the flower was both innovative and insightful. All of the organs that make up a flower look like they could be produced with only a few variations using the same basic developmental programmes that make leaves. Flowers, in fact, seem to be nothing more than bunches of slightly strange leaves bearing spore-producing structures.

2.2 The foliar theory in an evolutionary context

The observations of Goethe on the analogous nature of leaves and floral organs paved the way for later evolutionary interpretations of flower development. Where organs can be seen to represent variations on a theme, it is only a short step to blur the boundaries between them and imagine intermediate forms. This 'fuzzy' approach to plant morphology (Rutishauser and Isler 2001) fits perfectly with the idea, propounded by Darwin, that organisms were formed by gradual transitions between types. Goethe's work provided important data that supported the theory of evolution by natural selection, suggesting that the various organs of the flower evolved by gradual changes, as a result of mutations, from an ancestral leaf-like structure. As evolutionary theory became accepted as part of mainstream biology, Goethe's ideas were seamlessly incorporated into an evolutionary context, with his cautious 'analogous' converted to a more radical 'related to' or 'derived from'.

The idea that floral organs are only a few mutations different from leaves has been the foundation stone of modern molecular genetic analysis of flower development. Genetic analysis in the 1980s and 1990s identified lines of plants with mutations that caused interconversion of floral organs. Combination of these mutations into a single plant

line resulted in a plant which produced four whorls of leaves in place of the four whorls of floral organs, finally providing the molecular 'proof' for Goethe's foliar theory (Coen and Meyerowitz 1991). These mutant lines, and the genes in question, are discussed in more detail in Chapter 9. However, their interpretation, and indeed much of the impetus for their identification, relies heavily on the historical work of Goethe himself.

2.3 The transition to flowering

The history of scientific work on flower development may belong to comparative morphologists, but the history of work on the transition to flowering lies with plant physiologists. The transition to flowering consists of a major phase change in the plant's lifecycle. A plant that has previously produced only leaves and axillary meristems from its apical meristem must now switch to producing bracts and floral meristems, along with any changes to the general axis and pattern of growth. This switch can be thought of as a 'decision' on the part of the plant—it is not a random event, but occurs in response to one or several of a large set of potential stimuli, which vary from species to species. The transition to flowering can be broken down into two distinct processes, induction and evocation. Floral induction is the commitment of the plant to start producing flowers, which often happens as a result of processes occurring in the leaves. Once induced, a plant is set on the path to flowering, even though no flowers have yet been produced. The induced plant then evokes the production of flowers by triggering changes in gene expression in the apical meristem. Induction and evocation are usually considered separately for the simple reason that they often occur in distinct organs. It is possible to induce many plants to flower by providing appropriate stimuli to a single leaf, and it is possible to evoke an uninduced plant to flower by grafting a single induced leaf from another plant onto it. It is clear from this physical separation of processes that floral evocation requires the production of a signal by the induced organs, which is transported to the apical meristem and activates flower production. The mechanisms of induction and evocation, and the nature of the signal between them, have long been the subjects of physiological analysis.

2.4 Developmental explanations of floral induction

The variables that interact to determine when a plant is induced to flower can include developmental stimuli and environmental stimuli. For the majority of plants in temperate climates the environment is a key factor in the decision to flower, as producing vulnerable flowers and seeds in the depths of winter, for example, would not be a successful strategy. For this reason environmental stimuli are considered in much greater depth than developmental ones, both in this chapter and in the literature as a whole. Developmental factors are of more importance to plants living in environments with little annual variation. It is unusual, for instance, to find a rainforest species with a floral transition determined by factors such as day length or temperature, which are more or less constant in that environment.

The developmental factors which may influence the floral transition generally refer to the age and health of the plant. Some trees, for example, can only flower after a certain number of years of growth. This ensures that the tree has had time to build up a store of photosynthate with which to support any developing seeds and fruit. Similarly, some herbaceous and weedy species flower after the production of a certain number of leaves, again allowing for the development of sufficient photosynthetic tissue to support the nutritional needs of nonphotosynthetic flowers, seed, and fruit (King 1997). While the human observer might interpret these responses as implying a 'counting' mechanism within the plant, it is more likely that each leaf produces a certain amount of a stimulatory substance, and that the 'right' number of leaves are required to produce sufficient stimulus to cross a threshold, beyond which flowering responses are activated. The early literature on flower induction is primarily concerned with the search for this stimulatory substance, as discussed in Section 2.6.

It is likely that all plants, even those with a strong environmental stimulation of flowering, use developmental cues to a certain extent. The production of organs which do not contribute to the carbon balance of the plant is an expensive business, and one not undertaken unless the plant is sufficiently large and in sufficiently good health. The only exception to this general rule is that, in extreme cases of stress, flower induction may occur to ensure that seed is produced before a plant dies. This can often be observed in badly treated garden and house plants!

2.5 Environmental explanations of floral induction

The early plant physiologists established very quickly that the environmental variables which seemed to influence flowering were day length, temperature, and, to a lesser extent, water stress.

2.5.1 Day length

The most commonly used environmental variable is the length of time for which light is available, the day length. Almost all plants outside equatorial regions have flowering responses at least partially triggered by day length.

Plants that use light availability to tell them when to flower are called photoperiodic. Photoperiodic plants use day length, and the change in day length as the seasons progress, to predict which season and therefore which weather conditions will come next. In the same way, if one had no access to a calendar, it would be more reliable to predict when autumn would arrive by measuring the shortening days than it would be to base everything on the weather. A late hot spell in September, for instance, could hide how far advanced autumn was, but day length is an entirely reliable indicator.

That plants use day length to determine when to flower implies that there is an optimal time for flowering to occur. This optimal time is different between species, and within species it may differ in relation to geographical distribution. The optimal time to flower depends on a large number of factors, and is a reflection of the selective pressures that the plant experiences. It is essential that flowers are produced in sufficient time to set seed before winter weather conditions kill off developing tissues. The vulnerability of the developing seed and the harshness of the local climate

Figure 2.4 Responses of long and short day plants with varying critical night lengths. The long day plant will only flower in the light regime shown if its critical night length is longer than the period of darkness. The short day plant will only flower if its critical night length is shorter than the period of darkness.

will interact to determine how late any individual plant can afford to leave flowering. At the same time, if seed or fruit is to be dispersed by animals then it is essential for that seed or fruit to be mature while the animals are foraging. Again, this will depend on the animals used, whether they migrate, hibernate, or feed all winter, and how the local climate conditions affect the animals themselves. Flowers themselves may also rely on animals for pollination, and if this is the case flowers must be produced when the appropriate pollinators are active. Finally, competition between flowers of different species for pollinator attention has been hypothesized to drive some plant species to flower at a time slightly suboptimal with regard to the above factors, simply to avoid competition and reduced levels of pollination (see Chapter 17).

So photoperiodic plants are plants that use day length to predict when their optimal flowering time is approaching, and flower accordingly. In fact, there are two types of photoperiodic plants—long day plants and short day plants. This is a very important distinction in photoperiodic research, and one that the early plant physiologists had to establish in order to avoid much later confusion. A long day plant is a plant that flowers when the days are longer than a certain minimum length. The

minimum length itself will vary greatly between different species, but the key point is that long day plants flower only when the days exceed this length. In a temperate climate that means that the plant is likely to flower in late spring or early summer. Many common European garden and crop plants are long day plants, including species like wheat, lettuce, and sugar beet. Short day plants, on the other hand, are plants that require the days to be shorter than a certain maximum length in order to flower. Again, the critical length itself can vary widely, but it is the flowering when days are shorter than this that defines a short day plant. Short day plants are usually found in latitudes lower than those of long day plants. In a temperate climate they are represented by species that flower very early in spring, like the snowdrop and the crocus, and by autumn flowering species, such as chrysanthemum. (Figure 2.4 shows the responses of long and short day plants to an example light regime; note that it refers to critical night length, not day length, as explained below.)

Having established that plants appeared to flower in response to day length, it then took a number of simple experiments to show that plants were actually measuring the length of the night, not the day. This was shown by interrupting the 'day' experienced by plants in controlled growth conditions, by turning the lights off for a variety of periods. These interruptions had no effect whatsoever on flowering. However, interrupting the 'night' by turning the lights on for short periods had dramatic effects on flowering. The effect of flashes of light during the night was to persuade the plants that the nights were much shorter than they really were, and to induce early flowering in long day plants but to repress flowering in short day plants. These experiments also showed that red and far red light were the most effective at disturbing the plants' flowering responses (Borthwick *et al.* 1952).

It was a short step from establishing that plants measured night length to investigating where in the plant the light regime is sensed. Surprisingly, it has been shown quite conclusively that day length (night length) is measured in the leaves of the plant, not in the meristem, even though it is the meristem which actually undergoes the change from a vegetative to a reproductive state. Again, very simple physiological experiments provided this important breakthrough. Plants were kept in non-inductive conditions, with only single organs exposed to the correct day length. This could be achieved by placing a box around the organ in question, with a light in that box under separate control from the main lights of the room. These experiments showed that providing a single leaf with the appropriate day length was enough to cause flowering to occur. However, exposing just the apex of the plant had no effect at all. Although it may seem counter-intuitive to measure day length in the leaves and then send a signal to the meristem to cause flowering to occur, in fact there are a number of good reasons why leaves are the best place to sense the light regime. Leaves have a much bigger surface area than the apical meristem, and are adapted to maximise light capture. They also act as a fail safe mechanism—measuring light in many different leaves at once means that if some leaves are shaded and do not perceive the correct day length, other leaves are likely not to be. The plant thus has multiple chances to perceive day length, and multiple chances to flower.

These experiments provided a baseline for modern research into the floral transition. The early physiologists had established that most plants flowered in response to the light regime; that they

needed either days longer than a certain minimum (long day plants) or days shorter than a certain maximum (short day plants); that they appeared to measure night length rather than day length; and that they did that measuring in their leaves. Modern research, discussed in Chapter 5, has focused on the mechanism by which light is perceived, the clock which allows plants to measure day length, and the signalling that connects the two.

2.5.2 Temperature

The second most commonly used environmental stimulus to regulate flowering time is temperature. However, temperature is generally additional to photoperiod, and most often occurs in plants which live for several years and only flower in some of them. It provides information on the passing of years rather than seasons. Henbane, *Hyoscyamus niger*, for example, is a long day plant, but will not flower at all unless it has been exposed to cold. It is biennial and only flowers in its second year. Putting the photoperiod and cold requirement together, we can see that henbane, which germinates in the spring, cannot flower in its first summer, because it has not been induced by cold. It cannot flower in winter, or the next spring, despite having been induced by cold, because it is in the wrong day length. Only in the second summer, when the cold induction has happened and the day length is right, will the plant flower. The degree of cold needed to induce flowering and the amount of time it is needed for vary a lot between species according to the environments in which they occur. Treating a plant with cold to induce flowering is called vernalization, and it is used extensively in agriculture, where crops may be sown in the autumn but will not flower until early the following spring, induced by the winter cold. Like the sensing of photoperiod, temperature sensing has been shown to take place in the leaves of the plant, although there is also some evidence that apical meristems can be induced to flower by the chilling of them alone.

2.5.3 Water stress

Early physiological experiments also showed that various sorts of stress could induce flowering.

The most effective of these was water stress. Droughting plants, or even cutting off their roots, was frequently able to induce flowering. This seems to ensure that a plant unlikely to survive does at least set seed for the next generation; it is a phenomenon which has been studied less extensively than light- or temperature-induced flowering. One explanation for this is that drought stress is not generally conducive to the production of much seed or seed of very good quality, so this form of flower induction is unlikely to be of use in agriculture or horticulture.

2.6 The florigen problem

Once the early plant physiologists had established that it was leaves that sensed the light regime, they quickly realized that there must be a signal which travelled from the induced leaves to the apical meristem to activate the floral transition. The search for that signal was one of the most active areas of plant research for many years. We now know that multiple signals interact to evoke the flowering response in the meristem, and that the process is much more complex than anything previously imagined.

2.6.1 Florigen

It was initially hypothesized that the signal sent by the induced leaf to the apical meristem was likely to be a plant growth regulator (otherwise known as a plant hormone). There were a number of pieces of evidence to support this suggestion. Most compellingly, it could be shown that the signal travelled at about the same speed as the phloem stream (by analysing time from leaf induction to changes in enzyme activity in the meristem). Ring girdling experiments were then conducted, removing only the phloem from a ring around the stem of the plant (and thus blocking the phloem at that point). Ring-girdled plants no longer flowered, unless a leaf had been induced above the ring. This served to confirm the idea that the signal travelled in the phloem, the location of much of plant growth regulator transport (Zeevaart 1976).

Grafting experiments suggested to the early plant physiologists that the same signal could induce flowering in all species. They therefore concluded that the signal represented a novel plant growth regulator, present in all plants, and they named it 'florigen' (Chailakhyan 1936, in Hoffmann-Benning and Zeevaart 2003). The evidence in support of florigen was initially very convincing. Numerous reports were published in which a single induced leaf of species A was grafted onto the stem of a plant of species B (which was kept under non-inductive conditions), resulting in the appearance of flowers on plant B. The number of species to which this applied was large, and led to a concerted effort to identify the mysterious florigen. However, no novel plant growth regulator could be extracted from the phloem of these plants. And as research continued, it became clear that the same signal did not promote flowering in all species. One of the final nails in the florigen coffin was the discovery that grafting an induced leaf from tobacco onto the stem of a potato plant resulted not in the production of potato flowers but of potato tubers (discussed by Martinez-Garcia *et al.* 2002).

2.6.2 The anti-florigen theory

An alternative hypothesis to explain the transmission of a signal from the leaves to the meristem is known as the anti-florigen theory. This proposes that non-induced leaves emit a signal that travels to the apical meristem and prevents flowering. When the leaves are induced, it is the absence of this inhibitory message which allows the transition of the meristem. Initially, the anti-florigen hypothesis assumed a specific inhibitory signal which only prevented flowering, and was common to all species. It is now generally accepted that the anti-florigen hypothesis is a useful model which could apply to the release of any growth regulator or other signalling molecule, and which might vary in different species (Colasanti and Sundaresan 2000).

2.6.3 The source/sink resource allocation model

Another model to explain the evocation of flowering by induced leaves is the source/sink resource allocation model, which states that the meristem of a non-induced plant is not sufficiently supplied with the appropriate nutrients to change into a flowering meristem. Induction of the leaves changes the source/sink properties of various organs of the plant, resulting in a rush of nutrients to the meristem, which can then undergo the transition to flowering. While there is no specific evidence to support source/sink relations as the major determinant of flowering, it is certainly the case that extra energy, along with sufficient nitrogen and phosphate for all the RNA and protein synthesis associated with flowering, is required at the apex. The source/sink properties of plant organs clearly

do change during development, and it is very likely that there are many such changes at the transition to flowering, some of which are probably critical to the correct development of flowers (Heyer *et al.* 2004).

The various ideas and models to arise from the work of the early plant physiologists have been as important in understanding the data arising from modern molecular genetic analyses of the floral transition as the ideas of Goethe have been in understanding floral development. In Section II of this book, we look at the current state of knowledge of these processes.

THE MOLECULAR MECHANISMS OF FLOWERING: INDUCTION AND DEVELOPMENT

Induction of Flowering

CHAPTER 3

Flower Induction in *Arabidopsis thaliana*

Section II of this book explores our current state of knowledge concerning the induction, evocation, and development of flowers. In the twenty-first century, research into these processes is almost exclusively conducted through a molecular genetic approach. Such an approach uses the backwards-seeming logic of searching for a plant with a mutation which prevents a process occurring properly. By analysing the process in the mutant plant, and by identifying and analysing the mutated gene, it is possible to build up a picture of how a normal plant and a normal gene work. When molecular genetic analysis becomes the usual way of exploring a biological process, there is a tendency for the majority of work to focus on one or a few species. This is partly because information already gained in a species informs the next set of experiments, but also partly because there are some species simply better suited for molecular genetic analysis than others. Thus, the fruit fly, *Drosophila melanogaster*, is by far the best-studied animal, simply because a number of features of its biology make it an excellent model system. Similarly, the weedy annual plant *Arabidopsis thaliana* has become the most extensively studied plant species. This chapter introduces Arabidopsis and the reasons it has become the model of choice for the study of floral induction. It then goes on to look at the description and characterization of mutants with perturbations in floral induction, mutants which have allowed the development of models explaining the different ways in which the flowering response can be induced.

3.1 *Arabidopsis thaliana* as a model system for the study of flowering

3.1.1 Why Arabidopsis?

Arabidopsis thaliana is commonly referred to simply as Arabidopsis, although its traditional English names are thale cress and mouse ear cress. It is a weed native across Eurasia and East Africa, and a member of the Brassicaceae, an advanced family in the Rosid clade of the Eudicots. Arabidopsis was selected as a model species by plant geneticists as early as the 1950s, but it was not until the 1980s that it became the system of choice for the majority of plant scientists (Redei 1973). By December 2000, when its complete genome sequence was published (The Arabidopsis Genome Initiative, 2000), Arabidopsis had become so much the most usable plant model species that to work on anything else was highly unusual. In the post-genomic era, Arabidopsis continues to be the main focus, with many facilities and services available that allow experiments which would be unthinkable with any other species.

In the wild, Arabidopsis tolerates a very wide range of conditions. It is commonly found as a garden weed, in disturbed fields, on railway lines, and on walls. This unfussy approach to habitat was one of the first reasons Arabidopsis was identified as a useful plant model—it is perfectly capable of completing its lifecycle on sterile, chemically defined media in a Petri dish. Other advantages include its small size and fast lifecycle. The mature plant can be grown within 1 cm^2 of space, and in soil does not usually spread more than 8–10 cm in diameter and 12–15 cm in height (Fig. 3.1, see also Plate 4). Arabidopsis has a very short generation time—just 6 to 8 weeks—allowing several generations of a mutant to be studied within a relatively short period. It is self-fertile, which is very important for genetic analysis as it allows the simple generation of homozygotes. However, Arabidopsis can also be cross-pollinated to generate double and triple mutants. A single plant can produce over a thousand seeds, making segregation analysis straightforward. Its small genome is another great advantage. It has a haploid DNA complement of around 130,000 kilobase pairs—only fifteen times that of the bacterium *Escherichia coli*, five times that of yeast, and about

half the amount that *Drosophila* has. Mice, humans, and most other flowering plants have 30 to 40 times as much DNA as Arabidopsis, making their molecular analysis much more unwieldy (Redei 1973).

Being such a widespread plant, Arabidopsis is not entirely homogeneous. In its many different habitats it is found as a range of natural accessions (sometimes called ecotypes), each with significant differences in development, morphology, and behaviour. Even the commonly used laboratory lines have thousands of polymorphisms (Putterill *et al.* 2004), which may be morphological, easily identified by the naked eye, or molecular, requiring the use of DNA fingerprinting techniques to identify them. In both cases they provide useful tools

Figure 3.1 *Arabidopsis thaliana.* (a) Seedlings growing on an agar plate. (b) A mature vegetative plant. (c) During the reproductive phase of the lifecycle.

for the mapping of genes of interest, as we shall see below.

A final advantage of Arabidopsis as a model system is the remarkable ease with which transgenes can be introduced into its genome. Most plant species have not yet been tested for ease of transformation, and many that have were found to be resistant. Although it is relatively easy to introduce transgenes into some species, such as the garden snapdragon *Antirrhinum majus*, it is then incredibly difficult to regenerate new plants from their transgenic cells (Heidmann *et al.* 1998). However, an annually increasing number of plants are being reported as 'transformable', and the techniques used all rely on tissue culture for regeneration. In general, plants are transformed using the crown gall-inducing bacterium *Agrobacterium tumefaciens*, which has the ability to transfer a portion of a plasmid into the genome of a host plant cell. In modern genetic engineering, a modified version of both the bacterium and its plasmid are used. The transgene of interest is inserted into the modified plasmid, which is then transferred back into the bacterial cell (Zupan *et al.* 2000). A piece of plant tissue, typically a leaf disk or a section of root or hypocotyl, is infected with the *Agrobacterium*, and gene transfer occurs. It is then necessary to culture the transformed plant tissue on a variety of chemical media, to kill any remaining bacterial cells and potentially invading fungi, while supplying the nutrients and plant growth regulators which will allow regeneration of a new plant. Even in species amenable to this process, the tissue culture can take many months, and is labour intensive. However, Arabidopsis is routinely transformed using a much simpler and faster procedure, one which has not been found so far to work for any other plant. Flowering Arabidopsis plants are dipped into a culture of *Agrobacterium* containing the transgene in the modified plasmid. After a few minutes the plants are removed and returned to the greenhouse. The *Agrobacterium* inserts the transgene into the ovules within the flowers. When the plants set seed over the next few weeks, a proportion of those seed will contain the transgene, as a result of its presence in the egg from which they developed. Seed from the dipped plants is sown on a selective medium, which allows the identification of transgenic plants,

and they are simply picked off and used in further experiments. The speed and simplicity of this procedure has added greatly to the range and extent of experiments which can now be conducted using Arabidopsis (Clough and Bent 1998).

3.1.2 Generating mutants in Arabidopsis

The best genetic approach to a developmental process is to find mutants which prevent that process occurring normally. The rationale behind this method is that the study of the impaired function would give one insight into the genes and proteins involved in the normal process. Generating mutants in Arabidopsis is a simple process, and is usually done in one of three different ways. EMS mutagenesis uses ethyl methane sulfonate (EMS) to induce chemical damage to DNA. Typically, seed are soaked in EMS prior to germination. The chemical induces single base changes in the DNA, known as point mutations, by alkylating bases, primarily guanine. Alkylation causes mispairing of bases, with alkylated guanine pairing to thymine rather than cytosine, thus introducing single base pair changes when DNA is copied. A second approach uses X-rays to cause deletions of pieces of DNA or large-scale rearrangements of sections of chromosome. Thirdly, it is possible to use transformation technologies to insert foreign DNA randomly into the genome. Insertion of foreign DNA into a gene causes mutation of that gene, and has the added advantage of providing a tag or handle which can later be used for molecular cloning of the mutated gene. This approach, called T-DNA or insertion mutagenesis, is the method of choice for large-scale mutagenesis programmes. However, individual research groups focusing on a single developmental process might prefer EMS mutagenesis because it is simpler and generates new mutants more quickly.

These seed which are mutagenized, for example by soaking in EMS, are called the M0 generation (M for mutant). They germinate and the plants they produce are most likely to be heterozygous for any mutation. This is simply because it is vanishingly unlikely that damage to DNA will occur at precisely the same position on two homologous chromosomes. This heterozygous generation is

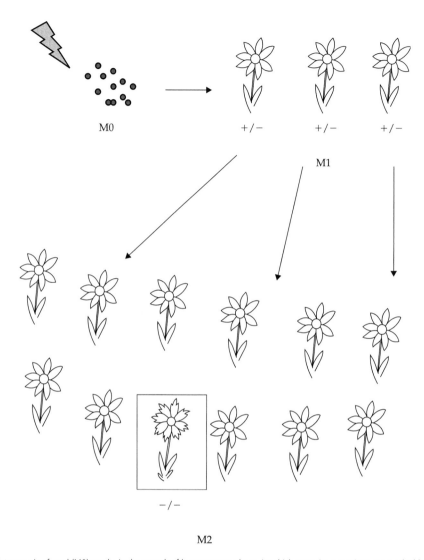

Figure 3.2 Mutagenesis of seed (M0) results in the growth of heterozygous plants, in which recessive mutations are masked by dominant wild type alleles (M1). Homozygous mutants are selected in the next generation (M2).

called the M1 generation. Since most mutations are recessive, M1 plants usually look perfectly normal. M1 plants are allowed to self-pollinate and set seed, and the seed grows into the plants of the M2 generation. In the M2 generation there will be segregation of the mutation, so one in four plants will be wild type, two in four will be heterozygous, and the final one in four will be homozygous for the mutation. Assuming this mutation causes a visible phenotype, the mutant plants will be easy to pick out from the rest (Fig. 3.2).

3.1.3 Complementation analysis—finding out how many genes have been mutagenized

When a mutagenesis programme goes well, a large number of mutants with interesting phenotypes might appear in the M2 generation. Usually only those mutants with a phenotype relevant to the process of interest are selected—so the 17 mutants (for example) which have problems with the induction of flowering are picked, but the dwarf plants and the yellow plants and the plants with strange

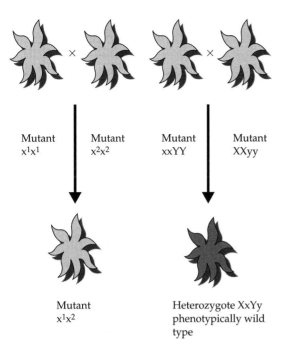

Figure 3.3 Complementation analysis determines whether two phenotypically similar mutants contain mutations in the same (left) or different (right) genes.

shaped leaves are ignored. However, before any analysis of these mutants is carried out, it is important to establish how many genes they represent. In theory, 17 mutants with the same phenotype might mean 17 different genes have been mutagenized, a different one in each plant, or that a single gene has been mutagenized 17 times. Or, more likely, one has managed something in between. The process of sorting out how many genes have been mutagenized is called complementation analysis.

Complementation analysis begins by making a cross between two mutant lines (illustrated in Fig. 3.3). Thus pollen from mutant 1 is used to pollinate flowers of mutant 2. Remember that both parents were homozygous for their mutation. So if the mutation in mutant 1 and the mutation in mutant 2 are in the same gene, X, the progeny of this cross will have one mutated allele of gene X from each parent. They will still not have a functional version of gene X, and will still look mutant. However, if the mutation in mutant 1 and the mutation in mutant 2 are in different genes, X and Y, the story is very different. In this case, mutant 1 will be

homozygous for a mutation in gene X, but homozygous for a wild type allele of gene Y. Mutant 2 will be homozygous for a mutant allele at gene Y, but also homozygous for a wild type allele at gene X. So in this case the progeny of the cross will have one wild type allele of gene X (from mutant 2) and one mutant allele (from mutant 1), and one wild type allele of gene Y (from mutant 1) and one mutant allele (from mutant 2). These plants will be heterozygous at both loci, so will have a functional copy of each gene. Since the functional copies are usually dominant, and the mutant copies recessive, these plants will appear wild type. Very simply, then, if the progeny of a cross between two mutant plants is also mutant, then the two mutations are in the same gene. However, if the progeny of a cross between two mutants is wild type, then the two mutations are in different genes as the mutants are able to **complement** one another.

3.1.4 Cloning genes following mutagenesis

Once mutants have been identified and characterized, it is useful to isolate the mutated gene and investigate its normal function. There are a number of different ways of going about this. If the mutant was generated by insertion mutagenesis, then it is possible to use the transgene as a tag with which to clone the gene. One way to do this is to make a radiolabelled probe that will bind the transgene, and use it to probe a Southern blot of the digested DNA of the mutant plant. The DNA fragment to which the probe binds can then be isolated from the gel and purified, and its sequence investigated. Alternatively, there exist a variety of PCR-based techniques which allow the amplification and isolation of a piece of DNA for which the sequence of one flanking region (i.e., the transgene) is known.

If the mutation was generated using EMS then it is necessary to map the position of the mutation within the genome. Once that has been done, it is a simple matter to analyse the DNA sequence within that region using the whole genome sequence available on public databases. Mutations are mapped onto the genome using the many polymorphisms, both morphological and molecular, that exist between different laboratory lines of Arabidopsis. Mapping populations are readily available, containing many polymorphisms at known map positions. These are crossed to the mutant lines of interest, and the resulting F1 generation self-pollinated to produce a segregating F2 population. Within this population the new mutation will segregate with some of the polymorphisms, but completely independently from others. This indicates that the mutation is on the same chromosome as those polymorphisms with which it segregates. Fine mapping of the mutation involves investigating how frequently there is crossing over between the mutated genes and other polymorphisms on the same chromosome. This approach allows one to pinpoint the location of the newly mutated gene between known polymorphisms. Investigation of the genome sequence within that area may indicate the presence of several genes, but it is usually a simple matter to find out which is responsible for the mutant phenotype. The usual approach is to take wild type copies of each of the candidate genes and insert them into the mutant plant, using the transformation protocol described earlier. The one which restores a normal phenotype to the mutant plant is the one which has been mutated.

Once cloned, the gene of interest can be analysed in a number of ways, including investigating its expression pattern within the plant, the consequences of changing that expression pattern using transgenic methods, and identifying other proteins with which its protein product interacts.

3.1.5 Reverse genetic approaches in Arabidopsis

The genetic approach described above has been responsible for enormous advances in understanding many processes, including floral induction. In recent years, and particularly following the publication of the genome sequence, it has become increasingly possible to use a reverse genetic approach as a counterpoint to the traditional genetic method. The reverse genetic approach begins with a gene, perhaps one with a sequence similar to another gene that has an interesting function, and progresses towards an understanding of its function. Over 200,000 insertion lines containing foreign DNA tags have been generated by a range of research groups and consortia, and the DNA

flanking the tags sequenced and made available on the Internet, making it a simple matter to search for tags which have inserted in or near a gene of interest (see http://signal.salk.edu/cgi-bin/tdnaexpress). The lines containing these tags can then be ordered and the plants grown for phenotypic characterization and further experimental analysis.

3.2 Flowering-time mutants

The molecular genetic approaches outlined above have generated enormous insights into the ways in which flowering is induced in Arabidopsis. In the chapters which follow we will focus on the results of this analysis, looking at the pathways that have been proposed to induce and repress flowering and the genes and proteins which work within them. The rest of this chapter focuses on the original mutants that began this process of discovery, and the ways in which they are categorized. Over 80 genes have been shown to influence floral induction to some degree, although many have effects on other developmental processes as well.

3.2.1 Arabidopsis is a vernalization-sensitive facultative long day plant

Arabidopsis is a useful model for the study of the floral transition, because it uses several different pathways and thus provides insight into a wide range of processes. To begin with, Arabidopsis is clearly a plant which has developmental control of its floral transition. Even when grown under optimal environmental conditions for flowering, Arabidopsis remains vegetative until it has produced a rosette of leaves—it will not flower immediately after germinating. In fact, this observation indicates that Arabidopsis has an autonomous pathway which represses the flowering response until a certain developmental stage has been reached. Secondly, Arabidopsis is a long day plant. Grown in a growth room with 16 hours daylight and 8 hours darkness every 24 hours, Arabidopsis plants will flower within 2–3 weeks of germination. However, grown in a growth room at the same temperature but with 12 hours daylight and 12 hours darkness, the plants will show no sign of flowering after the same time period has passed. They clearly measure day length (or night length) and respond to photoperiod by flowering only when the days are longer than a certain critical minimum. However, this photoperiodic response is not absolute. A plant entirely limited by photoperiod would never flower in the growth room with 12-hour days. Arabidopsis, on the other hand, continues to grow vegetatively in the short day room for several weeks after its counterparts in the long day room have flowered, and then it flowers itself. For this reason it is described as a facultative long day plant—long days promote flowering and cause it to occur early, but in their absence the plant will still flower eventually (Jack 2004). This shows that, besides its photoperiodic induction pathway, Arabidopsis must also contain an autonomous pathway which induces flowering eventually, irrespective of environmental variables. Finally, Arabidopsis is also sensitive to vernalization. Accessions from parts of the world with cold winters germinate in the spring or summer and remain vegetative until the next spring. The long days of their first summer do not induce flowering, because they require between one and three months of cold temperature to become responsive to day length. Accessions from warmer regions usually flower in the same year that they germinate, either through lack of this vernalization requirement or through especially great sensitivity to vernalization, ensuring that they flower early in their first spring, before full summer with its attendant droughts arrives (Stinchcombe *et al.* 2004). However, by extending molecular analysis into a range of accessions, it has been possible to use this one model species to investigate developmentally controlled, photoperiod-sensitive, and cold-induced flowering responses.

3.2.2 Classification of flowering-time mutants

When mutant plants are first described, they are classified and categorized according to their phenotype alone. Later, as molecular information on gene function is accrued, it sometimes becomes apparent that the initial classification was unhelpful. However, since only phenotypic data are available in the first instance, it continues to make sense to classify according to those data. Arabidopsis mutants in floral induction are traditionally described as either late flowering or early flowering (illustrated in Fig. 3.4). These classifications refer to

Figure 3.4 Flowering-time mutants are described according to how they behave under long and short day conditions. Wild type plants flower at an early time point under long day conditions, but at a later time point under short day conditions. Early flowering mutants flower at the early time point in both long and short days. Late flowering mutants flower at the later time point in both long and short days.

their time (or developmental stage) of flowering in different growth rooms. A late flowering Arabidopsis mutant is one whose behaviour in a long day growth room is similar to that of a wild type plant in a short day room—it flowers later than a wild type plant would do in the long day room. An early flowering Arabidopsis mutant, on the other hand, is one whose behaviour in a short day growth room is similar to that of a wild type plant in a long day room—it flowers earlier than a wild type plant would do in the short day room (Coupland 1995).

Because time of year, age of seed, and growth room temperature can all affect the rate of Arabidopsis growth and development, researchers often measure flowering time in terms of number of leaves rather than in number of days (Koorneef et al. 1991). Using this scheme, a wild type plant might flower after producing 6 leaves in a long day room, but not until it had produced 12 leaves in a short day room. A late flowering mutant is a plant which would not flower until it had produced 12 leaves *in the long day room*. An early flowering mutant, in contrast, is a plant which would flower after producing only 6 leaves *in the short day room*.

These classifications only refer to time of floral induction in a particular growth room—the late flowering mutant is defined by its behaviour in long days while the early flowering mutant is defined by its behaviour in short days. In fact, these categories can be broken down further, as shown in Table 3.1. Note that the number of leaves is simply an indication of relative flowering time. The absolute number of leaves formed before flowering depends on the individual mutant and may also be variable between plants carrying the same mutation.

This further classification of flowering-time mutants is determined by their behaviour in the *other* day length—late flowering mutants are divided up by their behaviour in short days and early flowering mutants are divided up by their behaviour in long days (Koorneef 1991).

Late flowering mutants in class 1 show the typical late flowering in long days and produce more rosette leaves than wild type plants in this situation. Their response to short days is an even more dramatic delay in flowering, producing even more rosette leaves before being induced to flower. These mutants are clearly still responsive to photoperiod. Their ability to measure day length is nicely demonstrated by the fact that they flower earlier in long days than in short days. However, compared to wild type plants they are always very slow to flower. It seems, then, that these mutants are lacking a promoter of flowering not related to photoperiod. They are perturbed in an autonomous promotion pathway, one which is always required to initiate flowering.

Late flowering mutants in class 2 also show the typical late flowering in long days, producing around 12 leaves before flowering. But when placed in the short day room these mutants still flower after producing 12 leaves. In contrast to class 1 mutants, class 2 mutants do not discriminate between different day lengths, but behave as though in short days however long the days are. This phenotype suggests an inability to measure day length, with the mutants always convinced that the days are short. The mutated genes in these plants might include those encoding the proteins which perceive light or measure time, so that no promotive long day signal is ever perceived, even if it is present.

Early flowering mutants can also be divided into two distinct classes. Early flowering mutants in class 1 show the typical response of flowering too early in the short day room—producing flowers with only around 6 leaves. If these plants are moved into the long day room, they flower even earlier—after making as few as 3 or 4 leaves. These mutants, like those in late flowering class 1, are clearly still sensitive to photoperiod—they flower later in short days than they do in long days. But they flower earlier than 'supposed' to, whatever the day length. These

Table 3.1 Approximate number of leaves produced before flowering by wild type and four mutant categories of Arabidopsis

	Long days	Short days
Wild type	6 leaves	12 leaves
Late flowering class 1	12 leaves	18 leaves
Late flowering class 2	12 leaves	12 leaves
Early flowering class 1	4 leaves	6 leaves
Early flowering class 2	6 leaves	6 leaves

mutants are probably lacking something which represses flowering until the plant has reached a certain developmental stage. So this mutant class should make it possible to identify the genes which provide a developmental brake to flowering, giving the plant a chance to build up a certain amount of photosynthetic tissue before it starts to expend energy on producing flowers and seed.

Early flowering mutants in class 2 also show the typical pattern of flowering with only 6 leaves in short days. However, unlike mutants in class 1, they will also flower in the long day room only after they have produced 6 leaves. These mutants appear to be unable to respond to different photoperiods, and perceive all days as being long days. In the same way as mutants in late flowering class 2, these mutants have some problem with the perception of light or time. However, whereas that problem made mutants in late flowering class 2 behave as though all days were short, mutants in early flowering class 2 behave as though all days were long. In fact, these mutations may turn out to be in similar genes, with the differences in phenotype related to whether the mutation makes the gene more or less active.

3.2.3 Epistatic analysis places flowering-time genes into pathways

The phenotypes of flowering-time mutants give some clues as to which pathways they might operate in—those regulated by light, developmental cues, or temperature. This is confirmed by conducting epistatic analysis to test whether two mutated genes operate in the same pathway or different ones. If they function in the same pathway then epistatic analysis also indicates in which order they are required within that pathway.

Epistatic analysis begins by producing a double mutant (aabb) by crossing two different mutant plants (aaBB and AAbb) and selfing the heterozygote progeny (AaBb). There are a number of possible phenotypes which might show up in this double mutant plant. If the double mutant (aabb) has the same phenotype as one of the parents (say aaBB) then the gene A is epistatic to the gene B. That means that gene A and gene B work in the same pathway, with gene A acting before gene B. The

logic behind this is that if the two genes work in the same pathway, and if the pathway is blocked at the earlier point (A), then the phenotypic effect of mutating a gene further down the pathway (B) will be invisible. Alternatively, the double mutant (aabb) may have the same phenotype as parent plant AAbb—indicating that gene B works before gene A in the pathway. It is also possible that the double mutant has the phenotypes of both parents. In this case we say there is no epistasy between the genes, which work in different pathways and do not interact. Epistatic analysis is most easily imagined with morphological mutants, where crossing a mutant with no roots to a mutant with short roots would generate a double mutant with no roots— showing that one needs to make a root before one worries about its length. On the other hand, crossing a mutant with no roots to a mutant with purple leaves is likely to generate a mutant with no roots and purple leaves—because the pathways are unrelated. However, the process can be used to investigate more complex processes such as floral induction. In these cases phenotype is not defined simply as late or early flowering, but factors such as the expression patterns of other known genes within the mutants are taken into account as well (Coupland 1995).

3.2.4 Arabidopsis contains one repression-of-flowering pathway and four promotion-of-flowering pathways

Epistatic analysis of the sort described above, using mutants including those above and other lines with perturbed responses to cold, has shown that at least five genetic pathways regulate flowering time in Arabidopsis (Simpson and Dean 2002). There is a single endogenous repression pathway which acts to prevent flowering. Its job is to delay flowering until the plant has reached a certain developmental stage and has the photosynthate to support developing flowers and seed. There are then four promotion pathways. Having multiple promotion pathways gives the plant plasticity, because it means that weak promotion in one pathway can be compensated for by strong promotion in another pathway, ensuring flowering always happens. Two of these promotion pathways respond to environmental signals while two of them are under endogenous control. The photoperiodic promotion pathway relies on photoreceptors and a circadian clock to transduce information regarding day length. The vernalization promotion pathway is activated by exposure to low temperatures. The ability of plants like Arabidopsis to flower under non-inductive conditions indicates that plants must contain an autonomous promotion pathway, which ensures flowering occurs if all else fails. The fourth promotion pathway in Arabidopsis is the gibberellin pathway, which acts independently of the other pathways and may serve to activate flowering in response to certain developmental triggers. These different pathways are discussed in the next four chapters.

CHAPTER 4

The Autonomous Pathways for Floral Inhibition and Induction

The autonomous pathways for floral inhibition and induction are central to the entire process of floral transition in Arabidopsis. The floral inhibition pathway is the central regulator of flowering, acting throughout the plant's life to ensure flowering does not occur until the appropriate environmental, developmental, and physiological cues have been received. Without this autonomous pathway, Arabidopsis plants could flower immediately on germinating, before sufficient photosynthate had been produced to support the developing flowers and seeds. The inhibition pathway, as we will discuss below, essentially operates through a single protein repressing the transcription of genes necessary for the floral transition to occur at the shoot apical meristem. Other proteins interact to enhance this basic repression mechanism. On the other hand, the autonomous induction pathway is equally essential to the life history of a weedy annual. Unlike long-lived perennial plants, Arabidopsis does not have the option to delay flowering indefinitely until perfect environmental conditions are perceived. Instead, the autonomous induction pathway acts antagonistically to the inhibition pathway to ensure that the plant flowers eventually, irrespective of a lack of inductive environmental signals. These two autonomous pathways are the central players in the control of Arabidopsis flowering, around which the environmentally induced pathways we will discuss in later chapters revolve.

4.1 The floral inhibition pathway

The floral inhibition (or repression) pathway is the central regulator of flowering time in Arabidopsis.

The repression caused by this pathway is constitutive, and can only be broken by signals which either downregulate or bypass it. In Chapters 5 and 6 we will consider how environmental signals achieve just that, but we must begin with a discussion of how the inhibition pathway itself works. The floral transition requires activation of a set of genes in the shoot apical meristem which convert the meristem from the vegetative state to the inflorescence state. The identities and functions of these floral meristem identity genes will be discussed in Chapter 8. In particular, it is essential that three genes described as flowering-time integrators (Chapter 7) are expressed. These genes, *FT*, *LEAFY* (*LFY*), and *AGAMOUS-LIKE20* (*AGL20*) (also known as *SUPPRESSION OF OVEREXPRESSION OF CONSTANS 1* (*SOC1*)), activate the floral meristem identity genes, and it is thought that all pathways affecting flowering time are integrated at these three genes (He and Amasino 2005). The inhibition pathway works by repressing transcription of the flowering-time integrators, thus ensuring that the floral meristem identity genes are not activated and that the meristem remains vegetative (Fig. 4.1). Repression of the flowering-time integrators is achieved primarily through the action of the FLOWERING LOCUS C (FLC) protein, reinforced by the activity of FRIGIDA (FRI). Dominant alleles at either locus result in plants that are extremely late flowering. Loss of function of either gene reduces the repression of flowering and causes an early flowering phenotype. In order for the plant to flower, signals promoting flowering must overcome repression of the flowering-time integrators by FLC. This can be achieved in either of two ways.

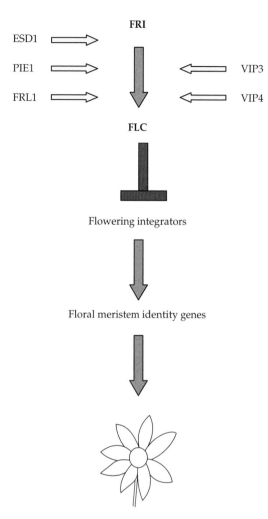

Figure 4.1 Diagram of the floral inhibition pathway. Expression of the flowering-time integrators, and thus the floral meristem identity genes, is prevented by FLC, itself enhanced by FRI and a number of less well-characterized proteins.

Proteins in the autonomous and vernalization promotion pathways promote flowering by downregulating *FLC* expression. These proteins directly target the *FLC* gene, and, through a number of routes, reduce its expression level to the point where it no longer represses the flowering-time integrators. The alternative approach is taken by proteins in the photoperiod and gibberellin-based promotion pathways. These proteins act to directly activate the flowering-time integrators, simply bypassing repression by FLC. This sets up a conflict

at the flowering-time integrators, with numerous promotive signals, plus an FLC-based inhibitory signal which varies in strength, converging on them and controlling their expression levels and subsequent functions.

4.1.1 *FLC*

The *FLC* gene was isolated using positional cloning and encodes a MADS box protein, a member of a large family of transcription factors found in all eukaryote genomes (Michaels and Amasino 1999). Since the central role of FLC is to repress the expression of the flowering-time integrators, it is likely that the protein acts as a negative regulator of transcription, perhaps binding to promoter sequences but failing to activate transcription. Initial reports focused on establishing the central role of FLC in floral repression. Loss of *FLC* function causes early flowering, while ectopic expression of FLC from the cauliflower mosaic virus 35S (CaMV 35S) promoter caused plants to flower extremely late, with some failing to ever flower at all, senescing and dieing after producing over 80 rosette leaves but never bolting (Michaels and Amasino 1999). Early experiments showed that the level of *FLC* transcript in a range of backgrounds correlated directly with the time those plants took to initiate flowering (Sheldon *et al.* 2000b). Later work confirmed that *FLC* transcript abundance correlated directly with FLC protein abundance across a range of genetic, developmental, and environmental variables, indicating that *FLC* regulation operates at the transcriptional level (Rouse *et al.* 2002).

More recently, analysis of FLC activity has centred around two aspects: the regulation of *FLC* expression both by proteins within the floral repression pathway and by proteins in floral promotion pathways; and the regulation by FLC of genes such as the flowering-time integrators. The activities of floral repressors which enhance FLC function are discussed below, while the regulation of *FLC* expression by floral promotion pathways is discussed in Section 4.2, as well as in Chapters 5 and 6. Chapters 7 and 8 are primarily concerned with the activities of the flowering-time integrators and the floral meristem identity genes. However, it is worth mentioning here the way in which FLC may repress the activity of those flowering-time integrators. Hepworth *et al.* (2002) focused on the role of FLC in repression of *SOC1*, and identified a key region of 351 base pairs in the *SOC1* promoter. This region was essential for repression of *SOC1* expression by FLC, but also necessary for activation of *SOC1* expression by CONSTANS (CO, a promoter of flowering operating in the photoperiod pathway; see Chapter 5). FLC was shown to bind to this 351 bp region *in vitro*, probably at a CarG box site identified within the sequence. CarG boxes are the binding motifs usually recognized by MADS box transcription factors. Hepworth *et al.* (2002) proposed that FLC and CO compete for binding of similar regions within the *SOC1* promoter, and that it is only when insufficient FLC protein is present that sufficient CO-driven activation can occur to cause *SOC1* expression and subsequent function.

4.1.2 *FRI*

The main enhancer of FLC activity, FRI, was identified in 2000 by Johanson *et al.* It had long been known that the requirement for vernalization in Arabidopsis was attributable to a single locus, *FRI*. Where a dominant *FRI* allele is present, an Arabidopsis ecotype will flower only after a period of cold treatment (i.e., it is late flowering without cold) but where a recessive *fri* allele is present (as is the case with all lab lines) the Arabidopsis ecotype does not require vernalization and flowers early. *FRI* was isolated by map-based cloning, and found to encode a 609-amino acid protein with little sequence similarity to proteins from other species. The protein contains two coiled-coil domains, one near the N terminus and one near the C terminus, and it is possible that these may be involved in transcriptional regulation of *FLC*. Analysis of *FRI* loci in Arabidopsis ecotypes confirmed that most early flowering varieties contained mutations within *FRI* (Johanson *et al.* 2000). This result was confirmed by the later analysis of Le Corre *et al.* (2002), who sequenced *FRI* from 25 Western European ecotypes of Arabidopsis and concluded that recent positive selection for early flowering had occurred within a set of isolated populations, resulting in variation at the *FRI* locus. It had

previously been shown that FRI is necessary for *FLC* expression (Michaels and Amasino 1999), but in 2001 Michaels and Amasino also demonstrated that, in the absence of a functional *FLC* allele, FRI has no effect on flowering time in long or short days. The role of FRI as an enhancer of FLC activity was thus clearly established.

4.1.3 *FRIGIDA LIKE 1 (FRL1)*

Since mutagenesis has traditionally been conducted in lab lines of Arabidopsis, which all contain non-functional *fri* alleles, it is only since the discovery of the role of FRI that investigation into proteins which interact with it has begun. Michaels *et al.* (2004) conducted a mutant screen for early flowering plants in an Arabidopsis line containing a dominant FRI allele in a Columbia background. They identified the *frl1* mutant, and confirmed that it had no phenotypic effect in a normal Columbia (i.e., *fri* mutant) background. The *FRL1* gene was isolated using a T-DNA insertion as a tag, and found to encode a protein 23% identical to FRI. However, the two proteins share no large blocks of conserved sequence, lending no clues to conserved domains which might explain their functions. Analysis of FRL1 function in a number of backgrounds confirmed that its only role is to facilitate the FRI-mediated expression of *FLC*. In an *frl1* mutant background, FRI has no effect on *FLC* transcription. Analysis of the Arabidopsis genome sequence has revealed a further 5 genes with significant sequence similarity to *FRI* and *FRL1*, suggesting that *FLC* regulation may be more complex than originally thought (Michaels *et al.* 2004).

4.1.4 *EARLY IN SHORT DAYS 1 (ESD1)*

The *esd1* mutant causes early flowering, as a result of a reduction of *FLC* expression (Martin-Trillo *et al.* 2006). Analysis of *esd1* in conjunction with other mutants suggests that the ESD1 protein is a strong inducer of *FLC* expression, but may also act on other genes involved in the repression of flowering, as the *esd1/fca* double mutant is earlier flowering than the *fca* mutant alone. Map-based cloning revealed that *ESD1* encodes ACTIN-RELATED PROTEIN 6 (AtARP6), which is required for modification of *FLC* chromatin. These modifications, which include methylation and histone acetylation, are necessary to achieve high levels of *FLC* expression and subsequent repression of flowering (Martin-Trillo *et al.* 2006).

4.1.5 *PHOTOPERIOD INDEPENDENT EARLY FLOWERING 1 (PIE1)*

PIE1 was first described from a screen to identify new early flowering mutants (Noh and Amasino 2003). The *pie1* mutant flowers early in short days compared to wild type, and earlier still in long days, confirming its role in a non-photoperiod-dependent floral repression pathway. The gene was isolated using a T-DNA insertion in the mutant as a tag, and the very large (2055-amino acid) protein it encodes found to resemble chromatin remodelling proteins from vertebrates with roles in the activation of gene expression. Two nuclear localization signals make it very likely that the protein is nuclear localized and plays a role in chromatin remodelling and gene expression. On the basis of this characterization, the authors investigated a potential role for PIE1 in activation of *FLC* expression. They discovered that PIE1 does regulate *FLC* expression in the shoot apical meristem, but not in other tissues, such as the root. This function is consistent with the expression pattern of *PIE1*, which is restricted to the shoot apical meristem. It appears that PIE1 is involved in the activation of *FLC* expression by FRI, and delays flowering time through this enhancement of FLC activity (Noh and Amasino 2003).

4.1.6 *VERNALIZATION INDEPENDENCE* Genes and the PAF1 Complex

In 2002 Zhang and van Nocker reported the results of a screen for mutants which flowered early and had reduced levels of *FLC* expression. They called these new mutant lines the *vernalization independence* lines (*vip*), and they represent a class of mutants with perturbations in genes involved in *FLC* upregulation. The *VIP4* gene was isolated first, and found to encode an extremely hydrophilic protein with similarity to animal and yeast proteins involved in forming multiprotein complexes.

VIP4 is expressed throughout the plant, but is not downregulated after vernalization, confirming that it plays a role in an autonomous enhancement of *FLC* (and thus repression of flowering) pathway (Zhang and van Nocker 2002).

The following year the *VIP3* gene was isolated, by positional cloning, and found to encode a protein consisting almost entirely of repeated WD (tryptophan and aspartate) motifs. Such a structure is indicative of a protein which acts as a scaffold for the assembly of multiprotein complexes. Ectopic expression of *VIP3* from the CaMV 35S promoter was insufficient to activate *FLC* expression, even though *FLC* transcript levels are reduced in the *vip3* mutant. This suggests that VIP3 acts with other proteins to activate *FLC* expression, perhaps in a complex for which it provides the base (Zhang *et al.* 2003).

In 2004 Oh *et al.* (2004) reported that *VIP6* encodes a protein with significant similarity to a yeast protein, CTR9, which forms part of the polymerase II-associated factor 1 (PAF1) complex. The PAF1 complex comprises five proteins which interact with RNA polymerase II to regulate gene expression through the modification of chromatin and the elongation of transcripts. In the same year He *et al.* (2004) also described a mutation in a gene they called *EARLY FLOWERING 8 (ELF8)*. Loss of function of *ELF8* resulted in loss of *FLC* transcript and early flowering. The *ELF8* locus is in fact the same locus as the *VIP6* locus, and both groups noted the similarity of the protein product to the CTR9 component of the yeast PAF1 complex. Indeed, Oh *et al.* (2004) also noted that the protein encoded by *VIP4*, previously described in yeast as forming multiprotein complexes, is now also known to associate within the PAF1 complex. Another component of this complex is encoded by the *EARLY FLOWERING 7 (ELF7)* locus. The *elf7* mutant flowers early in both long and short days and contains no detectable *FLC* transcript (He *et al.* 2004). The *ELF7* gene encodes a protein with great similarity to the PAF1 component of the yeast PAF1 complex. These data provide considerable support for the presence in Arabidopsis of a multiprotein complex involved in regulating transcription of *FLC* through modification of chromatin.

4.2 The autonomous induction pathway

The autonomous induction (or promotion) pathway was first formally described by Maarten Koornneef and colleagues, who conducted a mutant screen for flowering-time loci and then used epistatic analysis to describe the pathways within which these loci operated (Koornneef *et al.* 1991, 1998). Genes that confer autonomous promotion of flowering in Arabidopsis were identified by the isolation of late flowering mutants that retain photoperiod sensitivity and respond strongly to vernalization. That is, the mutants flowered later than wild type plants in long days, but later still in short days, and their flowering could be promoted by cold treatment. Such late flowering mutants were clearly not deficient in photoperiod-induced flowering, nor in vernalization-induced flowering, but in some general regulator of floral induction. Koornneef's original analysis identified four genes in this autonomous flowering pathway, *FCA, FY, FPA*, and *FVE*. However, a further three genes are also considered to act in this pathway, namely *LUMINIDEPENDENS* (LD) (Lee *et al.* 1994), *FLK* (Lim *et al.* 2004), and *FLD* (Sanda and Amasino 1996). The proteins encoded by all of these genes

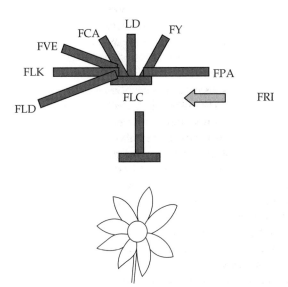

Figure 4.2 Diagram of the autonomous induction pathway. Proteins acting in the autonomous induction pathway repress FLC, eventually overcoming its repression of flowering.

have been shown to autonomously promote flowering in the same way—by repressing the FLC-based repression of flowering in the floral inhibition pathway (Fig. 4.2). Indeed, this repression appears to be the result of direct inhibition of FLC activity in all cases, as each protein operates, albeit in different ways, to prevent the accumulation of *FLC* mRNA in shoot apical meristem cells (Simpson 2004). Each mutant has increased levels of *FLC* transcript when compared to wild type plants (Michaels and Amasino 1999).

4.2.1 *FCA*

The best studied of the genes operating in the autonomous induction pathway is *FCA*. The *FCA* gene was isolated by positional cloning and encodes a 747-amino acid protein which interacts with RNA (Macknight *et al.* 1997). Specifically, FCA is classed as a plant-specific RNA-recognition motif-type RNA-binding-domain-containing protein, or RRM for short. In fact the protein contains two RRM domains, both in the N terminal region, and a C terminal WW domain, a region which is usually, and in this case has been confirmed to be, involved in protein–protein interactions. The WW domain is essential for FCA function (Simpson *et al.* 2003). The *FCA* transcript is found throughout the plant, although it is at highest abundance in areas of active cell division, including the root and shoot apical meristems (Simpson 2004). Within those cells the FCA protein is located in the nucleus, consistent with a role in RNA binding. Macknight *et al.* (1997) were able to confirm that FCA does bind RNA *in vitro*, with a preference for U- or G-rich sequences, but will not bind DNA. FCA is believed to down-regulate *FLC* by interfering with RNA processing, such that less *FLC* transcript survives and is available for translation. The *FCA* transcript has alternative polyadenylation sequences, which play an important role in its function (see below). This alternative polyadenylation is conserved in *FCA*-like sequences in *Brassica napus* and *Pisum sativum* (Macknight *et al.* 2002). Surprisingly, *FCA* transcript is found throughout the plant, and is present in particularly high levels not only in shoot meristems but also in the root apical meristem and root lateral meristems. Consistent with this expression pattern, *fca* mutant plants have shorter roots with 20% fewer lateral roots than wild type plants (Macknight *et al.* 2002). These data suggest that FCA plays other roles in plant development, besides that in promoting flowering. In 2006 Razem *et al.* demonstrated that FCA binds the plant growth regulator abscisic acid (ABA) and hypothesized that it behaves as an ABA receptor, explaining the pleiotropic phenotype of the *fca* mutant. FCA has long been known to activate responses over distances of some cells, as it has been shown not to act cell-autonomously. Furner *et al.* (1996) demonstrated that *fca* mutant sectors in a wild type plant flowered at the same time as wild type.

4.2.2 FY

The molecular identity of FY was established in 2003 as a result of a search for proteins that interacted with the WW domain of FCA. The gene encodes a protein with seven WD40 repeats and which shows significant sequence and structural similarity to a yeast protein which operates as part of a complex that cleaves and polyadenylates RNA (Simpson *et al.* 2003). This protein is very conserved throughout the angiosperms, with proteins from *Sorghum bicolor* and *Medicago truncatula* both showing very strong similarity. This is not surprising, as recent reports indicate that total loss of function of *FY* is lethal at the embryo stage (Henderson *et al.* 2006). Analysis of FY function indicated that it acts as a 3'-end RNA-processing factor, important at many stages of plant growth, and has two main functions in the induction of flowering. In conjunction with FCA it interferes with *FLC* RNA accumulation, and thus allows floral meristem conversion genes to be expressed. The importance of FY in this interaction is clear from transgenic experiments. Ectopic expression of *FCA* from the CaMV 35S promoter in Arabidopsis causes early flowering, but this construct has no effect in an *fy* mutant background (Simpson *et al.* 2003). Clearly FY protein is essential for the FCA-induced downregulation of *FLC*. At the same time, FY is necessary for regulation of FCA activity. The FCA/FY complex promotes premature polyadenylation within the third intron of *FCA* RNA, as opposed to the normal polyadenylation at the 3' end of the molecule. This results in the production of a truncated RNA molecule which encodes a short inactive protein (Quesada *et al.* 2003). This negative autoregulation of *FCA* was unexpected, and appears to be developmentally controlled, suggesting that at key times and places in the plant it may be possible to generate variable amounts of FCA protein. Certainly bypassing this autoregulation causes early flowering. The importance of FY in *FCA* autoregulation is seen by the great reduction of the amount of short *FCA* transcript present in the *fy* mutant background, compared to wild type (Simpson *et al.* 2003). Abscisic acid plays a key role in the association between FCA and FY, with ABA inhibiting FCA/FY binding and thus the formation of a functional complex, delaying the onset of flowering (Razem *et al.* 2006).

4.2.3 FPA

FPA is another gene encoding an RNA-binding protein, indicating that this mechanism of *FLC* regulation is a key feature of the autonomous promotion pathway. However, FPA does not interact directly with FCA or FY, and appears to regulate *FLC* independently. *FPA* was isolated by positional cloning and encodes another RRM-containing protein (Schomburg *et al.* 2001). Unlike FCA, FPA contains three RRMs and no apparent protein–protein interaction domain. The RRMs are all located in the N terminal region of what is a rather large protein (901 amino acids), and the structure and function of the rest of the protein have not yet been well characterized. *FPA* transcript is found throughout the plant at low levels, but is strongest in meristematic regions, particularly the inflorescence meristems (both apical and axillary). Ectopic expression of *FPA* in Arabidopsis using the CaMV 35S promoter caused plants to flower early in short days, and generated an almost day-neutral behaviour with respect to photoperiod (Schomburg *et al.* 2001).

4.2.4 FLK

The *FLK* gene was one of the last two genes in the autonomous promotion pathway to be identified, and was isolated, using the T-DNA in an insertion mutant as a tag, in 2004. The *flk* mutant flowers extremely late, and contains no detectable *FLK* transcript but a tenfold increase in *FLC* transcript compared to wild type plants. This elevated *FLC* expression results in a downregulation of the flowering-time integrators, *FT* and *SOC1*. The gene encodes a protein with three K-homology-type RNA-binding domains, very similar to those found in animal RNA-binding proteins (Lim *et al.* 2004). However, the rest of the protein bears no resemblance to anything found outside the plant kingdom. The protein is predominantly nuclear localized, in keeping with its function in RNA binding. It is clear that FLK functions to downregulate *FLC* through RNA interference, just as FCA/FY and FPA do, but these proteins are entirely dissimilar and presumably operate through different molecular mechanisms.

4.2.5 *FLD*

Although the four proteins described above all regulate FLC through RNA processing, that is not the only mechanism operating in the autonomous induction pathway. Two proteins, FLD and FVE, appear to regulate *FLC* through chromatin modification. The *fld* mutant is late flowering and has elevated levels of *FLC* transcript. However, in an *flc* mutant background the *fld* mutation has no effect on flowering time, confirming that the role of FLD is to suppress *FLC*. FLD was isolated using a map-based approach, and encodes a protein containing a domain involved in chromatin remodelling. The protein has significant sequence similarity to a human protein which functions as part of a histone deacetylase co-repressor complex (He *et al.* 2003). FLD has been shown to be necessary for the histone deacetylation of *FLC* chromatin, a process which plays an important role in reducing gene expression level, and histone H4 is hyperacetylated in the *FLC* region of the *fld* mutant. However, *FLD* expression is not sufficient to entirely eliminate *FLC* expression, merely to reduce it (He *et al.* 2003; Simpson 2004).

4.2.6 *FVE*

FVE also encodes a protein involved in chromatin structure and subsequent gene expression. Like

FLK, it was isolated in 2004, using a map-based approach, and has provided the last pieces of the autonomous induction pathway puzzle. The FVE protein contains six WD40 repeat domains, and shares sequence and structural elements with proteins from yeast and mammals that are present in complexes involved in chromatin modification. Like FLD, FVE has been shown to be involved in histone deacetylation of *FLC* chromatin, presumably modifying *FLC* transcript level in the process. *FLC* chromatin was found to have excess acetylated histones in the *fve* mutant, compared to wild type (Ausin *et al.* 2004).

4.2.7 *LUMINIDEPENDENS (LD)*

LD was the first gene of the autonomous induction pathway to be identified at the molecular level, although we still know little about how it functions. The *LD* gene was isolated using a T-DNA insert as a tag, and encodes a homeodomain protein, a structure usually associated with DNA binding and transcriptional regulation. The simplest role that can be hypothesiszd for LD from this sequence is a role in transcriptional repression of *FLC* (Lee *et al.* 1994). However, it has also been proposed that LD may function in RNA binding, rather than DNA binding. In rare cases the homeodomain motif has been shown to interact with RNA, and this mechanism would be consistent with the known importance of RNA processing in the autonomous pathway (Simpson 2004). The *LD* transcript is found throughout the plant, although it is most strongly expressed in shoot and root apical meristems and leaf primordia. An LD-GUS fusion protein was found to be nuclear localized, supporting a role for LD in transcriptional regulation (Aukerman *et al.* 1999).

4.3 Other endogenous factors which influence flowering time

While the two pathways described above are the central endogenous regulators of flowering time in Arabidopsis, many other factors, both directly genetic and otherwise, can also influence the floral transition. For example, in some species the floral transition will not occur until the plant is a certain size, irrespective of the nutrients it has received (Bernier and Perilleux 2005). Developmental factors such as these only really apply to longer lived plants, particularly perennials. However, even short-lived annuals like Arabidopsis undergo certain life stage transitions before flowering can occur. Loss of function of genes such as *EMBRYONIC FLOWER 1* (*EMF1*) eliminates rosette formation and transforms the apical meristem into an inflorescence meristem from the point of germination (Aubert *et al.* 2001). Although *EMF1* is not thought of as a flowering-time gene, because its role is clearly in controlling phase transition during shoot development, it is clear from the phenotype of the *emf1* mutant that other factors besides the inhibition and autonomous induction pathways are also necessary for appropriate development and accurate timing of flowering.

CHAPTER 5

The Photoperiodic Pathway of Floral Induction

PRESTON RIDGE LIBRARY
COLLIN COLLEGE
FRISCO, TX 75035

It is nearly a hundred years since the formal discovery that flowering time in many plant species was regulated by day length, and yet we have only recently begun to understand how this regulation occurs. In 1920 Garner and Allard published a categorization of plants into three types: those that flowered when days were longer than a critical period, those that flowered when days where shorter than a critical length, and those whose flowering did not appear to be affected by day length. The third group, day neutral plants, flower in response to triggers considered in Chapters 4 and 6. The remaining two groups, long day plants and short day plants, experience photoperiodic induction of flowering. In this chapter we will consider the photoperiodic regulation of flowering in Arabidopsis, which is a facultative long day plant (that is, one whose flowering is brought forward by long days, rather than one with an absolute requirement for them). Understanding the photoperiodic induction pathway requires an analysis of how light is perceived, how time is measured, and how the two signals are integrated to activate responses. In Chapter 7 we will contrast what we know of Arabidopsis floral induction with the systems in other species, particularly the short day plant rice.

5.1 Sensing daylight

Plants clearly have the ability to absorb light, because doing so is an essential component of photosynthesis. Leaves are adapted to function as solar panels, and thus make an ideal site for the perception of daylight. However, it is the presence of particular photoreceptors that allow plants to perceive the quality of daylight, and to convert that perception into signals. Extensive physiological work over many years has demonstrated that red light, far red light, and blue light are all important in photoperiodic induction, although their relative importance varies in different species and under different conditions.

5.1.1 Sensing red and far red light: phytochromes

Red (660 nm) and far red (735 nm) light are perceived by phytochromes, which are soluble cytosol proteins that form a dimer and have a tetrapyrrole chromophore covalently bonded to their N terminal photosensory domain (Fig. 5.1a). This chromophore, called phytochromobilin, selectively absorbs particular wavelengths of light, allowing the whole molecule (protein plus chromophore) to act as a photoreceptor and activate downstream responses. The C terminus of phytochrome contains two repeats of a motif called a PAS domain, which is believed to be involved in signal relay through protein–protein interactions with other molecules, including transcription factors (Yanovsky and Kay 2003). The C terminus also contains a histidine-kinase-related domain (HKRD) with serine/threonine kinase activity. The presence of this domain has led to the suggestion that phytochromes activate responses by phosphorylating their target proteins (Rubio and Deng 2005).

Traditionally, phytochrome function was described physiologically and was thought to depend on interconversion of the chromophore between two states according to the available light (Fig. 5.1b). The Pr form of the molecule was said to absorb red light, and be converted into the Pfr form.

(a)

(b)

Figure 5.1 Phytochromes. (a) The chromophore is bonded to the N terminal portion of the phytochrome protein. The PAS domains and the histidine kinase-related domain are towards the C terminus. (b) The chromophore of phytochrome takes one of two forms, depending on the light regime. These forms depend on the position of a single hydrogen atom (circled).

The Pfr form then absorbs far red light and is converted back to the Pr form. In the light, the transition between the two forms sets up a stable equilibrium, but in the dark Pfr declines progressively as the Pr form is more stable. The Pfr form was thought to be the active form, the one that causes things to happen in the plant, and so when it is absent, in the dark, the plant can no longer activate downstream light responses (Quail 2002). While this model still holds generally true for phytochrome function, more detailed analysis in Arabidopsis has revealed the presence of five different phytochrome molecules, each of which responds to light in subtly different ways. In particular, while the Pfr form of phytochrome has been shown to be the active form for four of the five molecules, it has been proposed that a short-lived intermediate generated during the switching between Pr and Pfr may be at least one active form

of phytochrome A (Shinomura *et al.* 2000; Wang and Deng 2003). Alternatively, it may be the case that continuous far red light is able to establish and maintain a small fraction of the phytochrome A population in the Pfr form over an unusually extended period (Quail 2002).

The five phytochrome proteins in Arabidopsis are encoded by five different genes. The genes are called *PHYTOCHROME A* to *E* (*PHYA–PHYE*), and the molecules they encode are referred to as phytochrome A, B, C, D, and E. A plant with a lesion in one of the *PHY* genes will still make the other four phytochromes and still be responsive to light. Analysis of the individual mutants has allowed dissection of the roles of the five different phytochromes, and it has become clear that the physiological model described above is oversimplified. Phytochromes B, D, and E all perceive red light, while phytochrome A distinguishes between far red light and darkness (Quail 2002). Some of the phytochrome molecules also seem to be more important than others in different responses by plants to light. Phytochrome A seems to be the most important photoreceptor in the photoperiodic regulation of Arabidopsis flowering.

The differences between the activities of the different phytochromes are attributable to differences in the protein structures and also to differences in their light lability. Phytochrome A is extremely light labile, being degraded within minutes of the light treatment needed to activate it (Nagy and Schafer 2002). It has been suggested that the phosphorylation status of the protein may influence both its lability and activity level. This lability may serve to ensure that plants do not respond with large-scale developmental changes (such as flowering) to very brief periods of illumination. Phytochrome A is also at its most active on exposure to far red light. In contrast, the other four phytochrome molecules are all light stable and are most active on exposure to red light (Yanovsky and Kay 2003). In addition, phytochrome A protein has a daily oscillation pattern when plants are grown in short days. In short day regimes phytochrome A protein levels peak in the night, but in long days they remain constant throughout the day (Mockler *et al.* 2003).

The importance of phytochrome A (and far red light) in the floral transition has been demonstrated using Arabidopsis plants exposed to varying light regimes. Wild type plants exposed to a normal short day (12 hours), followed by several hours of only far red light, flower early (at the same time as plants exposed to normal long days). However, *phyA* mutant plants flower late under this regime, as they are not able to perceive the far red light extension to the day. Exposure of wild type plants to short days followed by several hours of only red light results in late flowering (at the same time as plants exposed only to short days), indicating that far red light (and the phytochrome A which perceives it) is more important than red light (and the other phytochromes) in the flowering response. Similar experiments have shown that exposing plants to short days, but breaking up the long night with short periods of light (light breaks) results in early flowering. While all wavelengths of light are somewhat effective as light breaks, far red was shown to be the most effective of all (Goto *et al.* 1991; Putterill 2001). Far red light alone is insufficient to allow photosynthesis, but has been shown to be sufficient for flowering. Mockler *et al.* (2003) grew wild type and *phyA* mutant plants on nutrient media that met the plants' energy requirements in the absence of photosynthesis. Plants were kept in continuous far red light, and wild type plants flowered at the normal time, while *phyA* mutant plants were extremely late flowering. Clearly far red light and phytochrome A are essential for the floral transition.

The chromophores themselves are encoded by a different set of genes, called the *HY* genes. *HY1* and *HY2* are both necessary to make phytochromobilin (the chromophore, which absorbs light) and loss of function of either of these genes results in a plant lacking functional molecules of all the phytochromes because there are no chromophores to bond to the core proteins. These mutants are usually deficient in all phytochrome-mediated responses. Phytochromobilin is synthesized within the plastids and then released into the cytosol, where it is bonded to the main phytochrome proteins. *HY1* was isolated through positional cloning and shown to encode the enzyme heme oxygenase (Davis *et al.* 1999). Heme oxygenase converts the heme precursor of chromobilin into the intermediate biliverdin. *HY2*, isolated by map-based cloning,

encodes biliverdin reductase (otherwise known as phytochromobilin synthase), which catalyses the final step of chromophore synthesis (Kohchi *et al.* 2001).

The mode of action of active phytochromes is still not entirely understood, with both direct interaction with transcription factors and phosphorylation believed to be involved. However, it is clear that after the conversion of Pr to Pfr on exposure to light, all five phytochrome molecules move from the cytoplasm to the nucleus (Quail 2002). This transition is very rapid, occurring within only a few minutes of light treatment. Although it is not known how this transition occurs, it has been suggested that Pr is anchored somehow within the cytoplasm but that Pfr no longer interacts with the anchor and thus is able to move into the nucleus (Wang and Deng 2003).

The movement of phytochrome to the nucleus is likely to be crucial to its function. One known mode of rapid phytochrome action is through the interaction of its PAS domain with transcription factors such as the basic helix-loop-helix protein PIF3 (PHYTOCHROME INTERACTING 3). The active light-induced forms of PHYA and PHYB have been shown to bind PIF3 in yeast (although PHYB binds tenfold more strongly than PHYA; Ni *et al.* 1998; Zhu *et al.* 2000), and the PIF3 protein is localized to the nucleus. PIF3 is known to bind G-box motifs (CACGTG) in the promoters of other genes, and PHYB has been shown to bind to PIF3 after PIF3 is already bound to target promoters (Martinez-Garcia *et al.* 2000). These data suggest that PIF3 acts to recruit light-activated and nuclear-imported phytochrome proteins to the promoters of target genes, whose transcription is then directly activated. In support of this, an active PIF3 protein is necessary for PHYB-induced expression of CCA1 and LHY (genes active in the circadian clock, see below), genes which both contain G-box motifs in their promoters (Martinez-Garcia *et al.* 2000). Ectopic expression of *PIF3*, or its downregulation through antisense technology, results in perturbation of many photoresponses, including flowering time (Ni *et al.* 1998).

The influence of phytochromes on gene expression may operate through several different pathways and at several different rates. Microarray experiments have indicated that the expression profiles of 10% of all genes change in response to far red light perceived through phytochrome A. Of the genes that responded within one hour of treatment, 44% encoded transcription factors (Tepperman *et al.* 2001). It is unlikely that all these genes are regulated through a PIF3-mediated pathway, and, while some are likely to be activated as a result of downstream processes, it is also likely that some are regulated by, as yet unknown, rapid response systems.

Direct interactions with transcription factors are not the only way in which phytochrome proteins function. The active form of phytochrome A has also been shown to phosphorylate proteins, including autophosphorylation of itself. The active (Pfr) form of the molecule is over twice as effective as the Pr form at causing such phosphorylation (Fankhauser *et al.* 1999), confirming that light activation is important in this response. A particular target of phytochrome phosphorylation activity appears to be the PHYTOCHROME KINASE SUBSTRATE 1 protein (PKS1), initially isolated through its physical interaction with phytochrome A, but also shown to interact with phytochrome B. The PKS1 protein is localized to the plasma membrane (Lariguet *et al.* 2006), and its phosphorylation by active phytochromes confirms this pathway of phytochrome-mediated signalling (Fankhauser *et al.* 1999).

5.1.2 Sensing blue light: cryptochromes

Blue light is perceived through two entirely different sets of photoreceptors, the phototropins and the cryptochromes. The phototropins are not known to influence flowering time, but play roles in cellular responses such as chloroplast movement and stomatal aperture. However, the cryptochromes are very important for photoperiodic induction of flowering, as well as hypocotyl growth, and have been found in animals as well as plants. Sequence analysis suggests that the plant cryptochromes are ancient, and evolved shortly after the origin of eukaryotes. This ancestral cryptochrome was then lost in the animal lineage, but shortly after the animal–plant divergence the animals acquired a new cryptochrome, by modification of an existing photolyase (Cashmore *et al.* 1999).

There are two cryptochromes present in Arabidopsis, encoded by two different genes, *CRY1* and *CRY2*. *CRY1* was isolated first, from a T-DNA tagged line of the *cry1* mutant (originally known as *hy4*) (Ahmad and Cashmore 1993). The *CRY1* gene encodes a 681-amino acid protein with an N terminal domain very similar to that of the microbial enzyme DNA photolyase. DNA photolyases are enzymes activated by blue and UVA light to bind to and repair pyrimidine dimers caused by the exposure of DNA to UVB light (Eckardt 2003). However, the plant cryptochromes do not have photolyase activity. Instead, CRY1 associates with two different chromophores, flavin adenine dinucleotide (FAD) and pterin (5,10-methyltetrahydrofolate), which absorb blue light (Lin 2000). The C terminal domain of CRY1 (and CRY2) is also essential for protein function, although its role is less clear. However, there is speculation that the photochemical mechanism of signal capture and transfer by cryptochromes may involve a redox reaction (Gyula *et al.* 2003).

CRY2 (originally known as *FHA*) was isolated using the *CRY1* cDNA as a probe (Lin *et al.* 1998) and encodes a protein 51% identical to CRY1 in terms of amino acid sequence. However, there are clear differences between CRY1 and CRY2 function, probably as a result of their differential regulation. Both proteins are nuclear localized, but CRY2 has been shown to be the more crucial for the floral transition. CRY2 protein (but not *CRY2* mRNA) is downregulated in blue light, possibly as a result of light-induced conformational changes triggering protein degradation (Lin *et al.* 1998). The *cry2* mutant is late flowering in long days, while ectopic expression of *CRY2* (using the CaMV 35S promoter) results in early flowering in short days (Lin 2000).

Less is known about the signalling cascade downstream of cryptochromes than about that of phytochromes, but it is likely that there are some similarities. For example, Shalitin *et al.* (2002) showed that protein–protein interactions are important for CRY2 function, and that CRY2 undergoes a blue light-dependent phosphorylation, rather as PHYA does in far red light. They hypothesized that, in the dark, CRY2 is unphosphorylated, inactive, and stable. However, when blue light activates CRY2 it autophosphorylates, triggers responses, and activates its own degradation. The importance of phosphorylation in cryptochrome-mediated photoresponses was shown by an investigation of *cry2* mutant alleles that express full length (but apparently nonfunctional) CRY2 proteins. In all cases the absence of blue light-triggered responses was correlated with the absence of blue light-induced phosphorylation (Shalitin *et al.* 2003).

There is evidence that the C terminal domains of the cryptochromes cause constitutive responses which are downregulated by the N terminal domains unless they receive blue light illumination. Yang *et al.* (2000) showed that, in the absence of the N terminal domains, the C terminal domains of CRY1 and CRY2 are constitutively active, causing photomorphogenic responses (including early flowering) even in the dark. Presumably they are maintained in an inactive state by the N terminal domains in the dark, and only released into activity on blue light exposure. It has also been shown that CRY1 and CRY2 interact with COP1 (a downstream component of both phytochrome and cryptochrome signalling; see below) in the nucleus in the dark (Wang *et al.* 2001). These data have led to the suggestion that the perception of blue light by cryptochromes triggers degradation or deactivation of COP1, which eventually allows enhanced transcription of other genes (Gyula *et al.*, 2003).

The interaction of the cryptochromes and the various phytochromes ensures that day and night are correctly identified in all the different light regimes a plant may experience during its development. At least part of this interaction is physical, with phytochrome B and cryptochrome 2 physically interacting *in vitro* and co-localizing to speckles in the nucleus on exposure to light (Mas *et al.*, 2000).

5.1.3 Light-induced signals

The signal transduction pathways further downstream of the photoreceptors are still relatively poorly characterized, mostly because of their extreme complexity. However, we do know a considerable amount about one downstream pathway, operating through a protein called CONSTITUTIVE PHOTOMORPHOGENESIS 1 (COP1). A large class of mutants isolated from early genetic screens in Arabidopsis show similar or identical phenotypes, including anthocyanin accumulation, chloroplast development, epidermal cell differentiation, inhibition of hypocotyl elongation, and cotyledon opening, all in the dark (Yamamoto *et al.* 1998). These responses are normally restricted to light-grown plants, but are constitutive in this class of mutants. The proteins involved are referred to as COP/DET/FUS proteins, and they act as repressors of certain responses in the dark, but are themselves deactivated in the light. A large subset of these proteins are nuclear-localized components of a ubiquitin-protease pathway that targets degradation of downstream proteins, including a key transcription factor which activates expression of many light-regulated genes (reviewed in Quail 2002).

The best studied member of this group of genes is *COP1*, which encodes a protein with a RING-finger domain and WD40 repeats, characteristic of protein–protein interactions (Deng *et al.* 1992). The COP1 protein functions as an E3 ubiquitin-protein ligase, targeting proteins for degradation by binding them to ubiquitin (Quail 2002). Many of the other *COP/DET/FUS* genes encode subunits of an oligomeric complex involved in degradation of ubiquitinated proteins. So it appears that the role of COP1 is to target proteins for degradation, by ligating them to ubiquitin, and the other components of the pathway are involved in the actual degradation. The primary target of this complex is a transcription factor called HY5 (Oyama *et al.* 1997)). The *HY5* gene encodes a bZIP transcriptional activator that binds to the promoters of light responsive genes. However, HY5 protein is only present in the nucleus in the light, being virtually absent from the cell in the dark (a 15- to 20-fold difference in protein level has been measured; Osterlund *et al.* 2000). In the dark, COP1 has been shown to bind HY5, targeting it for degradation. However, in the light COP1 becomes significantly less abundant in the nucleus, allowing HY5 to function (Fig. 5.2). This light-induced change in COP1 nuclear abundance is mediated by the activities of at least PHYA, PHYB, CRY1, and CRY2, and requires the presence of a 110-amino acid domain in the COP1 protein called the cytoplasmic localization signal (CLS; Stacey *et al.* 1999). Analysis of protein function has shown that the CLS induces nuclear export of

(a) Events in the nucleus in the dark

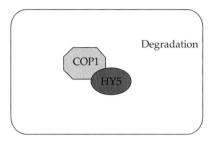

(b) Events in the nucleus in the light

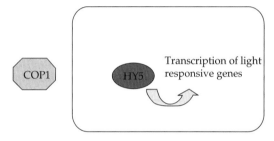

Figure 5.2 The COP1 protein mediates transcriptional responses to light. (a) In the dark COP1 binds to the HY5 transcriptional activator in the nucleus and targets it for degradation. (b) In the light COP1 is exported to the cytosol, allowing HY5 to activate transcription.

COP1, removing it from the nucleus in the light and preventing its activity (Subramanian *et al.* 2004).

Recent reports have suggested that a suite of proteins similar to HY5 may be the direct and immediate targets of COP1-regulated degradation. Yang *et al.* (2005) identified a basic helix-loop-helix transcription factor, HFR1, as physically interacting with COP1. HFR1 is short lived in darkness but stable in the light, and COP1 appears to be necessary for the degradation of HFR1. Other transcription factors may well be regulated through the same mechanisms.

5.2 Measuring time

As well as perceiving the quality of daylight available, photoperiodic induction of flowering depends on the ability of plants to measure time. All eukaryotes have biological clocks, although the first circadian rhythm recorded in any organism was the daily leaf movements of plants (leaves tend

to be fully open in the day but fold up to some extent at night; Barak *et al.* 2000). In order for such rhythms to be truly representative of a circadian clock, and not of tissue responses to immediate light or dark conditions, they must be maintained, if imperfectly, when the organism is kept in a constant light regime. Clock-based responses are also unaffected by changing temperature, unlike most biological processes. In all organisms where the clock has been analysed it is based on a transcription/translation feedback loop, although the actual proteins involved are different in plants and animals. The basic model is one in which a clock gene autoregulates itself at a speed which takes approximately 24 hours. The clock is entrained by daily light signals, to keep it running accurately. When organisms are kept in constant dark their clock usually drifts slightly away from the 24-hour optimum. In the model, light signals occurring at the transition from dark to light every morning ensure the transcription of a clock gene (although, since the clock is basically light independent, transcription should occur even in the absence of light). This is translated into a clock protein, which activates clock-controlled genes and is responsible for all the other clock-controlled responses. However, the protein also represses the transcription of its own gene, either directly or through activation of an inhibitor of its own transcription, or through repression of an activator. When protein levels fall below a certain threshold transcription can begin again.

The isolation of clock elements to test this model has involved extremely elegant genetic screens. For example, several clock genes were isolated using a screen which relied on the circadian expression patterns of certain genes. The promoter of one such gene, *CAB2*, was fused to the gene for firefly luciferase, and introduced into transgenic *Arabidopsis* plants. These transgenic plants were mutagenized and then the progeny monitored under timelapse video cameras. Wild type plants expressed luciferase (glowed) on a 24-hour cycle, but those with mutations in clock genes glowed on arrhythmic cycles (Millar *et al.* 1995). Several criteria which a gene must meet to be considered a core part of the clock were identified. The gene should have an expression pattern that cycles over 24 hours, it should control its own expression in a negative feedback loop, and, thirdly, by clamping the gene (fixing its expression at a high or low level) it should be possible to stop the clock.

The Arabidopsis circadian clock is based around the activities of three main proteins (Fig. 5.3). *CIR-CADIAN CLOCK ASSOCIATED 1* (*CCA1*) and *LATE ELONGATED HYPOCOTYL* (*LHY*) both encode transcription factors of the MYB family, although they are unusual in containing only a single copy of the MYB DNA-binding domain, whereas most plant MYB proteins contain two. Levels of transcript and protein of both *CCA1* and *LHY* circle on a circadian pattern, peaking at dawn. Ectopic expression of either gene from the CaMV 35S promoter downregulates the expression level of the other, and ruins the rhythmicity of the clock. The *cca1* mutant still retains rhythmic patterns of gene expression (Green and Tobin 1999), as does a loss-of-function *lhy* mutant (Mizoguchi *et al.* 2002), although in both mutants those rhythmic patterns deviate to some extent from wild type and both flower early in short days. However, a double mutant lacking functional copies of both genes is completely arrhythmic, flowering early in short days, showing random leaf movements, and having no rhythmic patterns of gene expression (Mizoguchi *et al.* 2002). These data suggest that the LHY and CCA1 proteins are partially redundant, overlapping in function incompletely.

The *TIMING OF CAB EXPRESSION 1* (*TOC1*) gene has an expression pattern that also cycles over 24 hours, but with a peak at dusk. Ectopic expression of *TOC1* causes arrhythmicity, while *toc1* mutants have shorter circadian patterns (typically 21 hours in constant light) and flower late in long days (Strayer *et al.* 2000). Expression levels of both *CCA1* and *LHY* are reduced in a *toc1* mutant background, compared to wild type. It has also been shown that the expression pattern of *EARLY FLOWERING 4* (*ELF4*) cycles with that of *TOC1*, but, since the gene encodes a protein of unknown function, it is not yet clear whether this is significant. *TOC1* was isolated by map-based cloning, and encodes a protein containing an N terminal motif very similar to the receiver domain of response regulators from two-component signal transduction systems (Strayer *et al.* 2000). However, the conserved aspartates that normally undergo

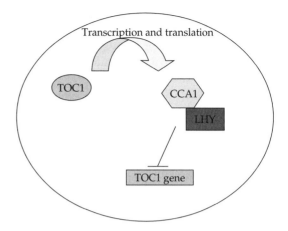

Figure 5.3 The circadian clock operates through a feedback loop based on three proteins. TOC1 activates expression of *CCA1* and *LHY*, which in turn repress the transcription of the *TOC1* gene. Several other proteins act to reinforce this autoregulatory loop.

phosphorylation in these systems are replaced by other amino acids in TOC1, so perhaps it has a different function. The TOC1 protein appears to be unable to undergo phosphorylation *in vitro*, leaving the function of the receiver-like domain unclear, and leading to the protein's current classification as a 'pseudo-response regulator' (Makino *et al.* 2000). The C terminus of TOC1 contains a domain very similar to that of the transcription factor CON-STANS (see below). Mas *et al.* (2003) showed that the gradual reduction in TOC1 protein levels after the dusk peak was caused not only by transcriptional repression by LHY and CCA1, but also by post-translational modification. The *ZEITLUPE* (*ZTL*) gene encodes a protein containing a number of domains, including an F-box motif, typical of proteins involved in ubiquitin ligase complexes (Somers *et al.* 2000). The *ztl* mutant has constitutive levels of TOC1 protein (although the mRNA expression pattern continues to cycle in a circadian manner). This excess of TOC1 protein is caused by the failure of TOC1 degradation in the dark, which requires a functional ZTL protein (Mas *et al.* 2003).

Putting these data together, a model has emerged in which TOC1 (and possibly several other proteins, including ELF4) upregulates expression of *LHY* and *CCA1*. The LHY and CCA1 proteins have been shown to bind to the promoter of the *TOC1* gene, where they repress its transcription. High constant levels of LHY or CCA1 (in transgenic plants) result in low constant levels of TOC1 protein. The *TOC1* promoter contains a 200 base pair region which is essential for rhythmicity, and it is to this that LHY and CCA1 have been shown to bind (Alabadi *et al.* 2001). The entire feedback loop, from TOC1 upregulation of *LHY* and *CCA1*, through LHY and CCA1 downregulation of *TOC1*, to the point where lack of TOC1 (assisted by ZTL-induced degradation in the dark) has resulted in sufficiently little LHY and CCA1 to allow *TOC1* transcription to begin again, takes approximately 24 hours to complete (Alabadi *et al.* 2001).

Correct rhythmicity of the clock is maintained by daily entrainment. Without exposure to a light burst each morning, the clock tends to run to slightly more than 24 hours. Somers *et al.* (1998b) showed that phytochrome A and phytochrome B are necessary to ensure red light signals entrain the clock, that cryptochrome 1 and phytochrome A are necessary for blue light entrainment, but that cryptochrome 2 does not seem to be involved in circadian clock input.

5.3 Integrating light and clock signals

An ability to perceive day length and an ability to follow a 24-hour pattern are not sufficient to explain the very precise photoperiodic induction requirements of many species. The two features must be integrated in such a way that the plant only flowers when the appropriate critical day length is achieved. Understanding the integration of light and clock signals has involved analysis of flowering-time mutants not perturbed in other circadian or light responses—that is, those not perturbed either in the clock itself or in the photoreceptors, both of which will alter flowering time.

It is now believed that signals from the circadian clock and the phytochromes are integrated by the *GIGANTEA (GI)* and *CONSTANS (CO)* genes. The *co* mutant was originally identified as late flowering in long days but flowering at the normal time in short days (Koornneef *et al.* 1991). *CO* was cloned using a map-based approach (Putterill *et al.* 1995). The gene encodes a protein with two zinc finger domains, commonly found in transcription factors, and also regions indicative of protein–protein interactions. It is expressed in both leaves and stems, at very low abundance. *CO* expression is regulated by the endogenous circadian clock on a 24-hour cycle that peaks during the night. So, in short days *CO* expression does not overlap with the period of daylight, but in longer days there is an overlap of *CO* expression and daylight in the evening (Suarez-Lopez *et al.* 2001). The CO protein directly causes floral induction by activating expression of *FT*, one of the flowering-time integrators (see Chapter 7) and a key necessary component of the floral transition. However, CO is only able to activate *FT* expression in daylight. This means that *CO* gene *expression* is regulated by the circadian clock, but CO protein *function* is controlled by the presence of daylight. Therefore CO protein can only activate *FT* expression if it is present during the evening, that is, in long days. In short days *CO* is never expressed at the time which would allow it to activate *FT*. In support of this model, ectopic expression of *CO* from the CaMV 35S promoter does cause increased *FT* expression, but only during the hours of daylight. Although *CO* transcript is equally abundant as a result of this constitutive promoter in both light and dark, it is only during the light period that *FT* expression is upregulated, causing the floral transition to occur (Valverde *et al.* 2004). This elegant model of integration has finally resolved the question of how photoperiod induces flowering, some 85 years after the phenomenon was formally described (Fig. 5.4, Yanovsky and Kay 2003).

The role of light in allowing CO to activate *FT* expression was investigated by Valverde *et al.* (2004). Firstly, they showed that blue light treatment increased *FT* transcript level in an ectopically expressing *CO* background, far red light had some effect (but less than blue), and red light did not enhance *FT* transcription. They also showed that the *cry1*, *cry2*, and *phyA* mutants had less *FT* transcript than wild type plants, confirming that both cryptochromes and phytochrome A are necessary to allow far red and blue light to initiate CO activation of *FT* expression. The key role that light plays in CO activity is in stabilizing the protein. The CO protein produced in the evening is stable, but CO protein produced during the hours of darkness is rapidly degraded. Application of proteasome inhibitors was shown to stabilize CO protein

Figure 5.4 The *CONSTANS* gene (above) is expressed in a circadian pattern. It is only able to activate *FT* expression (below) when its expression coincides with daylight, that is, in long days (right hand panel). Redrawn from Yanovsky and Kay (2003).

during dark periods, suggesting that the protein is normally degraded by the proteasome unless light signals (operating through the mechanisms described earlier) prevent this degradation (Valverde *et al.*, 2004).

A number of genes have been predicted to play roles upstream of *CONSTANS*, moderating *CO* expression and thus regulating the interaction between clock-induced *CO* expression and light-controlled CO activity. The *GI* gene, for example, is considered to act upstream of *CO* and to be important for promoting *CO* transcription (Putterill *et al.* 2004). The *gi* mutant flowers late in long days and has reduced levels of *CO* expression (Simpson 2003). *GI* was isolated by two groups, one using a T-DNA tagging approach and the other a

map-based approach. The gene encodes a large novel protein, which was initially predicted to contain multiple membrane-spanning domains (Fowler *et al.* 1999; Park *et al.* 1999). However, the GI protein was subsequently shown to be nuclear localized when fused to green fluorescent protein in both transient assays (Huq *et al.* 2000) and stable expression (Mizoguchi *et al.* 2005), leaving doubt as to its method of functioning. However the protein functions, it is clear that the gene's expression pattern is circadian-regulated, peaking 8–10 hours after dawn, and that ectopic expression of *CO* is able to bypass the *gi* mutant phenotype (Fowler *et al.* 1999). These data strongly support a role for GI upstream of *CO*, enhancing its transcription. However, the observation that circadian rhythmicity of gene expression patterns is altered in plants ectopically expressing *GI* or containing a *gi* mutant allele suggests that GI also feeds back into the central circadian oscillator (Mizoguchi *et al.* 2005). In contrast, the *EARLY FLOWERING 3* (*ELF3*) gene is also predicted to act upstream of *CO*, but in this case as a negative regulator, ensuring *CO* expression is not too active. In support of this hypothesis, the *elf3* mutant has increased levels of *CO* transcript (Putterill *et al.* 2004). The ELF3 protein has also been predicted to play a role in light entrainment of the circadian clock, as the *elf3* mutant retains rhythmicity in constant dark but loses it in constant light (Hicks *et al.* 1996). Positional cloning of *ELF3* revealed that it encodes a putative transcriptional regulator, with a potential nuclear targeting signal, suggesting that it might regulate *CO* expression through a direct interaction with the *CO* promoter (Hicks *et al.* 2001). It is likely that other genes will also be shown to act as intermediates between *CO* expression and CO function.

CHAPTER 6

The Vernalization Pathway of Floral Induction and the Role of Gibberellin

Day length is not the only environmental signal used by plants to optimize the timing of flower production. Temperature is also a factor, and particularly vernalization, exposure to a prolonged period of cold. This is a sensible adaptation of plants in temperate climates to prevent the risk of autumn germination leading to flowering in winter, a time when pollinators are scarce and freezing temperatures may inhibit seed production. At the same time, the plant growth regulator gibberellin has been shown to be a powerful promoter of flowering in many species. For some time there was speculation that the vernalization-induced flower induction pathway operated through gibberellins, but we now know, in Arabidopsis at least, that vernalization and gibberellin represent two independent pathways which can both induce flowering. Chandler *et al.* (2000) showed that vernalization was not affected in mutant backgrounds which failed to make or perceive gibberellins, as well as those defective in abscisic acid and phytochrome B signalling. In this chapter we will consider the two pathways separately, before looking at the evidence that they overlap. In Chapter 7 we will integrate all the flowering-time pathways discussed in Chapters 4, 5, and 6 to produce a holistic model of the Arabidopsis floral transition.

6.1 The vernalization promotion pathway

Photoperiodic induction is the most commonly cited and best studied example of the environmental regulation of flowering. However, flowering is likely to be accelerated by any condition that reliably indicates the passage of winter and the onset of more favourable conditions for reproduction. Although day length is one such reliable indicator, the end of a long period of cold weather is another good sign that winter has passed. Many plants flower in response to a period of between one and three months of exposure to temperatures between 1 and 10°C. The precise cold requirements vary from species to species and from plant accession to plant accession, and can be extremely specific. This strategy is adopted by many wild species, and has also been bred into many of our crop varieties, to produce lines which will overwinter in the vegetative state and flower in the spring or summer (Fig. 6.1, Simpson and Dean 2002). Such cold-induced induction of flowering is known as vernalization (from the Latin *vernus*, meaning 'of the spring'; Sung and Amasino 2004b).

Vernalization involves two separate processes. First, the plant must perceive the prolonged period of cold temperature. Then, it must 'remember' that perception in order to induce flowering later in the spring or summer. Very few plants flower immediately on release from the cold period, so it is clear that the 'memory' of winter is retained over many weeks and months. Analysis of vernalization in Arabidopsis has been broken down into analysis of cold perception and analysis of the maintenance of the vernalized state. While the mechanism by which a plant 'remembered' cold seemed initially the most challenging problem to solve, it has been the perception of vernalization that has proved more difficult to analyse using conventional genetic approaches.

Winter wheat

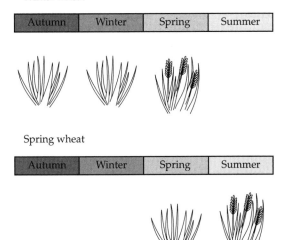

Spring wheat

Figure 6.1 Vernalization sensitivity is used in agriculture to increase the number of harvests possible in a year. Vernalization-sensitive winter wheat (top) is sown in autumn, germinates and remains vegetative in the winter, and then flowers and sets seed in spring. Vernalization-insensitive spring wheat (below) is sown in spring and germinates, grows and produces a crop in the summer, without the requirement for a period of cold.

It has long been known that prolonged periods of cold are perceived by cells in the shoot apical meristem and root apical meristem of Arabidopsis. This perception can be passed onto new daughter cells through mitotic divisions, long after the cold period has passed, but cannot pass through meiosis. This is a logical system that ensures that the next generation of plants starts with a fresh slate and no 'memory' of the winter their parents experienced. Extensive experiments with *in vitro* regeneration of various vernalized tissues of *Lunaria biennis* (honesty, a member of the Brassicaceae) revealed that only tissues containing actively dividing cells regenerated into plants that retained a memory of vernalization (Wellensiek 1962). It seems necessary for cells to be dividing, or at least undergoing DNA replication, for vernalization to be successful (Sung and Amasino 2004b). Since the apical meristems are the source of the most active cell division within the plant, it is hardly surprising that vernalization seems to operate through these tissues.

Initial analysis of cold-induced flowering in Arabidopsis established that a requirement for vernalization segregated as a monogenic trait, which mapped to the *FRIGIDA* (*FRI*) locus (Simpson and Dean 2002). *FRI* was introduced in Chapter 4. The main function of the protein is to enhance expression of *FLC*, the main repressor of flowering in Arabidopsis. FLC represses flowering by downregulating the flowering-time integrator genes, such as *FT*. Having established that many Arabidopsis plants with a summer annual habit (i.e., those that do not need a cold period to induce flowering) contain mutations in *FRI* or *FLC*, attention focused on the mechanism of vernalization interaction with these genes (Simpson and Dean 2002). The first intron of *FLC* is necessary for maintenance of *FLC* repression after cold treatment, but other regions of the gene are needed for the initial downregulation, confirming the hypothesis that cold repression and maintenance of that repression are two different processes (Sheldon *et al*. 2002). Vernalization results in a reduction in *FLC* transcript levels in the plant, and this reduced expression is maintained after plants are transferred to warmer environments. Increasing periods of cold exposure result in increasing downregulation of *FLC*, a quantitative effect which appears to be related to the extent of the initial downregulation of *FLC*, rather than the security of maintenance of the repressed state (Sheldon *et al*. 2006). This maintenance of the *FLC* repressed state, and the temporal separation between the timing of cold treatment and the onset of flowering, led to the hypothesis that vernalization operated through epigenetic control of *FLC* expression.

In support of the suggestion of an epigenetic method of response to cold treatment, prolonged growth in cold temperatures was shown to reduce methylation level across the genome. It was further shown that genome-wide demethylation, whether induced by chemical means, through the use of transgenes, or in mutant plants, promoted earlier flowering in Arabidopsis (Finnegan *et al*. 1998; Sheldon *et al*. 2000a). These data seemed to contradict the idea that vernalization induced an epigenetic repression of *FLC* transcription, until Sheldon *et al*. (1999) showed that plants expressing an antisense *METHYLTRANSFERASE 1* gene, and with only 15% of the wild type level of genomic methylation, flowered early and had reduced *FLC* transcript levels. Since demethylation usually allows transcription, rather than represses it, the authors concluded that methylation inhibits a repressor of *FLC* expression, and that such methylation is reduced by venalization. More recently, Finnegan *et al.* (2005) conducted a detailed analysis of the effects of demethylation and vernalization on the *FLC* locus. They showed that both processes resulted in reduced acetylation of histones H3 and H4 in the promoter of *FLC* and in neighbouring genes. However, DNA demethylation also resulted in repression of other MADS box genes with roles in flowering, lending support to the idea that methylation inhibits a repressor of MADS box genes in general, and may therefore affect flower induction in a rather indirect way.

Vernalization is conceptually and mechanistically different from cold acclimation, where short periods of cold temperatures induce responses that prepare plants for freezing temperatures. It is essential that these two processes are discrete, to ensure that plants are not tricked into flowering by short cold spells in autumn. However, both cold acclimation and vernalization must occur in temperatures above freezing, as both require metabolic activity severely inhibited by subzero temperatures (Sung and Amasino 2004b). Given the rapid temperature changes experienced by plants in many habitats, cold acclimation can be induced by only a short period of chilling, with one day sufficient for Arabidopsis. In contrast, vernalization of Arabidopsis is optimal after 30–40 days of continuous cold, a period of time sufficient to overcome the FRI-induced enhancement of *FLC* expression (Lee and Amasino 1995). Cold acclimation may operate through changes in membrane fluidity in different temperatures, altered calcium ion fluxes, and protein phosphorylation, although none of these processes are sufficiently stable to induce the memory of winter that is part of vernalization (Sung and Amasino 2004b). Further evidence for the mechanistic differences between cold acclimation and vernalization comes from the perfectly normal expression of and response to cold acclimation-induced genes shown to occur in mutants compromised in vernalization (*vrn1* and *vrn2*, see below). Liu *et al.* (2002) also showed that ectopic expression of *CBF1*, the first transcription factor gene whose

expression is upregulated upon cold acclimation, has no effect on *FLC* transcript levels or on flowering time. Vernalization and cold acclimation do share some elements of a common pathway, even if the majority of their actions seem to be separate. The *HIGH EXPRESSION OF OSMOTICALLY RESPONSIVE GENES 1* (*HOS1*) gene is involved in both processes, with *hos1* mutants showing reduced cold acclimation and thus reduced freezing tolerance, reduced *CBF* expression, early flowering, and reduced *FLC* transcript levels (Lee *et al.* 2001). The *HOS1* gene encodes a RING-finger protein with domains suggestive of a role in ubiquitin-mediated protein degradation. This protein is translocated from the cytoplasm to the nucleus in response to exposure to cold (Lee *et al.* 2001).

Analysis of the memory aspect of vernalization has been enhanced by the identification of two mutants, the *vernalization* (*vrn*) mutants, in which flowering is not induced in response to a prolonged period of cold. The *vrn1* mutant experiences reduction in *FLC* expression during its exposure to cold, but on return to warmer conditions *FLC* transcript level recovers and flowering is not induced (Levy *et al.* 2002). This suggests that the mutant can perceive cold but then fails to retain a memory of it. *VRN1* was isolated by map-based cloning and encodes a protein containing two B3 DNA-binding domains and a nuclear localization signal, that is, a protein with several hallmarks of a transcription factor. GFP fusion experiments confirmed that VRN1 is nuclear localized, supporting the idea that it is involved in transcriptional regulation. Levy *et al.* (2002) showed that the expression levels of *VRN1* mRNA did not change on exposure to cold temperatures, confirming that the gene plays a role in maintenance of the vernalized state, rather than induction of it. Ectopic expression of *VRN1* from the CaMV 35S promoter induced early flowering without vernalization, and even earlier flowering still with a cold treatment. Finally, Levy *et al.* (2002) showed that VRN1 protein binds DNA, including *FLC*, *in vitro*, and proposed that VRN1 binds to the *FLC* locus *in vivo*, targeting complexes (possibly including VRN2, see below) to it which cause methylation and subsequent inactivation of the gene.

The *vrn2* mutant, which is vernalization-insensitive, also experiences cold-dependent

Plate 1.

Plate 2.

Plate 3.

Plate 4.

Plate 5.

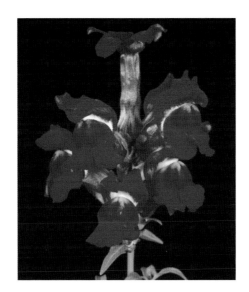

Plate 6.

(a)

(b)

(c)

(d)

(e)

(f)

Plate 7.

(a)

(b)

(c)

Plate 8.

(a)

(b)

Plate 9.

reduction in *FLC* transcript levels, but return of *FLC* expression on increasing temperature (Gendall *et al.* 2001). *VRN2* was isolated by map-based cloning and encodes a protein with a nuclear localization signal and several features indicative of a member of the polycomb group of genes. Polycomb group genes regulate gene transcription through modifying chromatin structure, and are active in both plant and animal development (Gendall *et al.* 2001; Chanvivattana *et al.* 2004). Just like *VRN1*, *VRN2* expression pattern is not affected by vernalization, and the protein is nuclear localized at all times. VRN2 is therefore predicted to function by maintaining the chromatin of the *FLC* gene (to which it may be targeted by VRN1) in a state incompatible with transcription. VRN2 has also been shown to interact physically with other polycomb group proteins, which may function in a complex to inhibit *FLC* expression (Chanvivattana *et al.* 2004).

The observation that *VRN1* and *VRN2* are expressed even in the absence of vernalization indicated that they could not be responsible for the cold-induced repression of *FLC* expression, only its maintenance following the action of some other protein. In 2004, Sung and Amasino reported the identification of VERNALIZATION INSENSITIVE 3 (VIN3) as the solution to this mystery. In the *vin3* mutant repression of *FLC* expression does not occur in response to vernalization, and there is thus no cold effect for the plant to remember. *VIN3* expression is only induced on exposure to a long period of cold temperature, and, as *VIN3* transcript levels increase in the plant, *FLC* transcript levels decrease. *VIN3* expression ceases after the plant is returned to warmer temperatures, becoming undetectable within 3 days of warmth, at which point it is assumed that VRN1 and VRN2 take over the maintenance of the induced state. *VIN3* was isolated by map-based cloning and encodes a plant homeodomain-finger-containing protein (PHD protein), a structure usually involved in chromatin remodelling (Sung and Amasino 2004a). Indeed, during vernalization, acetylation levels of *FLC* chromatin decrease in response to VIN3 activity, and this deacetylation is followed by methylation. In the *vin3* mutant neither deacetylation nor methylation of *FLC* occurs, whereas the *vrn1* and *vrn2* mutant

lines experience deacetylation of *FLC* but reduced methylation (Sung and Amasino 2004a). It has recently been shown that the methylation occurs at lysines 9 and 27 on histone H3. In the *vrn1* mutant methylation of lysine 9 is lost, while in the *vrn2* background neither lysine is methylated (Bastow *et al.* 2004).

Like VRN1 and VRN2, LIKE HETEROCHROMATIN PROTEIN 1 (LHP1) is also necessary to maintain repression of *FLC* after exposure to cold. LHP1 is very similar to animal and yeast proteins that maintain chromatin in particular states. LHP1 protein is enriched in the chromatin of *FLC* after exposure to cold, and remains in position after the plant is returned to warm temperatures. The involvement of LHP1 as well as polycomb group proteins in the flower induction process confirms that the epigenetic control of *FLC* operates through mechanisms very similar to those present in animals and fungi (Mylne *et al.* 2006; Sung *et al.* 2006).

These experiments have allowed us to build up a complete picture of the vernalization induction of flowering in Arabidopsis. During a prolonged cold treatment, *VIN3* expression is initiated, and the VIN3 protein causes deacetylation of *FLC* chromatin. VIN3 protein is associated with VRN2 in a multiprotein complex which may target *FLC* chromatin for repression (Wood *et al.* 2006). VRN1 and VRN2, along with LHP1, are responsible for methylation of the modified *FLC* gene, preventing its transcription, and this methylation is stable on return to warmer temperatures, providing the 'memory' of winter (see gene expression patterns in Fig. 6.2). In the absence of *FLC* expression the flowering-time integrators can be expressed and flowering can be induced.

It is worth noting that a second, poorly characterized, vernalization induction pathway operates in Arabidopsis. A different group of polycomb proteins have been shown to repress the activity of a different MADS box gene, *AGL19*. AGL19 induces expression of floral meristem identity genes (Chapter 8), bypassing both FLC and the flowering-time integrators it represses. Although this pathway has been little studied, recent data suggest that VIN3 may play a role in maintaining the repressed state of *AGL19* (Schoenrock *et al.* 2006).

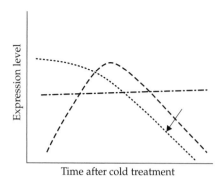

Figure 6.2 Graph showing expression levels of various genes following vernalization. *FLC* transcript levels (dotted line) start high, repressing flowering. *VIN3* (dashed line) is induced by cold treatment, and represses *FLC* expression, before expression levels decline again. *VRN1* and *VRN2* (dots and dashes) expression is unaffected by vernalization. However, the VRN1 and VRN2 proteins (arrow) act to maintain the repressed state of the *FLC* locus.

6.2 The gibberellin promotion pathway

Almost all of the classic plant growth regulators have been linked with the floral transition. There is evidence from a range of species that, under appropriate conditions, each of gibberellin, auxin, cytokinin, and brassinosteroids can influence flowering time. This is hardly surprising, as levels of each of the major plant growth regulators (PGRs) are affected by environmental conditions, including nutrient and water availability and light quality. General changes to plant development as a result of those environmental stimuli may include a direct effect on flower induction, or an indirect one operating through changes in source/sink relationships of various tissues. However, within Arabidopsis at least there is evidence to suggest that gibberellins directly promote flower induction, and that that promotion operates through a pathway genetically distinct from the autonomous, photoperiodic, and vernalization induction pathways discussed previously.

Gibberellins are tetracyclic diterpenoids, and 126 different forms of gibberellin have been described from plants, fungi, and bacteria (Gomi and Matsuoka 2003). They are synthesized in differentiating chloroplasts, and the biosynthetic pathway leading to their production has been characterized through biochemical and molecular genetic methods. The primary signal that high gibberellin levels send to the plant is one of good conditions. In

high-quality light, and with sufficient water and nutrients available, young leaves develop rapidly and more chloroplasts differentiate, leading to increased gibberellin levels. Gibberellins are perceived by soluble receptors (AtGID1a, AtGID1b, and AtGID1c in Arabidopsis) which also interact with DELLA proteins (Nakajima *et al.* 2006). DELLA proteins are negative regulators of signalling, which are repressed in the presence of gibberellin, allowing signalling to occur.

The first piece of evidence in support of a gibberellin induction pathway is the observation that treating Arabidopsis plants with exogenous gibberellins promotes earlier flowering. In fact, application of gibberellin seems to promote the floral transition in all rosette plants investigated, but has a different effect in non-rosette plants. Plants with elongated stems at the vegetative stage seem to experience further stem elongation when treated with gibberellins, but do not flower early (Corbesier and Coupland 2005). The molecular mechanism of action of exogenous gibberellin is as yet unclear, but there is evidence that gibberellin application may induce expression of *LEAFY* (a flowering-time integrator and floral meristem identity gene; see Chapters 7 and 8). Blazquez *et al.* (1998) showed that growth on 1% sucrose enhanced expression of a GUS reporter gene fused to the promoter of *LEAFY*. Although treatment with gibberellin alone had no obvious effect on reporter gene activity, the combination of gibberellin and 1% sucrose resulted in much stronger (more than double) reporter gene expression than was caused by sucrose alone. Eriksson *et al.* (2006) further showed that GA_4 is the active gibberellin which induces upregulation of *LEAFY* and therefore flowering, and that its concentration at the shoot apical meristem increases shortly before the floral transition occurs. It has also been shown that gibberellin promotes flowering in short days through a GAMYB-dependent activation of *LEAFY*. The gibberellin responsive MYB transcription factor, AtMYB33, is necessary to allow gibberellin to activate *LEAFY* expression, and is presumably responsible for direct transcription of *LEAFY* in response to gibberellin signals. Appropriate *LFY* expression is dependent upon a MYB recognition motif within the *LFY* promoter and this motif is bound by AtMYB33 (Blazquez *et al.* 1998; Blazquez and

Weigel 2000; Gocal *et al.* 2001). It is also known that DELLA proteins act as the molecular switches through which gibberellin signalling occurs, and so are presumably important in the flowering response. DELLA proteins are named for a group of amino acids which make up a conserved motif within their sequence, and they act as repressors of growth. DELLA proteins bind to gibberellin and its receptors, and are then degraded (Gomi and Matsuoka 2003; Nakajima *et al.* 2006). Achard *et al.* (2006) showed that DELLA proteins delay the onset of flower production under high salt conditions, supporting a role for gibberellin in the general induction of flowering through the degradation of these repressors.

Whatever the details of gibberellin signalling, it is clear that gibberellin does not promote flowering through repression of *FLC*, unlike the autonomous and vernalization induction pathways (Fig. 6.3). *FLC* transcript levels are unchanged in response to gibberellin application and subsequent early flowering (Sheldon *et al.* 2000b), indicating that gibberellin operates downstream of FLC. In conjunction with the observed role of gibberellin in activating *LEAFY* expression, this has led to the proposition that gibberellin is able to bypass the autonomous inhibition pathway and induce flowering by direct activation of flowering-time integrators or even floral meristem identity genes. Such a direct response might be seen as a sensible response of a short-lived plant to the information that high levels of gibberellin provide—namely that the young leaves are vigorous and growing, and that there is plenty of light for photosynthesis. Since good conditions are by no means guaranteed in the life of a short-lived annual, it may be beneficial to be able to respond rapidly to such conditions when they arise.

Further evidence for the role of gibberellins in flower induction, and particularly of their independence from the other floral induction pathways, comes from mutant studies. The *spindly* (*spy*) mutant has constitutively active gibberellin signalling, and flowers early (Jacobsen and Olszewski 1993). The mutant is also resistant to inhibitors of gibberellin synthesis, and the gene encodes an enzyme involved in post-translational protein modification, which acts as a negative regulator of gibberellin signalling (Jacobsen *et al.* 1996;

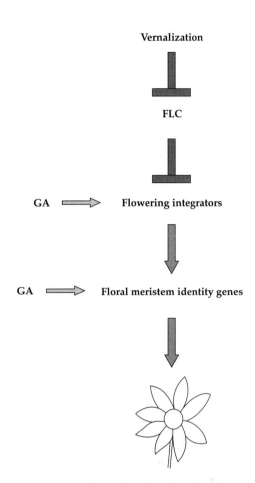

Figure 6.3 Despite early hypotheses that gibberellin and the vernalization promotion pathway were parts of the same process, it is now clear that they operate at different points in the pathway that represses flowering. Vernalization represses the activity of the floral repressor FLC, while gibberellin bypasses FLC and directly activates expression of the flowering integrators and floral meristem identity genes.

Thornton *et al.* 1999). Chandler *et al.* (2000) analysed double mutant plants containing lesions in the autonomous pathway promoter *FCA* and lesions in gibberellin synthesis or perception. Plants that are *fca* mutants have a strong vernalization requirement, as the autonomous promotion pathway is perturbed and *FLC* expression remains high. Double mutants between *fca* and *ga1*, and between *fca* and *gai*, were produced and analysed. The *GA1* gene encodes ent kaurane synthase, the enzyme required for the first committed step of gibberellin synthesis, and so *ga1* mutants produce no gibberellin. The *ga1* mutant is late flowering in long days, and rarely flowers at all in short days. However, the double mutants were still responsive to vernalization and photoperiod, confirming that gibberellin induces flowering through a separate pathway. The *GAI* gene is required for the perception of gibberellin, and *gai* mutants can grow on toxic levels of gibberellin. Again, *gai/fca* double mutants were responsive to vernalization and photoperiod, indicating that gibberellin promotes flowering through an independent pathway. While much remains to be discovered about the way in which gibberellin promotes the floral transition in Arabidopsis, these genetic studies make it clear that the gibberellin induction pathway must be thought of as distinct from the autonomous, photoperiodic, and vernalization pathways, and operates at a different level, bypassing FLC.

6.3 Does gibberellin act in the vernalization promotion pathway as well as independently?

The proposal that vernalization may promote flowering through gibberellin is a long-standing one in the literature, and should be considered distinct from analysis of the gibberellin promotion pathway. Evidence in favour of a role for gibberellin in the vernalization pathway was provided by Hazebroek *et al.* (1993), who observed an increase in the activity of an enzyme of gibberellin synthesis following vernalization. Using shoot apical meristems of *Thlaspi arvense* (pennycress, Brassicaceae, closely related to Arabidopsis) they measured an increase in the activity of ent kaurenoic acid hydroxylase after a prolonged cold treatment. This increase resulted in increased synthesis of a gibberellin, GA_9, albeit a less biologically active one than many. A model was proposed in which vernalization resulted in methylation changes to genes encoding enzymes of gibberellin biosynthesis (including ent kaurenoic acid hydroxylase), allowing enhanced gibberellin production and thus early flowering (Sheldon *et al.* 2000a). However, there is little other evidence to support a role for gibberellin in the vernalization promotion pathway, and the genetic data in Arabidopsis indicate quite clearly that the pathways function independently of one another.

CHAPTER 7

Integrating the *Arabidopsis thaliana* Flower Induction Pathways and Assessing the Extent to Which the Model Is Ubiquitous

In the previous three chapters we have discussed the various pathways which regulate the floral transition in Arabidopsis. Chapter 4 considered the floral inhibition pathway, which prevents the plant flowering until it has grown vegetatively and acquired sufficient photosynthate to fuel seed production. It also considered the autonomous induction pathway which ensures that the plant will flower eventually, even if it does not experience inductive environmental conditions. Chapter 5 discussed the photoperiodic promotion pathway, which encourages the floral transition in long days, indicative of summer and good growth conditions. In Chapter 6 we considered the influence of vernalization on flowering, a pathway which operates in ecotypes inhabiting more northerly latitudes, where floral induction occurs after winter has passed. We also discussed the ability of gibberellin, the plant growth regulator produced by young leaves in response to good conditions, to bring forward the floral transition. An average Arabidopsis plant will experience inhibitory and promotive signals from each of these pathways to a greater or lesser extent. In the first part of this chapter we will investigate how the pathways are brought together to result in a single response, through the activities of the flowering-time integrator genes. In the second section we will look at the evidence that the same or similar pathways operate in other species, with particular emphasis on the short day plant rice, and Gregor Mendel's famous genetic model, the garden pea.

7.1 Integrating the flowering-time pathways

The various flowering-time pathways can be thought of as converging on a set of genes, known as the flowering-time integrators. It is tempting to think of the pathways converging on the activity of *FLC*, the floral repressor, but in fact FLC is situated immediately upstream of the point of convergence. As Fig. 7.1 shows, while several of the pathways do moderate *FLC* activity, others bypass it, and so the true point of convergence is at the genes which FLC itself downregulates. The identification of these genes, whose function or expression is regulated by more than one of the input pathways, was described by Simpson and Dean (2002) as 'a key step forward in our understanding of flowering time control'. In summary, the autonomous inhibition pathway culminates in FLC repression of the flowering-time integrators. The autonomous promotion pathway results in downregulation of *FLC*, through RNA binding and chromatin remodelling, and thus allows activation of the flowering-time integrators. The vernalization promotion pathway also operates through downregulation of *FLC* by VIN3, and its maintenance in a non-transcribed state by VRN1 and VRN2. This also allows induction of the flowering-time integrator genes. The photoperiodic promotion pathway bypasses FLC, and in response to circadian clock-regulated *CONSTANS (CO)* expression coinciding with the daylight-activated forms of phytochromes and

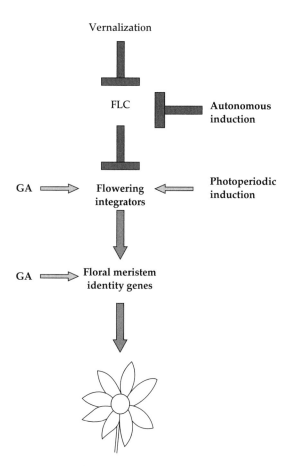

Figure 7.1 The four flowering induction pathways converge on the flowering-time integrators, either directly or through repression of FLC.

cryptochromes, the flowering-time integrator genes are directly activated. Gibberellin is also able to directly induce expression of flowering-time integrators, bypassing the FLC block. It can be seen, then, that the pathways all converge on the activity of a small subset of flowering-time integrator genes, either directly or through FLC. At any one time point these genes will be experiencing some degree of inhibition as a result of FLC action (varying in its strength according to the relative strength of signal from the autonomous inhibition and promotion pathways and the vernalization pathway) and some degree of activation as a result of gibberellin-induced signals and/or CO activity. When the activation signals are stronger than the FLC-based inhibition, the flowering-time

integrators will be expressed and the floral transition will occur.

7.1.1 Function of flowering-time integrators

It can be seen that the flowering-time integrators are critical to the Arabidopsis floral transition. There are three main genes described within this category, and we will discuss each in turn below. However, it is useful to note first that all act, either directly or indirectly, as transcriptional regulators, and they promote flowering by activating the expression of genes which operate at the meristem to commit it to the reproductive state. The activity of each of the flowering-time integrators is based around the expression of a subset of these 'floral meristem identity genes' (FMI genes). The FMI genes will be discussed in detail in Chapter 8, when we consider the physiological and molecular changes that occur at the meristem once the plant is committed to the reproductive state. Not only is the regulatory activity of each flowering-time integrator slightly different, the output pathways that lead into them vary slightly as well, as can be seen in Fig. 7.2 (Jack 2004).

7.1.1.1 FLOWERING LOCUS T (FT)

The *ft* mutant was one of the original late flowering lines identified by Koornneef *et al.* (1991). The mutant has normal expression of *LEAFY (LFY)*, but a double mutant between *lfy* and *ft* has substantially reduced expression of the floral meristem identity gene *APETALA1 (AP1)*. FT was isolated by two independent research groups in 1999. Kardailsky *et al.* (1999) generated activation tagged lines of plants, with an enhancer of transcription transformed randomly into the genome. One of the early flowering lines they selected contained an insertion that mapped to the same place as the *FT* locus, and they were able to use the insertion to isolate the gene. Kobayashi *et al.* (1999) also isolated the *FT* locus, this time using a T-DNA insertion line. The *FT* locus encodes a protein related to membrane-associated mammalian proteins that can bind hydrophobic ligands. It is predicted to play a role in signal transduction, and is over 50% identical at the amino acid level to the protein encoded by the floral meristem identity gene *TERMINAL*

Figure 7.2 The different floral induction pathways are integrated slightly differently by SOC1, FT, and LFY. Only SOC1 integrates all 4 pathways. LFY responds directly to GA, as well as indirectly to everything that regulates SOC1. FT integrates all pathways bar GA, to which it is not responsive (redrawn from Jack 2004).

FLOWER 1. Indeed, Hanzawa *et al.* (2005) showed that only a single amino acid controlled the difference between the protein functioning as FT or as TFL1, providing evidence for extreme evolutionary lability of this class of protein. The single amino acid is part of a region which may act as a molecular surface for interactions with other molecules, including the promoters of target genes (Ahn *et al.* 2006).

Kardailsky *et al.* (1999) conducted an analysis of the effects of ectopic expression of *FT* from the CaMV 35S promoter, and concluded that *FT*

expression was sufficient to induce flowering. The transgenic plants showed early and excessive expression of *AP1*, although only in the floral primordia, suggesting that *AP1* activation requires the presence of another protein not present in vegetative tissues. The transgenic plants also contained higher levels of *LFY* transcript than wild type, a surprising observation in light of the normal *LFY* expression seen in the *ft* mutant. Ectopic expression of *FT* was sufficient to rescue a *co* mutant, confirming the hypothesis that FT acts downstream of CO, integrating the floral induction pathways. Strikingly, the combination of ectopic expression of *FT* and ectopic expression of *LFY* resulted in plants that entirely bypassed the vegetative stage of development. These double transgenic lines produced a single terminal flower on germination, and produced no leaves at all, only the cotyledons formed during embryogenesis (Kardailsky *et al.* 1999).

FT expression itself is induced by the photoperiodic induction pathway, and repressed by FLC (and therefore indirectly induced by the vernalization and autonomous promotion pathways). The gibberellin promotion pathway does not seem to operate through FT. Evidence for the photoperiodic induction of *FT* was provided by Kobayashi *et al.* (1999), who fused a rat glucocorticoid receptor to *CO* and ectopically expressed the construct in Arabidopsis. Induction of CO activity by treatment with a steroid caused rapid induction of *FT* expression within 12 hours. As discussed in Chapter 5, CO is only able to induce *FT* expression in daylight, providing the photoperiodic control to the system (Valverde *et al.* 2004). At the same time *FT* receives inhibitory signals from FLC. This was best demonstrated by Hepworth *et al.* (2002), who showed that ectopic expression of *FLC* resulted in a total absence of *FT* transcript in transgenic plants. More recently, Helliwell *et al.* (2006) have shown that FLC protein binds to the first intron of *FT* as part of a multiprotein complex. This complex is assumed to function to inhibit *FT* transcription.

One of the most satisfying features of the *FT* gene is the recent demonstration that its protein can move from induced leaves to the shoot apical meristem, where it evokes the flowering response. This movement has been shown by a number of authors using a number of different techniques.

Corbesier *et al.* (2007) made transgenic Arabidopsis plants expressing an FT:GFP fusion protein from a promoter expressed only in phloem companion cells, and observed that, while mRNA was detected only in the companion cells, GFP was detectable in the vasculature, moving towards the shoot apical meristem. In a reverse experiment, Mathieu *et al.* (2007) fused FT to three copies of YFP (yellow fluorescent protein), making a protein too large to be trafficked between cells. Expression of this construct from the CaMV 35S promoter induced early flowering, as did expression of it from a shoot apical meristem specific promoter, but expression from a companion cell specific promoter did not induce early flowering, suggesting that immobilising the protein in the companion cells inhibited its normal function. Jaeger and Wigge (2007) further showed the role of FT protein movement in flower induction by attaching a nuclear localisation signal to the protein. Expression of this construct from the CaMV 35S promoter was sufficient to induce flowering, but expression from a companion cell specific promoter was not. Previous experiments by Huang *et al.* (2005) and others had shown that FT activates its own expression in the shoot apical meristem in a positive feedback loop, and that, once activated, expression of endogenous *FT* is stable. These data may finally shed some light on the identity of the mysterious "florigen" beloved of the early plant physiologists. *FT* is transcribed and translated in leaves in response to appropriate signals from the various flowering time pathways. The protein produced is mobile, travelling through the phloem to reach the shoot apical meristem. It should therefore also be graft-transmissible. Once the FT protein reaches the meristem it then induces expression of *FT* in those cells, and this expression is stable. These results show us how the evocation of flowering in the meristem can be caused by a mobile signal, and how such evocation can be stable once it occurs.

One further protein is required to allow FT to activate expression of the floral meristem identity genes. Recent reports have identified a bZIP transcription factor which interacts with the FT protein and is necessary for the floral transition to occur. *FD* was initially identified by Koorneef *et al.* (1991)

as necessary for the floral transition, but has been little studied since. In 2005, two groups reported that *FD* is expressed at the shoot apical meristem prior to the floral transition, encodes a bZIP transcription factor that interacts physically with FT protein, and is necessary for the induction of *AP1* expression by FT (Abe *et al.* 2005; Wigge *et al.* 2005). *FD* expression is upregulated once *FT* protein arrives at the meristem, an upregulation previously prevented by FLC (Searle *et al.* 2006). *FD* provides further support for the idea of a meristem 'waiting' for the arrival of *FT* mRNA and ready to be stably converted to the flowering state.

7.1.1.2 SUPPRESSION OF OVEREXPRESSION OF CONSTANS 1 (SOC1)/AGAMOUS-LIKE 20 (AGL20)

Like *FT*, *SOC1* was described by several different groups in the same year. It was given two different names, according to the method by which it was isolated. It will be referred to as *SOC1* throughout this work, as that is the name used most frequently in the literature. *SOC1* was analysed by Borner *et al.* (2000) because it encodes a MADS box transcription factor with an expression pattern indicative of a role in the floral transition. In particular, *SOC1* expression is induced in the shoot apical meristem as the floral transition occurs. However, *SOC1* expression is reduced in mutant lines with lesions in genes active in the photoperiodic and autonomous induction pathways. Borner *et al.* (2000) screened for and isolated an insertion mutant with an insertion in the *SOC1* locus, and described the phenotypic consequences of mutation of the gene. Loss of function mutants are late flowering, while ectopic expression of *SOC1* results in an early flowering phenotype. Ectopic expression of *SOC1* is sufficient to overcome mutations in the long day and autonomous flowering pathways (Mouradev *et al.* 2002). Expression of *SOC1* was also found to be regulated by gibberellin.

Also in 2000, Lee *et al.* identified *SOC1* from a screen for supressors of *FRI* activity. Using a transgenic approach involving the random insertion of transcriptional enhancers into the genome, they established that ectopic expression of *SOC1* was sufficient to induce early flowering. These authors showed that *SOC1* expression was positively regulated by the vernalization, autonomous, and photoperiodic flower induction pathways.

In the same year Samach *et al.* (2000) also isolated *SOC1* as a gene with a rapid increase in expression level following induction of CO activity using a steroid-activated form of CO in transgenic plants. *SOC1* transcript increased in abundance threefold within four hours of steroid treatment, and the authors then showed that *SOC1* was expressed at 10 times higher levels in plants ectopically expressing *CO* than in wild type plants.

SOC1 expression is induced by the photoperiodic pathway and by gibberellin, and is repressed by FLC (and therefore indirectly induced by the autonomous and vernalization promotion pathways). Unlike *FT*, then, *SOC1* is a true integrator of all of the flowering-time pathways. Gibberellin has been shown to directly enhance *SOC1* expression (Borner *et al.* 2000), while the photoperiodic pathway activates *SOC1* through the actions of CO (Samach *et al.* 2000). It is not yet known exactly how gibberellin achieves *SOC1* upregulation, but analysis of the *SOC1* promoter has indicated that overlapping promoter elements respond to FLC and CO signals, allowing repression and activation to occur at the same time, creating the necessary conflict at the flowering-time integrator (Hepworth *et al.* 2002). Within the *SOC1* promoter a MADS box binding motif was identified, and FLC was shown to bind this site *in vitro*. Mutation of this site *in planta* prevented FLC repression of *SOC1* expression. This particular region of the *SOC1* promoter was not bound by CO, but a nearby region responded to CO activity (Hepworth *et al.* 2002). It has also been shown that removal of FLC-based repression of *SOC1* is insufficient to result in strong expression of the gene. Moon *et al.* (2003) showed that vernalization (and thus reduction in FLC activity) was not sufficient to induce *SOC1* expression in a gibberellin-deficient mutant line (*ga1*, deficient in the first enzyme of gibberellin biosynthesis) grown in short days. Therefore, without a photoperiodic induction signal, gibberellin is necessary to induce *SOC1* expression.

7.1.1.3 *LEAFY* (*LFY*)

The third flowering-time integrator, *LFY*, is also described as a meristem identity gene, in part because it is believed to act downstream of SOC1. However, it can also respond directly to gibberellin signalling, and so is thought of as a flowering-time integrator too (Jack 2004). Through SOC1 *LFY* integrates signals from the photoperiodic, vernalization, and autonomous promotion pathways, but its expression is also directly induced by the application of gibberellin. Blazquez *et al.* (1998) showed that treatment with gibberellin alone had no effect on the activity of a reporter gene fused to the *LFY* promoter, but that the combination of gibberellin and 1% sucrose resulted in much stronger (more than double) reporter gene expression than was caused by sucrose alone. It has also been shown that gibberellin promotes flowering in short days through a GAMYB-dependent activation of *LFY*. The gibberellin responsive MYB transcription factor, AtMYB33, is necessary to allow gibberellin to activate *LFY* expression, and is presumably responsible for direct transcription of *LEAFY* in response to gibberellin signals. Appropriate *LFY* expression is dependent upon a MYB recognition motif within the *LFY* promoter and this motif is bound by AtMYB33 (Blazquez *et al.* 1998; Blazquez and Weigel 2000; Gocal *et al.* 2001). A separate promoter element is responsible for *LFY* induction in response to long days (i.e., the photoperiodic promotion pathway operating through SOC1; Blazquez and Weigel 2000).

The activity of *LFY* as a floral meristem identity gene will be discussed in more detail in Chapter 8.

These three flowering-time integrators may not represent the complete picture, and others may remain to be discovered. Evidence for the presence of other integrating genes comes from genetic analysis. Moon *et al.* (2005) showed that a triple mutant between *soc1*, *ft*, and *lfy* was still able to flower under long day conditions, although it flowered extremely late and the flowers it produced were developmentally abnormal. Further genetic screens for enhancers of the flowering-time integrator mutant phenotypes may identify additional integrators. However, in the meantime the flowering-time integrators provide a satisfying mechanism by which the signals from the different flowering-time pathways can be integrated and translated into the developmental response of flower production.

7.2 The Arabidopsis flowering-time model in other species

It is all too easy to forget that model species are only intended to provide the beginning of the answers to questions, rather than the full story. While we now have a good picture of how the floral transition is controlled in Arabidopsis, that does not mean that we understand the control of flowering in all plants. The next phase of flowering-time research will be to investigate the extent to which the Arabidopsis flowering-time model applies to other species, and what innovations other species have evolved to fit their floral transition to their environment. One particular question that must be addressed is the extent to which the long day photoperiodic induction pathway seen in Arabidopsis operates to control flowering in short day plants (i.e., those that flower when the days are shorter than a certain maximum length). It will also be important to assess the extent to which each of the other Arabidopsis pathways functions to induce flowering in day neutral plants. The best studied model aside from Arabidopsis is currently rice, which has been analysed both because of its economic importance and because it is a short day plant. We will also consider flowering time in pea, and touch briefly on other species being studied.

7.2.1 Flowering time in rice

Rice is one of the world's most important food crops, providing staple carbohydrate and protein for half the world's population. It is also a short day plant, which synchronizes its flowering and reproduction with the rainy season in its usual habitats (Putterill *et al.* 2004). Although it may seem strange for a wind-pollinated plant to flower under wet conditions, seed set is enhanced when sufficient water is available, and damp conditions are important for germination of rice seed. However, rice has no vernalization requirement, as it is adapted to life in warm environments. It is therefore of immense economic importance to establish how the short-day-induced flowering of rice occurs, since this is the only significant environmentally controlled promotive pathway it contains.

Analysis of the rice floral transition has primarily been conducted through mapping of quantitative trait loci, and, to a lesser extent, using a candidate gene approach to investigate the function of orthologues of Arabidopsis genes.

The first gene involved in the rice floral transition to be studied at the molecular level was isolated in 2000. The *PHOTOPERIODIC SENSITIVITY 5* (*SE5*) gene encodes a protein related to HY1 of Arabidopsis, and is needed for the synthesis of the chromophores on all phytochrome molecules (Izawa *et al.* 2000). The *se5* mutant flowers early in long days, at the same time as wild type plants in short days. These data confirm that phytochrome-induced signals in long days normally delay flowering in rice, whereas in Arabidopsis they activate it. However, analysis of the rice genome sequence suggests that the light receptors themselves differ slightly from those in Arabidopsis. Specifically, while orthologues of *PHYA*, *PHYB*, and *PHYC* are apparent in the rice genome, there is no sign of *PHYD* or *PHYE* orthologues. At the same time, three genes encoding cryptochromes appear to be present, compared to the two in Arabidopsis (Izawa *et al.* 2003). This suggests that detailed analysis of day length measurement in rice will reveal small differences to the ways in which different wavelengths of light are perceived in Arabidopsis.

Analysis of the rice genome has revealed that genes with very similar sequences to *CCA1* and *TOC1* are present, suggesting that the rice circadian clock may function through a mechanism essentially similar to the Arabidopsis one. The rice *CCA1* orthologue was shown to have a circadian-regulated expression pattern similar to that of Arabidopsis *CCA1* (Izawa *et al.* 2002).

Many of the other components of the Arabidopsis photoperiodic induction pathway also seem to be present in rice, although functioning in slightly different ways. A rice orthologue of *GIGANTEA*, *OsGI*, was identified by analysis of genes differentially regulated in wild type and *se5* mutant plants (Hayama *et al.* 2002). The protein encoded by *OSGI* is very similar to that encoded by *GI*, and it appears that the proteins fulfil similar, if opposite, roles. Ectopic expression of *OsGI*, using the CaMV 35S promoter, delays flowering, where ectopic *GI* expression in Arabidopsis induces flowering. In the transgenic rice plants, expression levels of *Hd1* (the *CO* orthologue, see below) were increased, suggesting that *OsGI* functions to upregulate *Hd1*, just as *GI* functions to upregulate *CO*. However, the *FT* orthologue, *Hd3a*, was expressed at lower levels in these transgenic plants, whereas ectopic expression of *GI* and enhanced *CO* expression increases *FT* transcript levels in Arabidopsis. These data led to the conclusion that the same pathway components are present in rice as in Arabidopsis, but that the central role of *CO* in inducing *FT* expression seems to be reversed in rice, with enhanced *Hd1* expression resulting in decreased *Hd3a* activity (Hayama *et al.* 2003; Simpson 2003).

Loci involved in flower induction in rice are traditionally named *Heading date* loci (*Hd*), and two

(a)

(b)

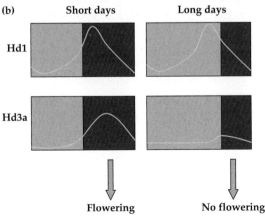

Figure 7.3 (a) A rice plant in flower (foreground). (b) The *Hd1* gene (above) is expressed in a circadian pattern. It is only able to activate *Hd3a* expression (below) when its expression coincides with darkness, that is, in short days (left panel). Compare to Figure 5.4.

of the *Hd* loci encode the rice orthologues of *CO* and *FT*. Yano *et al.* (2000) isolated *Hd1* by map-based cloning and found that it encodes the rice *CO* orthologue, a protein very similar in sequence to CO itself. *Hd1* expression is controlled by the circadian clock in the same way that *CO* expression is, with maximum transcript occurring at night in short days, but in the later afternoon/early evening in long days. *Hd3a* encodes a protein with strong sequence similarity to FT, the crucial flowering-time integrator in Arabidopsis. As discussed in Chapter 5, in Arabidopsis *CO* expression in the light causes activation of *FT* expression, but *CO* expression in the dark (i.e., in short days) does not induce *FT* activity. However, in rice the situation appears to be entirely the opposite (compare Fig. 7.3 to Fig. 5.4). In the dark, *Hd1* expression induces *Hd3a* expression, which then fluctuates on a circadian cycle similar to that of *Hd1*, and induces flowering. But when *Hd1* expression occurs in the light (i.e., in long days) it does not induce *Hd3a* expression and flowering does not occur (Izawa *et al.* 2002; Kojima *et al.* 2002). Indeed, exposure to a single 10-minute burst of light during the night is sufficient to suppress *Hd3a* expression and delay flowering (Ishikawa *et al.* 2005). It is not currently known how the molecular mechanism of CO/Hd1 function in the light and dark differs in rice from that in Arabidopsis.

Another rice orthologue of an Arabidopsis flowering-time integrator was recently isolated, and shown to play a similar role. The *OsSOC1* gene (also known as *OsMADS50)* was isolated by a screen for MADS box genes with similarity to Arabidopsis *FLC* (Tadege *et al.* 2003), and found to map to the known flowering-time locus *Hd9*. In fact the OsSOC1 protein has greater sequence similarity to SOC1 than to FLC, and a number of experiments have indicated that it functions as a SOC1 orthologue. Expression of *OsSOC1* increases at the floral transition, as would be expected for a flowering-time integrator, while ectopic expression of OsSOC1 in Arabidopsis complements the *soc1* mutant. Ectopic expression of *OsSOC1* in a wild type Arabidopsis background induces early flowering. In contrast, suppression of *OsSOC1* by RNAi methods causes late flowering in rice (Lee *et al.* 2004). Since Arabidopsis *SOC1* expression is down-

PRESTON RIDGE LIBRARY
COLLIN COLLEGE
FRISCO, TX 75035

regulated by FLC, Tadege *et al.* (2003) investigated potential FLC/OsSOC1 interactions in rice. They were able to identify a CarG box in the *OsSOC1* promoter, which appeared suitable for FLC binding, but could not identify an *FLC* orthologue by library screening, searching through EST collections or analysis of the rice genome sequence. However, ectopic expression of Arabidopsis *FLC* in rice delays flowering by 2–4 weeks, and also delays upregulation of *OsSOC1*. These data suggest that another MADS box protein, recognizing similar binding motifs to FLC, may be involved in *OsSOC1* downregulation (Andersen *et al.* 2004).

Overall, the photoperiodic induction pathway in rice seems remarkably similar to that of Arabidopsis, suggesting that the evolution of long and short day plants has involved tinkering with the basic mechanisms of photoperiodic induction, rather than the development of two distinct systems. However, since rice is not sensitive to vernalization it is perhaps not surprising that no clear orthologues of *FLC*, *FRI*, *VRN1*, or *VRN2* have been identified in the rice genome, suggesting that no equivalent vernalization-promotion pathway is present (Izawa *et al.* 2003).

7.2.2 Flowering time in other cereals

It is anticipated that the photoperiodic control of flower induction in other cereals will turn out to be similar to the pathway identified in rice. However, many cereals have an additional vernalization requirement, which rice does not share. For example, both wheat and barley need around 60 days of cold to induce flowering. In both species spring varieties, which are planted in the spring and flower immediately without the need for vernalization, have been bred. The difference between these spring varieties and traditional vernalization-requiring winter varieties has been mapped to two major QTLs, named *VERNALIZATION1* and *VERNALIZATION2* (*VRN1* and *VRN2*). These loci should not be confused with the Arabidopsis *VRN1* and *VRN2* loci, to which they are not necessarily related (Andersen *et al.* 2004).

The *VRN1* loci have been identified in both wheat and barley, and renamed *TmAP1* (or *WAP1* or *TaVRT-1*) and *BM5*, respectively (Danyluk

et al. 2003; Trevaskis *et al.* 2003; Yan *et al.* 2003). Both genes encode orthologues of the Arabidopsis floral meristem identity gene *APETALA1* (*AP1*). Transcripts of both genes are upregulated by exposure to cold in winter varieties of wheat and barely, but are constitutively expressed at high levels in spring varieties. These findings were unexpected, as they appear to suggest that a key component of the Arabidopsis floral meristem identity system is involved in a much earlier stage of floral induction in cereals, acting to induce the floral transition in response to vernalization. The identification of a CArG box in the promoter of wheat *TmAP1*, and the observation that spring varieties usually have a deletion of part of the promoter in this region, suggests that another MADS box protein, perhaps more related to FLC, acts upstream of this AP1-like factor (Yan *et al.* 2003).

Wheat *VRN2* was isolated in 2004, using a map-based approach (Yan *et al.* 2004), and shown to encode a protein with many similarities to Arabidopsis CO. In contrast to the behaviour of *VRN1*, the *VRN2* gene is a dominant repressor of flowering, which is downregulated by vernalization. No *VRN2* transcript is detectable after vernalization has occurred. Loss of function of *VRN2* results in the production of spring lines of wheat with no requirement for cold induction. Again, it was surprising that a protein with similarity to another key Arabidopsis protein was involved in this process, but that protein was not the expected direct orthologue (i.e., FLC). It appears that, unlike photoperiodic induction, vernalization has evolved multiple times and that each evolutionary event has involved the recruitment of similar but distinct genes from the flowering-time pathways.

7.2.3 Flowering time in pea

The rationale for studying flowering time in pea is based on its unique suitability for both genetic and physiological studies. Although it is a long day plant, like Arabidopsis, it has the added benefit of being easy to graft, thus allowing researchers to investigate whether something missing from a mutant line is mobile. Mobile signals will travel from wild type root stocks into mutant shoots grafted to them, and rescue the mutant. Immobile signals will not. Extensive genetic stocks are also available in pea, which has been used since the time of Mendel for the study of inheritance.

A number of flowering-time mutants have been described in pea, and these are reviewed in Weller *et al.* (1997). Analysis of these mutants has led to the identification of a number of flowering-time pathways, some very similar to those of Arabidopsis. For example, the *gigas* (*gi*) mutant flowers late in long days, but later still in short days, indicating that pea contains an autonomous floral induction pathway, operating in both long and short days. Grafting a *gi* shoot onto a wild type root revealed that the signal produced by the *GI* gene is mobile, as the mutant shoot flowered at the normal time following this graft (Beveridge and Murfet 1996). So pea contains an autonomous floral induction pathway which operates through a mobile signalling molecule.

Similarly, the *STERILE NODES* (*SNE*), *DIE NEU-TRALIS* (*DNE*), and *PHOTOPERIOD* (*PPD*) genes are all required for inhibition of flowering in short days. Mutation in any of these genes causes early flowering in short days, but does not affect flowering time in long days. Grafting experiments with wild type root stocks and shoots of these mutants revealed that all three genes induce transduction of a mobile, graft-transmissible signal. Pea therefore also contains a floral repression pathway, operating only in short days, which functions through the transmission of a signal from vegetative tissue to the apex (Weller *et al.* 1997).

Overall, the genetic and physiological analysis of floral induction in pea has led to the suggestion that, as with Arabidopsis, a combination of promoting and inhibiting signals arrive at the apical meristem, and when the promotive signals are stronger than the inhibitory ones the floral transition occurs (Corbesier and Coupland, 2005).

7.2.4 Flowering time in other plants

Most crop plants are annuals, like Arabidopsis. But for many plant species, a perennial lifecycle is the norm. In these species flowering is followed by a return to vegetative growth and the cycle is repeated, often for many years. Perennials are usually polycarpous, with some meristems committing to the flowering state each year while others remain vegetative. Polycarpy requires an extra level of control, allowing some meristems to undergo the floral transition but preventing others from doing so. This is believed to be achieved trough a range of competence or juvenile/adult mechanisms, with only competent (adult) meristems undergoing the floral transition (Battey and Tooke 2002). The number of meristems undergoing the floral transition may also be determined by extra environmental factors. For example, in the wallflower (*Cheiranthus cheiri*) variations in the temperature perceived by the apical meristem result in the production of varying numbers of axillary meristems competent to adopt the floral fate, as opposed to those constrained to the vegetative fate (Battey and Tooke 2002). In this way total numbers of flowers and total reproductive output can be adapted to the varying conditions of different years, a sensible system for a species which has many opportunities to flower and should optimize maximum flowering to coincide with optimal environmental conditions.

Development of Flowers

CHAPTER 8

Changes at the Shoot Apical Meristem in Response to Floral Induction

Section IIA of this book was concerned with the multiple environmental and developmental stimuli that combine to induce the transition from the vegetative state to the floral state. Section IIB turns to look at the consequences of that transition. In this chapter we will consider the changes that occur at the shoot apical meristem once the decision to flower has been taken by the plant. These changes involve the expression of a succession of floral meristem identity genes, which convert the meristem to the reproductive state. Floral meristem identity genes can be defined as those genes that specify the floral fate of lateral meristems arising from a reproductive shoot apical meristem. At the same time, it is important that apical meristem indeterminacy is maintained, to allow multiple flowers to be produced. Later chapters in this section will analyse the development of the floral organs from a floral meristem, and the development of the all-important gametes within the reproductive organs themselves.

8.1 Physiological changes at the shoot apical meristem

Once floral induction has occurred, and the plant is committed to flowering, many physiological changes occur at the shoot apical meristem (SAM) to prepare it for the developmental transition it must undergo. Without these physiological changes, the floral transition will not usually occur successfully. An increase in the rate of RNA synthesis is one of the earliest events that occur in the SAM, in order to support the transcription of novel genes. The rate of protein synthesis also increases. Both changes are predictable, as approximately 50% of the transcripts in a typical floral organ are not found in leaves of the same species, highlighting the need for the transcription and translation of flower-specific genes. It has been shown that application of chemical inhibitors of RNA or protein synthesis to the SAM of *Sinapis alba* (white mustard, a close relative of Arabidopsis in the Brassicaceae) is sufficient to prevent the flowering response, even when the plant has been exposed to the usual inductive conditions.

RNA and protein synthesis are not the only metabolic processes which occur as the meristem undergoes its transition to the reproductive state. Rises in sucrose and ATP levels have also been recorded in *S. alba*, presumably to support the increased rate of RNA and protein turnover. Increases in invertase activity and mitochondrion number have also been correlated with the change in meristem state, and these will also support enhanced metabolic activity.

Cell cycle is another important process affected by the change in meristem state. The rate of cell division within the meristem has been shown to increase following floral induction in several species. For example, in *S. alba* exposure to one long day is sufficient to induce the photoperiodic floral promotion pathway and leads to the plant flowering. The single long day also results in a mitotic wave across the SAM, caused by a shortening of the G2 phase of cycling cells and a return to cycling by noncycling G2 cells. G2 is the cell cycle phase between S (DNA synthesis) and M (mitosis), and is where many meristematic cells 'rest' between cell cycles. This shortening of the G2 phase is followed by a shortening of G1 and S phases, which together speed up cell cycling and also synchronizes it so that all cells in the SAM are cycling together. Meristematic cell cycle length is reduced from approximately 86 hours to around 32 hours as a result of these changes, although it is not clear why the more rapid division of cells is necessary for a successful floral transition.

8.2 Shoot apical meristem anatomy

One of the key changes that occur at the SAM in response to the floral transition is an alteration in the expression patterns of certain genes. Before we begin a discussion of floral meristem identity genes and their expression patterns, it is necessary to briefly consider the anatomy of the SAM.

The Arabidopsis shoot apical meristem is a typical, if small, example (Fig. 8.1). It consists of

(a)

(b)

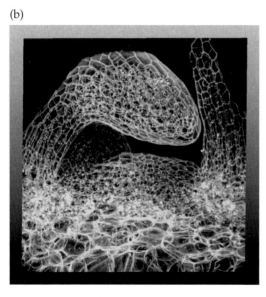

Figure 8.1 Anatomy of the shoot apical meristem. (a) The SAM is made up of three cell layers, L1, L2, and L3. It can also be divided into a central zone (CZ), peripheral zones (PZ), and rib meristem (RM), with lateral meristems arising at the sides. (b) Three-dimensional reconstruction of an Arabidopsis SAM. The tissue was fixed, stained using a propidium iodide pseudo-Schiff reaction to label cell walls, mounted in a chloral hydrate-based medium, and imaged using a Leica SP scanning laser confocal microscope. A series of optical sections was projected using Amira software (Mercury Computer Systems). Image kindly supplied by John Runions and Jim Haseloff (University of Cambridge).

around 100 cells, and that number remains constant despite continuous cell division and cell differentiation throughout the life of the plant. Extensive analysis of shoot apical meristem function and maintenance in Arabidopsis has identified a group of genes key to its functioning. The *CLAVATA1 (CLV1)* and *CLAVATA2 (CLV2)* loci encode a pair of disulphide linked receptors, while the *CLAVATA3 (CLV3)* locus encodes a small ligand. Interaction of these three proteins triggers cell differentiation in response to excess cell division. The *WUSHEL* and *KNOX* genes encode transcription factors necessary for the maintenance of cell division. Together these groups of genes act as a feedback loop which serves to keep meristem size constant under all environmental conditions.

The SAM is composed of three cell layers. The outermost layer is called the L1, and is a single cell thick. These cells divide anticlinally to generate the epidermis of the plant. Below the L1 layer is the L2, usually 2–3 cells thick, and again composed of cells which predominantly divide anticlinally. The bulk of plant tissue is produced from the L3 layer, which may be many cell layers thick. Cells in the L3 divide periclinally and anticlinally. Since these cell layers are maintained throughout the life of the SAM and the plant it produces, all cells derived from a single layer are clonally related to one another. This can be seen in chimaeric mutants, where, for example, the L1 cells may be lacking a gene required for pigment synthesis but the L2 and L3 cell layers are wild type.

Besides its layers, the SAM can also be divided into three functional zones. In the very centre of the SAM is the central zone, characterized by cells with a low rate of division relative to the rest of the SAM. These cells are thought of as a back-up population, which can enter into more rapid mitosis should cell proliferation in the rest of the SAM not meet the needs of various developmental programmes. The peripheral zone encircles the central zone like a doughnut. The peripheral zone is where the majority of cell division occurs, and on the edges of the peripheral zone cells are recruited to various developmental programmes and become part of leaves or other plant organs. At the base of the meristem is the region known as the rib meristem. Periclinal cell divisions within the rib meristem generate the stem

of the plant, and ensure the SAM stays at the top of the developing structure.

8.3 Gene expression patterns in the shoot apical meristem

Gene expression patterns in the SAM can be analysed using a number of techniques. One option is to fuse the promoter of a gene of interest to a reporter gene, such as GUS (β-glucoronidase, stains blue) or GFP (green fluorescent protein, fluoresces under appropriate illumination) and transfer the construct into a transgenic plant. GFP in particular allows the monitoring of expression patterns and their changes in a living plant, although there can be difficulties associated with the length of time for which the reporter lasts, making accurate analysis of temporal changes in expression pattern difficult. These techniques are also only possible in a system where it is easy to isolate promoter sequence and plant transformation is routine. They are often used in Arabidopsis, but less frequently in other species. The alternative is to use *in situ* hybridization to investigate expression pattern. In this technique, plant tissue is fixed, dehydrated, and embedded, as for normal microscopy, and then sectioned. Thin sections of tissue are placed on microscope slides, and then probed with the antisense strand of a transcript of interest, usually fused to a coloured marker. Coloured stain can then be observed through a microscope in cells expressing the gene of interest. An advantage of *in situ* hybridization is that it is possible to probe serial sections with different genes, giving an accurate impression of the expression patterns of multiple genes at a particular time point.

Genes may have one of three possible expression patterns with regard to the floral transition at the meristem. Some genes have fixed patterns of expression, which do not change as the meristem undergoes the transition. These genes are likely to be very important in general meristem function, and may play roles in maintaining meristem activity. Genes such as the *CLV* genes mentioned earlier may fall into this category.

Other genes may show modified patterns of expression. These genes are expressed in the SAM at both the vegetative and reproductive stages, but the layers or zones in which they are expressed varies as the SAM undergoes the floral transition. Such genes are likely to be involved in processes requiring modification when the meristem changes state, such as the control of phyllotaxy (the positioning of lateral organs). One example of an Arabidopsis gene with a modified expression pattern is *CDC2*, which encodes a protein kinase that controls the G1 to S phase cell cycle transition, and so is essential to mitosis. The gene is expressed throughout the meristem in the vegetative state, but when the SAM is converted to a single terminal flower its expression is concentrated in the rib meristem region. This pattern corresponds to the shift in maximum mitotic activity to this region in the floral meristem, to build the carpels and receptacle of the mature flower.

The final category of genes contains those with stage-specific expression in the apical meristem. Such genes are only expressed following the floral transition, or before the floral transition. These are the key genes whose expression is necessary to convert the meristem through to the floral state, and it is these genes which we call the floral meristem identity genes.

8.4 Floral meristem identity genes act downstream of the flowering-time integrators

In Chapter 7 three key Arabidopsis flowering-time integrator genes were introduced. The *FT* gene encodes a membrane-associated protein which integrates signals from the photoperiodic induction pathway (through CO), and the autonomous promotion, inhibition, and vernalization pathways (all through FLC). The *FT* protein moves from the leaves to the shoot apical meristem. The *SOC1* gene encodes a MADS box transcription factor repressed by FLC (and thus integrates the autonomous promotion, inhibition, and vernalization pathways) but activated by long days (through CO) and by gibberellin. The third flowering-time integrator, *LEAFY*, primarily acts as a floral meristem identity gene whose expression is induced by SOC1. However, it can also be directly activated by gibberellin, causing it to be classed as a flowering-time integrator as well. The three flowering-time

integrators activate the expression of the floral meristem identity genes, which are essential for conversion of the meristem to the reproductive state. These genes are not usually expressed prior to the floral transition, and are therefore genetically downstream of the flowering-time integrators.

8.5 Floral meristem identity genes

In this section we will consider the function of each of the major floral meristem identity genes in turn, looking at both the Arabidopsis genes and their orthologues in *Antirrhinum majus*. Meristem conversion has been studied almost as extensively in Antirrhinum as in Arabidopsis, and the combination of data from both species has allowed important advances in our understanding of the subject. A key characteristic of floral meristem identity is that many of the genes are involved in mutual regulatory interactions. These interactions make any discussion of floral meristem identity more complex, but they serve in nature to provide buffering against environmental noise and ensure meristem conversion occurs correctly and does not revert once it has occurred (Blazquez *et al.* 2006). While *LEAFY*, *APETALA1*, and *CAULIFLOWER* are considered to be the three most crucial floral meristem identity genes, we will also discuss the roles of *TERMINAL FLOWER 1*, *APETALA2*, *UNUSUAL FLORAL ORGANS*, and *AGAMOUS*, all of which are necessary for the development of normal, determinate floral meristems on the flanks of a normal, indeterminate inflorescence meristem.

8.5.1 *LEAFY (LFY)*

The central role of *LFY* in the conversion of the meristem is a major unifying principle of plant development, with *LFY* orthologues identified in many species from diverse parts of the plant kingdom (Jack 2004). Interestingly, *LFY* appears to be a plant-specific gene, and it has recently been shown that its role in the moss *Physcomitrella patens* is to ensure cell division within the developing embryo (Maizel *et al.* 2005; Tanahashi *et al.* 2005). The LFY protein is both necessary and sufficient to specify a floral meristem fate in many (but not all) developmental contexts, and the protein has a dual role in integrating signals from the floral inductive pathways (see Chapter 7) and acing as a floral meristem identity gene by activating downstream genes controlling floral organ development.

LFY is the best characterized gene involved in conversion of the meristem to the floral state. Plants that are *lfy* mutants produce inflorescence shoots where their flowers should be, subtended by bracts. Some flowers do develop in the mutant lines but they are abnormal and have characteristics of secondary inflorescences, including indeterminacy. Thus the basic conversion of the meristem to the reproductive state has occurred in these lines, but the flowers themselves cannot be properly produced without LFY activity. LFY was isolated using the Antirrhinum orthologue (FLORICAULA, FLO) as a probe in the genomic region to which LFY had been mapped (Weigel *et al.* 1992). The Antirrhinum FLO gene had been cloned first from a transposon tagged mutant line (Coen *et al.* 1990). The Arabidopsis and Antirrhinum proteins are 70% identical at the amino acid sequence level, and the *flo* mutant has a phenotype similar to that of the *lfy* mutant, with bracts subtending ectopic inflorescence shoots (Fig. 8.2a, see Plate 5). However, the *flo* mutant is more extreme in phenotype than even a null *lfy* mutant, as the conversion of flowers to inflorescence meristems is complete in Antirrhinum. Unfortunately neither protein has real sequence similarity to previously described proteins with known biochemical functions, although they do have motifs which suggest that they might be novel transcription factors. This hypothesis is consistent with their role in the activation of downstream genes involved in the development of floral organs. The LFY protein has been shown to be localized to the nucleus, which adds weight to the transcription factor idea.

LFY is expressed strongly in young floral organ primordia and floral meristems, but not in inflorescence meristems, consistent with its role in specifying floral identity (Weigel *et al.* 1992). Strangely, while FLO is also expressed in floral meristems and bract primordia, it is only expressed in the organ primordia which will become sepals, petals, and carpels, but is absent from the developing stamen primordia (Coen *et al.* 1990). Both genes are initially expressed in the inflorescence meristem at the floral

Figure 8.2 Early floral meristem identity mutants. (a) The Antirrhinum *flo* mutant (left) has inflorescence shoots produced in place of the flowers found in the axils of wild type bracts (right). Image kindly supplied by Enrico Coen (John Innes Centre). (b) The Arabidopsis *ap1* mutant is slightly better converted to the floral form, with indeterminate floral structures arising from the meristem.

transition, but rapidly resolve outwards to a region where the lateral meristems and bracts are specified.

Experiments to investigate the putative function of LFY in vegetative meristems demonstrated that the LFY protein is a developmental switch sufficient to convert all lateral meristems into solitary flowers, and even to convert the shoot apical meristem into a single flower, although this latter conversion takes some developmental time to complete (Weigel and Nilsson 1995). Ectopic expression of LFY in transgenic Arabidopsis plants using the CaMV 35S promoter results in the conversion of all lateral or axillary meristems to flowers. Thus the axillary meristems in the axils of the rosette leaves form precocious flowers, instead of remaining dormant and forming branches later in development.

Similarly the lateral meristems in the axils of the cauline leaves also form single flowers. Eventually, the SAM is also converted to a terminal flower. Strikingly, the same construct was shown to have the same dramatic effects on meristem identity in other species, including tobacco. Ectopic expression of *LFY* in aspen resulted in the production of flowers in the axils of leaves, and the early conversion of the SAM to a terminal flower. While aspen trees ordinarily take around 20 years to flower, these conversions all occurred within a few months (Weigel and Nilsson 1995). Similarly, ectopic expression of *LFY* in citrus trees has also been shown to reduce their time to flowering, from over 6 years to within 1 year (Pena *et al*. 2001).

The immediate cellular consequences of LFY activity are not yet well described, although we do know that LFY is involved in the activation of expression of *AP1* and *CAL* (see below). Recent reports suggest that it also activates the expression of a number of other genes, including *LATE MERISTEM IDENTITY 1 (LMI1)*, which encodes a homeodomain leucine-zipper transcription factor. LMI1 is required for activation of *CAL* expression, and in its absence the meristem defects of weak *lfy* mutant alleles are substantially worsened (Saddic *et al*. 2006). It seems likely that LFY has such a strong effect on meristem identity because it activates transcription of a suite of other genes, each of which is crucial for the proper development of a floral meristem.

8.5.2 *APETALA1 (AP1)*

Plants with mutations in *AP1* have phenotypes very similar to that of the *lfy* mutant, suggesting that *AP1* is also necessary for the floral identity of the meristem. The phenotype is, if anything, slightly less dramatic, with a fuller conversion of the inflorescence meristem to the floral form (Fig. 8.2b, see Plate 5). Sepals do form in *ap1* mutants, but they are leaf like, and they form secondary flowers within their axils. The first whorl organs of these secondary flowers may also form tertiary flowers in their axils (Mandel *et al*. 1992). *AP1* encodes a transcription factor which is a member of the MADS box family, and was isolated by a combination of mapping and a candidate gene approach (Mandel *et al*. 1992).

MADS box transcription factors, such as SOC1 and FLC (Chapters 4 and 7), are characterized by a region of fifty amino acids containing a DNA-binding and dimerization domain. The MADS box family is very large, with over 40 members apparent in the Arabidopsis genome. *AP1* is expressed in the floral meristems, from the first time point at which they are visible on the flanks of the inflorescence meristem, and remains expressed in petal and sepal primordia during flower development (Mandel *et al*. 1992).

The Antirrhinum orthologue of *AP1* is *SQUAMOSA (SQUA)*, which encodes a MADS box transcription factor with 68% sequence identity to AP1 (Huijser *et al*. 1992). The *squa* mutant is more extreme than the *ap1* mutant, but slightly less dramatic than *flo*. Most floral meristems are converted to inflorescence meristems, in a reiterating pattern, but the plant does occasionally produce anomalous flowers. *SQUA* is also expressed in the wild type floral meristem, but not the inflorescence meristem, and remains expressed in all developing floral organs except the stamens.

In both Arabidopsis and Antirrhinum, double mutant phenotypes between *lfy/ap1* and *flo/squa* are more extreme phenotypically than either single mutant. The *lfy/ap1* double mutant is very similar phenotypically to the *flo* single mutant, in that the conversion of floral meristems to inflorescence meristems is essentially complete.

Genetic analyses have indicated that *AP1* functions downstream of *LFY* in the conversion of the meristem to the floral state. Ectopic expression of *LFY* does not confer any phenotypic change in an *ap1* mutant background, implying that LFY works through *AP1* (Weigel and Nilsson 1995). In contrast, ectopic expression of *AP1* from the same CaMV 35S promoter does induce early flowering and conversion of meristems to the floral state, even in a *lfy* mutant background. The flowers produced by these plants are abnormal, with perturbations in organ identity and development. These data confirm that *AP1* functions downstream of *LFY*, but that the pathway is not entirely linear, with *LFY* necessary for the activation of certain floral organ identity genes not regulated by AP1 (Mandel and Yanofsky 1995). Nonetheless, ectopic expression of *AP1* in citrus was able to induce early flowering, just as ectopic expression of *LFY* was (Pena *et al.* 2001). Further evidence for *LFY* acting upstream of *AP1* comes from work by Wagner *et al.* (1999), who showed that inducible activation of *LFY* (achieved by fusing the cDNA to a glucocorticoid receptor and then treating transgenic plants with a steroid so that the protein moves into the nucleus and can function) resulted in an almost immediate upregulation of *AP1* expression. Furthermore, the *AP1* promoter contains a region bound by LFY *in vitro*, suggesting that the regulation of *AP1* expression by LFY is direct and that LFY does indeed act as a transcriptional regulator (Parcy *et al.* 1998).

Matters are slightly less clear-cut in Antirrhinum, where the *squa* mutant contains normal amounts of *flo* transcript and vice versa (Huijser *et al.* 1992), and the single mutants are so much more extreme than in Arabidopsis that the double mutant suggests much less synergy between the genes. It has been suggested that in Antirrhinum the FLO and SQUA proteins act together in the induction of downstream genes, rather than in a more linear pathway.

8.5.3 *CAULIFLOWER (CAL)*

The *CAL* gene is probably the least well understood of the floral meristem identity genes, and to date no orthologue has been identified in Antirrhinum, so no comparative analysis has been possible. *CAL* was isolated in a search for genes with sequence similarity to *AP1* and encodes a MADS box transcription factor 76% identical in amino acid sequence to AP1 (Kempin *et al.* 1995). The *cal* mutant has no discernible phenotypic differences from a wild type plant, but the *ap1/cal* double mutant plant has a very striking phenotype. Where flowers should be, the plant continues to proliferate inflorescence meristem, from which no organs develop. This results in the production of heads of meristematic tissue several millimetres across and resembling miniature vegetable cauliflowers. CAL is therefore clearly necessary for the correct transition of the inflorescence meristem to the floral meristem and the production of flowers. However, it is also apparent that *AP1* can compensate for the loss of *CAL*, since the *cal* single mutant is phenotypically wild type. The partial redundancy

between these genes has made determining the function of *CAL* more difficult.

CAL mRNA accumulates in young floral meristems, but does not persist in any floral organ during development, unlike *AP1*. This difference suggests that *CAL* cannot compensate for the loss of *AP1* in its later activities, such as the activation of particular floral organ identity genes (Kempin *et al.* 1995). One clear function of *CAL* is to enhance *AP1* activation of *LFY* expression. Although genetic and molecular data agree that *AP1* acts downstream of *LFY*, in the *ap1/cal* double mutant *LFY* expression levels are significantly reduced, especially in the early stages of its expression (Bowman *et al.* 1993). This has led to the conclusion that the AP1/CAL duo enhances *LFY* expression, perhaps operating as a positive feedback loop to ensure the full conversion of the meristem to the floral state.

A *CAL* orthologue has been isolated from *Brassica oleracea*, and it has been shown that the variety grown for vegetable cauliflowers (*B. oleracea var. botrytis*) contains a mutation within the *CAL* coding sequence, resulting in the production of a truncated 150-amino acid protein rather than the wild type 255-amino acid version (Kempin *et al.* 1995).

8.5.4 *TERMINAL FLOWER 1 (TFL1)*

The *TFL1* gene is required for a less obvious aspect of meristem conversion, namely the maintenance of the apical meristem in the inflorescence state, and therefore indeterminate. When *LFY* is first expressed, its transcript can be found right up into the apex of the plant, but it is shortly afterwards restricted to the flanks of the apical meristem, where the lateral floral meristems arise. Without this restriction, LFY would convert the apical meristem through to the floral state and the plant's growth would stop. The TFL1 protein is involved in repressing the floral nature of the apical meristem, and maintaining it in the inflorescence state and indeterminate. In *tfl1* mutants, up to five ordinary flowers are made and then the apical meristem itself becomes floral, producing a single terminal flower. In the *tfl1* mutant the *LFY* and *AP1* genes are expressed in apical regions of the meristem as well as in the usual peripheral regions, indicating that a role of TFL1 is to repress *LFY* and *AP1* expression in the apex. This appears to work through a feedback loop. Before the vegetative to reproductive transition, *TFL1* is expressed in the apical meristem but at low levels and not in the apex itself. It is upregulated in the apex itself following the floral transition, probably by LFY, and then represses *LFY* and *AP1* expression in the apical meristem (Bradley *et al.* 1997). Conversely, in the *lfy* mutant *TFL1* is expressed in the ectopic shoots formed where flowers should be, produced from the axillary meristems on the flanks of the apical meristem (Ratcliffe *et al.* 1999). So, in the absence of the LFY-induced enhancement of *TFL1* expression in the SAM itself, and exclusion from lateral meristems, the gene is expressed in lateral meristems which remain indeterminate. Further evidence for the feedback loop comes from transgenic experiments. Ectopic expression of *LFY* or *AP1* from the CaMV 35S promoter results in a reduction in *TFL1* transcript levels in the apical meristem (Liljegren *et al.* 1999), while ectopic expression of *TFL1* from the same promoter results in reduced levels of *AP1* and *LFY* transcript (Ratcliffe *et al.* 1998). This feedback loop is clearly essential to maintain the apical meristem in the indeterminate state while allowing determinate floral meristems to form on its flanks.

The Antirrhinum orthologue of *TFL1* is *CENTRORADIALIS* (*CEN*). *In situ* hybridisation experiments have revealed that *CEN* is expressed 2 to 3 days after *FLO*, and not at all in *flo* mutants. So in Antirrhinum it also appears that *FLO* expression induces *CEN* expression, and the CEN protein then excludes *FLO* expression from the apical region of the meristem, keeping it in the inflorescence rather than the floral form.

The *CEN* gene was cloned before *TFL1*, using a transposon tagged mutant line of Antirrhinum (Fig. 8.3, also Plate 6; Bradley *et al.* 1996). Using this sequence as a probe, *TFL1* was isolated the following year (Bradley *et al.* 1997), and, concurrently, was isolated from a T-DNA insertion mutant by another group (Ohshima *et al.* 1997). The two proteins are 70% identical, and there are also very similar proteins present in the rice genome. The biochemical nature of TFL1 and CEN is not yet clear, but they have most similarity to animal phosphatidylethanolamine-binding proteins, which associate with membrane protein complexes. This

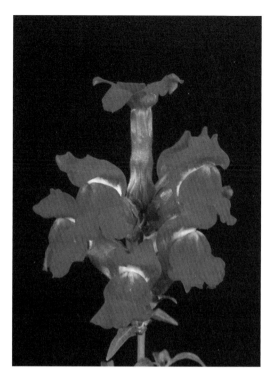

Figure 8.3 Maintaining indeterminacy. The *cen* mutant in Antirrhinum experiences conversion of the indeterminate inflorescence meristem to a determinate floral meristem. Image kindly supplied by Enrico Coen (John Innes Centre).

suggests that they are involved in signal transduction at the cell membrane. TFL1 is also 50% identical to the flowering-time integrator FT, although the proteins play opposite roles in the conversion of the meristem (Kardailsky *et al.* 1999). As discussed in Chapter 7, Hanzawa *et al.* (2005) showed that only a single amino acid controlled the difference between the protein functioning as FT or as TFL1. The single amino acid is part of a region which may act as a surface for interactions with other molecules (Ahn *et al.* 2006).

Indeterminate inflorescences and determinate inflorescences are widely scattered throughout the plant phylogenetic tree, suggesting either that indeterminacy evolved multiple times, or that it was lost multiple times. The presence of the same mechanism to maintain an indeterminate meristem, operating through the same proteins, in two species as phylogenetically distant as Antirrhinum and Arabidopsis, suggests that indeterminate inflorescences may be the ancestral state with indeterminacy lost in those species with determinate inflorescence meristems (Bradley *et al.* 1997).

8.5.5 *APETALA2 (AP2)*

The *AP2* gene has two roles in flower development. It is required for the maintenance of the floral identity of the meristem, and so is classed as a floral meristem identity gene, but it is also involved in sepal and petal development, so is also classed as an organ identity gene (Chapter 9). In this section we will focus only on the role of *AP2* in specifying floral meristem identity. In long days *ap2* mutant plants are phenotypically normal with regard to meristem identity, but in short days they resemble *ap1* mutants, with sepals converted to leaves and indeterminate flowers developing inside these leaves, often in a reiterating pattern (Bowman *et al.* 1993). So without signals from the photoperiodic induction pathway, *ap2* mutant plants fail to convert their inflorescence meristems into floral meristems. This suggests that the AP2 protein is required for the conversion of inflorescence meristems into floral meristems, but only when the plant is flowering in response to the autonomous or gibberellin promotion pathway, not the photoperiodic promotion pathway.

AP2 encodes a protein with very little similarity to anything that has been functionally characterized. However, it has a domain that includes a nuclear localization signal, suggesting that it might be a transcription factor. It also contains a domain initially thought to be involved in protein–protein interactions. This domain is 68 amino acids long, and is repeated directly, giving a central 136-amino acid domain, which has come to be called the AP2 domain (Jofuku *et al.* 1994). Recent evidence suggests that the AP2 domain functions as a DNA-binding domain, supporting the idea that AP2 is a transcriptional regulator. The AP2 domain contains a region predicted to form an amphipathic alpha helix, and tobacco AP2 domains have been shown to bind DNA (Ohme-Takagi and Shinshi 1995). Genes encoding over 150 proteins with AP2-like domains have now been identified in the Arabidopsis genome, and the functions of most are unknown (Wessler 2005). However, there is

evidence that AP2 itself may regulate the activity of some of these genes, as *ap2* mutant plants have altered expression patterns of some *AP2*-like genes (Okamuro *et al.* 1997). It is also becoming clear that AP2 has roles in some parts of vegetative plant development, perhaps through the activation of *AP2*-like genes. These roles will be considered more fully in Chapter 9. The AP2 domain was initially thought to be plant-specific, but recent evidence points to its existence in cyanobacteria, bacteriophages, and ciliates (Wessler 2005). The cyanobacterium AP2-like protein has been shown to bind DNA *in vitro* (Magnani *et al.* 2004). It has been suggested that the plant AP2-like proteins were derived by lateral gene transfer from a prokaryote, as they do not appear to be present in any other eukaryote (besides a single ciliate) and they contain few or no introns, a sign of a recent prokaryotic ancestor (Magnani *et al.* 2004).

AP2 is expressed continuously in both inflorescence and floral meristems, and in all four types of floral organs, as well as in the stem and leaves (Jofuku *et al.* 1994). However, this expression pattern does not reflect the true activity of AP2, as recent analysis has shown that it is translationally repressed in certain tissues by a microRNA (Chen 2004). This repression only really affects the role of AP2 in floral organ identity, so will be discussed in more detail in Chapter 9.

Antirrhinum orthologues of *AP2* were not identified through mutant screens, but by isolation of *AP2*-like sequences and characterization of their function. Two *AP2*-like genes have been identified, and are called *LIPLESS1 (LIP1)* and *LIP2*. Plants with mutations in either gene are phenotypically wild type. A double mutant between both genes has defects in floral organ identity but none in meristem conversion, suggesting that *AP2* and its Antirrhinum orthologues have divergent roles in floral meristem identity (Keck *et al.* 2003).

8.5.6 *UNUSUAL FLORAL ORGANS (UFO)*

The *UFO* gene has an effect on meristem identity in both long and short days. When *ufo* mutant plants are grown in long days the flowers have increased numbers of organs in all whorls, suggesting that the determinacy of the floral meristem is disrupted.

When *ufo* mutants are grown in short days the flowers are replaced by shoots which are intermediate in appearance between flowers and inflorescences, again suggesting that the determinacy of the meristem, and specifically its conversion from inflorescence to floral, is affected by this gene. *UFO* seems to be necessary, particularly in short days, to ensure the full floral identity of the lateral meristems, although its role is not as important as that of earlier floral meristem identity genes such as *LFY*, without which no conversion occurs at all (Levin and Meyerowitz 1995; Wilkinson and Haughn 1995).

The Antirrhinum orthologue of *UFO* is *FIMBRIATA* (*FIM*), and the *fim* mutant is very similar to the *ufo* mutant, although the phenotype is less extreme. The *fim* mutant shows defects in floral organ development and reduced determinacy of the floral meristems, but does not have reversion of floral meristems to an inflorescence fate in the way that the *ufo* mutant has. *FIM* was isolated from a transposon tagged mutant line of Antirrhinum and found to encode a protein with very little similarity to proteins then in the available databases (Simon *et al.* 1994). *UFO* was isolated using *FIM* as a probe, and the two proteins were found to be 60% identical (Ingram *et al.* 1995). However, later analyses suggested that the FIM and UFO proteins contain an F-box motif, a domain present in cell cycle-associated cyclin proteins and some transcription factors, and which is known to provide substrate specificity to certain types of ubiquitin ligases. FIM/UFO may therefore act to target proteins for degradation. FIM has been shown to physically bind three members of a small family of proteins called the FAPs (FIM associated proteins), which are related to human and yeast proteins involved in the targeting of proteins for degradation, particularly during cell cycle progression (Ingram *et al.* 1997). It has also been shown that the phenotypic consequences of ectopic expression of *UFO* from the CaMV 35S promoter are suppressed by mutations in the COP9 signalosome, a complex that selects proteins for degradation. The role of UFO and FIM therefore seems to be in the targeted degradation of particular proteins (Wang *et al.* 2003).

Both *UFO* and *FIM* are expressed in the floral meristems of the two species, fitting with their role as specifying the determinate floral nature of the meristem. *UFO* and *FIM* are expressed later in the development of the floral meristem than some other meristem identity genes, including *LFY*, but clearly earlier than the organ identity genes. Expression of both *UFO* and *FIM* initiates in the centre of the floral meristem and then resolves into a ring, leaving a clear area at the centre of the apex (Simon *et al.* 1994; Ingram *et al.* 1995). Neither gene is expressed in the apical meristem itself, only in the lateral floral meristems. Later in development expression is restricted further to a few cells at the base of the sepal and petal primordia. The expression pattern of the *FAP* genes overlaps that of *FIM* itself, adding weight to the idea that they interact *in vivo* (Ingram *et al.* 1997).

FIM acts downstream of FLO and SQUA, with a complete absence of *FIM* transcript in the meristems of *flo* mutants, and a great delay in its timing of expression in *squa* mutants (Simon *et al.* 1994). UFO has also been shown to act downstream of LFY. In a *lfy* mutant background there is no phenotypic difference between *ufo* mutants, plants wild type at the *UFO* locus, and plants ectopically expressing *UFO* (Lee *et al.* 1997). These data have led to the suggestion that LFY and UFO work together to activate downstream genes involved in floral organ development.

In contrast, UFO is also known to act upstream of AP1. Hepworth *et al.* (2006) showed that the floral meristems arising from a *ufo* mutant inflorescence meristem were initially bract-like as a result of the limitation of *LFY* and *AP1* transcript to a very small number of cells. Expression of *UFO* from the *AP1* promoter was unable to complement this phenotype, suggesting that UFO is required before AP1 is active.

8.5.7 *AGAMOUS* (*AG*)

The final floral meristem identity gene, *AG*, is very similar to *AP2* in that it is better known for its role in floral organ identity, but does have a limited role in ensuring the complete determinacy of the floral meristem. Flowers that are *ag* mutant are indeterminate. They produce a whorl of sepals, two whorls of petals, and then start again, producing indefinite numbers of whorls of sepals and petals. Under short day conditions the floral meristem of an *ag* mutant may revert to an inflorescence meristem, producing an inflorescence shoot from within the flower. AG therefore plays a role in the determinacy of the floral meristem and also a role in the maintenance of the floral state of the meristem, after the switch has occurred as a result of the activities of the other proteins (Yanofsky *et al.* 1990).

The *AG* gene was isolated from a T-DNA tagged line, and encodes a MADS box transcription factor. The gene is expressed slightly later than *LFY* and *AP1*, and appears in the central region of the floral meristem, and then later in the developing stamen and carpel primordia in whorls 3 and 4 (Yanofsky *et al.* 1990).

Unusually, although the 5′ promoter region of *AG* is important for the regulation of its expression, so are some intragenic regions. Sieburth and Meyerowitz (1997) used a GUS reporter gene fused to various elements of the *AG* genomic sequence to demonstrate that the large (3 kilobases) second intron is necessary for the correct spatial expression of *AG*. A reporter gene expressed solely from the *AG* promoter was expressed in vegetative tissues as well as in the floral meristem, indicating that the second intron contains sequence which represses vegetative expression of *AG*.

AG expression is known to be activated by LFY (Parcy *et al.* 1998), and the LFY protein has been shown to bind *in vitro* to the second intron of *AG* (Busch *et al.* 1999). The subsequent restriction of *AG* transcript to the third and fourth whorls of the developing flowers is the result of the combination of LFY and WUSCHEL (WUS) activity. *WUS* encodes a homeodomain transcription factor necessary to specify cellular proliferation in meristems and, without which, shoot apical meristems terminate prematurely (Laux *et al.* 1996). The WUS protein has also been shown to bind to the second intron of *AG*, but only in the presence of LFY, and acts in a negative feedback loop (Lohmann *et al.* 2001). WUS activity induces *AG* expression in the correct primordia, but AG then downregulates *WUS* expression in those cells, causing them to stop proliferating and causing the meristem to assume a determinate state. In the *ag* mutant the lack of this downregulation of *WUS* results in the maintenance of the indeterminate state of the floral meristem (Lenhard *et al.* 2001).

This feedback loop does not occur in a vegetative meristem due to the lack of LFY activity.

Ectopic expression of *AG* in Arabidopsis alters flower morphology but also the numbers of organs in whorls and numbers of whorls, which confirms the role of AG in inducing the determinate meristem state.

The Antirrhinum orthologue of *AG* is *PLENA* (*PLE*), which was isolated from a transposon insertion line and encodes a protein 64% identical to AG (Bradley *et al.* 1993). The *ple* mutant phenotype is less extreme than *ag*, showing altered floral organ development and some indeterminacy, but not the reversion of floral meristems to the inflorescence state. The role of PLE seems to be more focused on floral organ identity and meristem determinacy than on maintenance of the floral meristem state. However, the regulation of the two genes is very similar, with a transposon insertion in the second intron of *PLE* causing ectopic expression of the gene in vegetative tissues. Differences in protein function may be attributable to the observation that PLE and AG represent members of different ancient lineages, both derived originally from the same MADS box transcription factor but having experienced extensive divergence in the intervening time (Kramer *et al.* 2004; Causier *et al.* 2005).

To sum up, the conversion of the lateral meristems from inflorescence to floral following the transition of the apical meristem from vegetative to reproductive requires the activity of a suite of floral meristem identity genes. The timing of the expression of these genes suggests that they function sequentially. During normal floral development *LFY* is activated before the formation of a visible floral meristem. It is required for the vegetative to inflorescence transition, because mutants have leaf-like structures, and also for the inflorescence to floral transition, because mutants produce inflorescences where their flowers should be. *LFY* is then excluded from the very apex of the inflorescence meristem by *TFL1*, and marks the position of future floral meristems on the sides of the meristematic dome. *AP1* is also expressed in the newly marked floral meristems, along with *CAL*. These genes are then excluded from the central part of the floral meristem as *AG* is expressed there. *AG* maintains the determinate state of the meristem, preventing it from reverting to indeterminate by downregulating *WUS*, the regulator of cell proliferation. *UFO* is also expressed centrally, and then moves to a ring around the centre. Its interactions with *AP1* may help to confirm the determinacy of the meristem. *AP2* expression is also required throughout the meristem to maintain its determinacy. So the flowering-time integrators, acting in response to the signals from the various floral promotion and inhibition pathways, activate the floral meristem identity genes, with *LFY*, then *AP1*, then *AG*, gradually committing the meristem further and further towards the floral state.

CHAPTER 9

Development of the Floral Organs

In Chapter 8 we discussed the floral meristem identity genes, which act sequentially to commit the lateral meristems on the flanks of an inflorescence meristem to the floral state. This involves inhibiting the indeterminacy of the meristems, so that they do not continue to proliferate, by downregulating WUS, the homeodomain transcription factor which maintains cell division within the meristem. It also involves activating the expression of the genes required for the specification of particular floral organs, the floral organ identity genes. It is becoming clear that flower development is a cascade, with flowering-time integrators activating floral meristem identity genes which activate floral organ identity genes. The floral organ identity genes themselves encode transcription factors, and are predicted to activate the many structural genes required to form the new organs, as well as further regulatory genes controlling processes such as pigment synthesis. The expression patterns of the floral organ identity genes determine the positions of developing organs, and those expression patterns are themselves determined in part by cadastral genes (genes whose function is to specify the domain of expression of other genes). Floral organ identity genes act together as transcriptional complexes to ensure expression of the correct structural genes in each whorl of the developing flower.

9.1 The original ABC model of flower development

The ABC model of flower development was formally proposed in 1991, and was based on concurrent genetic studies in Arabidopsis and Antirrhinum (Coen and Meyerowitz 1991). The model has been integral to our modern understanding of flower development, and has been shown to apply to all angiosperm species tested. The model rests on Goethe's foliar theory of the flower (Chapter 2), in assuming that all floral organs are modified leaves, and that without the expression of key floral organ identity genes, the organs in the flower would develop as leaves.

Angiosperm flowers usually have four whorls. A whorl is the domain in which a single type of floral organ is produced, and may consist of one or more concentric circles (if one or more rings of each organ are produced). In some species the whorl boundaries are blurred, and in dioecious species one whorl of reproductive organs is absent from the adult stage. Some basal angiosperms produce flowers with a spiral arrangement of their floral organs, but even in these cases it is usually possible to distinguish between the spirals which make up each of the four sets of organs. According to the original ABC model, in the outer whorl (called whorl 1), only A function genes are expressed on top of the usual leaf developmental programme genes. In the presence of A function genes, the primordia undergo modifications to the basic leaf developmental programme which result in them developing into sepals. In the second whorl in, whorl 2, both A and B function genes are expressed on top of the standard leaf developmental genes. In the presence of both A and B function genes, the primordia undergo more dramatic modifications that result in them developing into petals. In the third whorl (whorl 3) the A function genes are no longer expressed. The B function genes remain active, and C function genes are now also expressed. In the presence of both B and C function genes the primordia are modified even more dramatically and develop into stamens. Finally, in the innermost whorl (whorl 4) the B function genes are downregulated and only C function genes are expressed.

In the presence of C function genes alone the primordia are modified and develop into carpels. This model is shown diagramatically in Fig. 9.1a.

There are two key points about the model which must be grasped in order to understand its simplicity. Firstly, and most importantly, the model states that organ identity is a function of the ABC genes expressed in developing primordia, not a function of the position of the whorl of primordia within a flower. The organs do not develop directly as a result of the position of their primordia, but only as a result of the expression of the genes. For example, the model predicts that loss of B function gene expression in whorl 2 would result in whorl 2 organs developing as sepals. This is because the gene expression pattern in that whorl would be A alone, and A alone specifies sepal identity, even though the position of the primordia (in whorl 2) would lead you to predict the development of

petals. The second key point about the model is that it assumes that the boundaries of A and C function gene activity are set by mutual antagonism, rather than by reference to external factors. A and C function do not overlap because each downregulates the other, and so they maintain a strict boundary. However, in the absence of either function (in a mutant, for instance) the expression of the other function would extend all the way across the meristem. So, in a mutant with a deletion of the C function genes, A function gene expression would be expected to invade whorls 3 and 4, with consequent effects on the identity of the organs in those whorls.

The model was derived from analysis of floral homeotic mutants with the wrong organs developing in particular whorls. The genetic screens which identified these mutants were conducted independently in Antirrhinum and Arabidopsis, by Enrico Coen's group in the John Innes Centre in Norwich, England, and Elliot Meyerowitz's group in CalTech, California. The Arabidopsis work was based on a variety of mutant collections, generated by EMS and X rays, while the Antirrhinum work focused on transposon mutagenized plants. However, both groups identified mutants that fell into the same three classes, which they named A, B, and C, and which are described in Fig. 9.1b.

Mutants in class A had carpels instead of sepals in whorl 1, stamens instead of petals in whorl 2, normal stamens in whorl 3, and normal carpels in whorl 4. Mutants in class B had normal sepals in whorl 1, sepals instead of petals in whorl 2, carpels instead of stamens in whorl 3, and normal carpels in whorl 4. Class C mutants had normal sepals in whorl 1, normal petals in whorl 2, petals instead of stamens in whorl 3, and sepals instead of carpels in whorl 4. While various mutants which showed abnormal development of particular organs were found, these were the only three classes of homeotic mutants observed; that is, only these three categories of mutants experienced complete conversion of one type of organ into another.

The ABC model was produced from these mutants by assuming that each mutant class carried a mutation in a particular gene (named A, B, or C). Each gene was postulated to be normally expressed (in a wild type plant) in the whorls phenotypically disturbed in the mutant. So genes in class A were

(a)

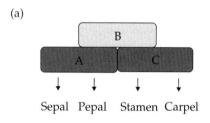

Sepal Pepal Stamen Carpel

(b)

	Whorl 1	Whorl 2	Whorl 3	Whorl 4
Wild type	Sepal	Petal	Stamen	Carpel
Mutant A	Carpel	Stamen	Stamen	Carpel
Mutant B	Sepal	Sepal	Carpel	Carpel
Mutant C	Sepal	Petal	Petal	Sepal

Figure 9.1 The original ABC model. (a) Diagram of the expression patterns of three gene functions in four whorls, generating 4 wild type organs. (b) Table describing the organs found in the four different whorls of plants with mutations in A, B, and C function genes.

predicted to be active in whorls 1 and 2, genes in class B are active in whorls 2 and 3, and genes in class C are active in whorls 3 and 4. These expression patterns produce the model described above, and are sufficient to explain the phenotype of the B function mutant (loss of B function leads to an expression pattern of A, A, C, C, which the model predicts would generate sepal, sepal, carpel, carpel, precisely the phenotype observed in a B function mutant). To explain the phenotypes of the A and C function mutants it is necessary to postulate the mutual repression of the A and C function genes, so that loss of one causes the other to spread across the meristem. A C function mutant would then have the expression pattern A, AB, AB, A which would lead to a phenotype of sepal, petal, petal, sepal. The reverse applies to the A function mutant. Finally, the model was tested using combinations of mutants. Coen and Meyerowitz (1991) used the model to predict the phenotypes of double mutant combinations in Arabidopsis, and then generated those mutants through genetic crosses. All double mutants carried the predicted phenotype, confirming the validity of the model. A triple mutant containing neither A, B, nor C gene activity produced four whorls of leaves where its floral organs should be, lending support to Goethe's foliar theory.

9.1.1 A function genes

The concept of A function is the part of the ABC model which has stood up least well to molecular analysis. In Arabidopsis there are two A function genes, although both play other roles in plant development. In Antirrhinum the original A function mutant was discovered to be a gain-of-function mutation in a C function gene. A search for orthologues of the Arabidopsis A function genes identified two sequences, but mutation of both only resulted in perturbations to whorl 1 and 2 organ development, rather than true homeotic conversion. In Antirrhinum, then, it has been postulated that A function simply represents absence of C function, reinforced by the activities of genes determining correct sepal and petal development.

The *APETALA2 (AP2)* gene introduced in Chapter 8 acts as an A function gene, besides playing its role in floral meristem identity. The *ap2*

mutant has carpels/stamens/stamens/carpels in its four whorls, the expected phenotype of an A function mutant if C function has spread across the meristem (Bowman *et al.* 1991). Indeed, *in situ* hybridization has revealed that the C function gene, *AGAMOUS (AG)*, is expressed in whorls 1 and 2 in an *ap2* mutant background, confirming this aspect of the model (Drews *et al.* 1991). The *AP2* gene encodes the only non-MADS box protein in the ABC model, a protein predicted to function as a transcription factor and containing the core AP2 domain which has since been identified in many other proteins (Jofuku *et al.* 1994). *AP2* is expressed throughout the plant and is required for activation of other genes involved in diverse aspects of plant development. It is also directly involved in other aspects of plant development. Recent reports indicate that correct AP2 function is necessary for maintenance of the stem cell population in the vegetative shoot apical meristem of Arabidopsis (Wuerschum *et al.* 2006). The expression of *AP2* throughout the plant, and particularly in all four whorls of the flower, initially seemed to contradict the ABC model, which predicted a whorl 1- and 2-specific expression pattern (Jofuku *et al.* 1994). However, it has recently been shown that *AP2* is translationally repressed by a microRNA that is only active in whorls 3 and 4, leaving a functional AP2 protein present only in whorls 1 and 2 (Chen 2004). Therefore, within the flower *AP2* meets all the criteria of an A function gene, being active only in whorls 1 and 2, repressing C function expression, and being necessary for correct sepal and petal development. It is only when the other roles of *AP2* are considered, including the specification of floral meristem identity and diverse roles in vegetative development, that it becomes clear that classic A function is an oversimplification of the role of *AP2*.

The second A function gene in Arabidopsis is *APETALA1 (AP1)*, which was also introduced in Chapter 8. The *ap1* mutant flower has the classic carpel/stamen/stamen/carpel phenotype (shown in Fig. 9.2a), although the important role of the gene in floral meristem identity may mask this phenotype except in long days (Mandel *et al.* 1992). *AP1* encodes a MADS box transcription factor, and is the first of the ABC genes to be expressed in the floral meristem, because of its essential role

Figure 9.2 The original ABC mutants. (a) *apetala1* (A function in Arabidopsis). (b) *ovulata* (A function in Antirrhinum) on the right compared to wild type (left). (c) *apetala3* (B function in Arabidopsis). (d) *deficiens* (B function in Antirrhinum) on the right compared to wild type on the left. (e) *agamous* (C function in Arabidopsis). (f) *plena* (C function in Antirrhinum) on the right compared to wild type on the left. Antirrhinum images kindly supplied by Enrico Coen (John Innes Centre). Image (e) provided by Ian Furner (University of Cambridge). See also Plate 7.

in establishing the floral identity of that meristem (Mandel *et al.* 1992). The initial expression of *AP1* is in all four floral whorls, but it is repressed by the activity of the C function gene, and confined to whorls 1 and 2 later in flower development

(Gustafson-Brown *et al.* 1994). It has recently been shown that this restriction of *AP1* transcript is also a direct result of the activity of the B function genes (*APETALA3* and *PISTILLATA*, see below), whose protein products bind to the *AP1* promoter

(Sundstrom *et al.* 2006). Unlike AP2, AP1 has no role in the repression of C function genes. In *ap1* mutant flowers the C function gene *AGAMOUS* is not expressed ectopically in whorls 1 and 2, indicating that AP1 is not involved in the repression of *AG* activity (Weigel and Meyerowitz 1993). *AP1* therefore meets some of the criteria of an A function gene, being restricted in its expression pattern to whorls 1 and 2 by C function gene activity (and B function activity) and being necessary to specify correct sepal and petal development, but it does not itself repress C function activity, indicating that the different aspects of A function activity can be separated.

The situation in Antirrhinum is even more complex. The original Antirrhinum A function mutant, *ovulata*, has the same phenotype as *ap1* and *ap2* mutant flowers (Fig. 9.2b; Carpenter and Coen 1990), but molecular analysis revealed that this phenotype results from ectopic expression of the Antirrhinum C function gene *PLENA (PLE)*, as a result of a transposon insertion within an intron of *PLE* (Bradley *et al.* 1993). No other A function mutants have been identified from Antirrhinum mutagenesis programmes, although multiple examples of the B and C function mutants have been obtained. To investigate the possibility that genetic redundancy between several A function genes was responsible for the absence of a mutant, Keck *et al.* (2003) isolated two *AP2*-like genes from Antirrhinum and investigated their function. The genes were named *LIPLESS 1* and *LIPLESS 2* (*LIP1* and *LIP2*), and encode proteins which are 72% identical to one another and each 65% identical to AP2. A reverse genetic approach was used to identify transposon insertions in each gene, but both mutants were found to be phenotypically wild type. However, the double mutant created by crossing them did show phenotypic abnormalities. In the *lip1/lip2* double mutant sepals were converted to leaf-like organs and petals were much reduced with no lips and underdeveloped lobes. Stamens were also shorter than wild type, and the ovules were carpelloid in structure, resulting in female sterility. Both genes were found to be expressed in sepal primordia early in development, and then later in developing petals and carpels. The C function gene *PLENA* was expressed normally in whorls 3 and 4 in this double mutant, and its expression did not extend into whorls 1 and 2. These data suggest that *LIP1* and *LIP2* act together to specify some aspects of A function in Antirrhinum. They are clearly necessary for correct sepal and petal development, and are expressed in whorls 1 and 2. However, unlike AP2 they play no role in repressing the activity of C function genes, and so loss of their function does not result in homeotic conversion of organs. The *AP1* orthologue in Antirrhinum is *SQUAMOSA (SQUA)*, which was introduced in Chapter 8. However, the *squa* mutant does not have an A function mutant phenotype, instead displaying an extreme conversion of floral meristems to inflorescence meristems. This suggests that SQUA is involved primarily in the establishment of floral meristem identity, and does not play much of a role in organ morphogenesis, unlike AP1. A function appears to be even more complex in Antirrhinum than in Arabidopsis.

9.1.2 B function genes

B function has been the simplest of the ABC functions to understand. Molecular analysis has revealed that there are two B function genes, which both encode MADS box transcription factors, and that these proteins act as a heterodimer. The system is conserved between Arabidopsis and Antirrhinum, and indeed genes from one species can act in place of the equivalent gene from the other species. The B function genes do not seem to play any other role in plant development, and their expression profile is restricted to the second and third whorl of the developing flower. The only area of B function activity which still requires significant analysis is the regulation of their expression within such tightly defined boundaries, as this regulation does not appear to occur as a result of their interactions with other ABC genes.

The first B function gene in Arabidopsis is *APETALA3 (AP3)*. The *ap3* mutant has the classic B function phenotype of sepals/sepals/carpels/carpels (Fig. 9.2c). *AP3* was cloned using the Antirrhinum B function gene *DEFICIENS (DEF)* as a probe (Jack *et al.* 1992). *DEF* had previously been isolated by subtractive hybridisation using *def* mutant and wild type tissue (Sommer *et al.* 1990).

The *def* mutant also has the classic B function phenotype (Fig. 9.2d). The proteins encoded by *AP3* and *DEF* are 61% identical and both encode MADS box transcription factors. *In situ* hybridization revealed that both *AP3* and *DEF* are expressed in the second and third whorl organ primordia from a very early stage, right through to development of the mature petals and stamens (Jack *et al.* 1992; Schwarz-Sommer *et al.* 1992). In fact, analysis of a temperature-sensitive *def* mutant line, in which *DEF* activity can be repressed at various stages of development by transferring plants to different temperature regimes, revealed that DEF activity is necessary throughout petal development. Loss of *DEF* expression in later stages of petal development results in a reversion to more sepaloid characters, such as green streaks at the edges of the tissue (Zachgo *et al.* 1995). While transgenic experiments are rarely conducted in Antirrhinum, which is very difficult to transform, ectopic expression of *AP3* in Arabidopsis using the CaMV 35S promoter resulted in the conversion of the carpels to stamens, confirming that B and C activity is sufficient to induce stamen formation. In the fourth whorl of these transgenic plants expression of the other Arabidopsis B function gene, *PISTILLATA* (*PI*), was activated, suggesting that *AP3* and *PI* regulate each other's expression.

The *PI* gene also encodes a MADS box transcription factor, and was isolated through its similarity to the second B function gene of Antirrhinum, *GLOBOSA* (*GLO*) (Goto and Meyerowitz 1994). *GLO* had been previously isolated in a search for novel MADS box transcription factors in Antirrhinum (Trobner *et al.* 1992). The two proteins are 58% identical at the amino acid level, and mutation of either gene results in plants phenotypically identical to *def* and *ap3* mutants. Both genes are also expressed throughout the petals and stamens from early development to maturity (Trobner *et al.* 1992; Goto and Meyerowitz 1994). The apparent identity of the *ap3* and *pi* (or *def* and *glo*) mutant phenotypes was surprising in view of the differences in the gene sequences. Outside the MADS box domain, the proteins encoded by the two B function genes in each species are extremely dissimilar. However, molecular analysis established that DEF and GLO function as a heterodimer, and only activate transcription of downstream genes once they have

heterodimerized. Since the heterodimer is absent from both mutants, they are phenotypically identical (Trobner *et al.* 1992). The same has proved to be true of AP3 and PI, which also function as a heterodimer. It has also been shown that the two B function genes in each species are responsible for the maintenance of each other's transcription. Although the transcription of *DEF* and *GLO* is initially regulated by other factors, once they are both active they take over the regulation of one another, and the same is true of *AP3* and *PI* (Schwarz-Sommer *et al.* 1992; Goto and Meyerowitz 1994).

The activation of B function gene expression is primarily achieved through the actions of floral meristem identity genes. In Arabidopsis LEAFY has been shown to induce *AP3* expression. The *lfy* mutant has reduced levels of *AP3* expression and a reduced domain of *AP3* activity (Weigel and Meyerowitz 1993). The UFO protein is also necessary for *AP3* expression, as the *ufo* mutant contains reduced levels of *AP3* transcript and fewer petals and stamens (Levin and Meyerowitz 1995). Transgenic experiments have shown that LFY and UFO together are sufficient to activate *AP3* expression. Although ectopic expression of *LFY* and *UFO* from the CaMV 35S promoter results in plants which die before flowering, it was possible to show that the transgenic seedlings expressed *AP3* by adding an *AP3* promoter fused to a reporter gene to the experiment. In the transgenic seedlings reporter gene activity was observed, indicating that *AP3* expression had occurred (Parcy *et al.* 1998). It is also apparent that B function gene expression is repressed during vegetative growth by the activity of DELLA proteins. Gibberellin directly inactivates DELLA proteins, and thus causes flower development by allowing expression of *AP3* and *PI* (Yu *et al.* 2004). This direct activation of B function genes (and also *AGAMOUS*, the C function gene) is a secondary route through which gibberellins induce flowering, alongside their role in activating floral meristem identity genes (discussed in Chapter 6).

One current area of fruitful research into B function is the analysis of downstream genes. Until recently, subtractive and differential methods had been used to identify a small number of genes acting downstream of the B function genes. However, high throughput techniques have recently been applied to this problem, and have revealed that a striking number of genes appear to be direct or indirect targets of B function activity. In Arabidopsis microarray analysis has indicated that 200–1100 genes depend on the activity of *AP3* and *PI* for their expression, the number varying with developmental stage (Zik and Irish 2003; Wellmer *et al.* 2004). In Antirrhinum macroarrays have been used to explore genes differentially expressed between sepal and petal development, and at different stages of petal development using the temperature-sensitive *def-101* allele (Bey *et al.*, 2004). Over 500 genes were found to be differentially regulated during petal development, and of those upregulated, the most frequent groups were of genes involved in cell cycle, cell growth, or secondary metabolism, all essential features of petal development.

The striking similarity of the sequences and mode of action of the B function genes in Arabidopsis and Antirrhinum led Irish and Yamamoto (1995) to investigate whether the proteins they encode could cross species boundaries. They created a construct containing the *DEF* cDNA fused to the *AP3* promoter and transferred it into Arabidopsis plants mutant for *AP3*. The construct was able to rescue a weak mutant phenotype completely, and a strong mutant phenotype partially, confirming that B function has been conserved to a very high degree throughout angiosperm evolution.

9.1.3 C function genes

C function is relatively simple, in that it is controlled by a single gene in each species analysed. That gene is also involved in floral meristem identity, maintaining the determinate nature of the meristem, and so, like A function genes, C function genes are classified as both floral meristem identity genes and floral organ identity genes. The C function mutants have a sepal/petal/petal/sepal phenotype (Figs. 9.2e and f), although the indeterminacy of the floral meristem in the mutants makes it hard to follow the development of the whorls, as extra whorls of organs develop within the first four. The *AGAMOUS (AG)* gene was isolated from a T-DNA tagged line of Arabidopsis and encodes a MADS box transcription factor (Yanofsky *et al.* 1990). The Antirrhinum orthologue, *PLENA (PLE)*,

was isolated from a mutant line containing a transposon insertion in the gene, and the proteins encoded by *PLE* and *AG* were shown to be 64% identical (Bradley *et al.* 1993). More recently it has become clear that *AG* and *PLE* are non-orthologous genes derived from a duplication in a common ancestor. This duplication led to two lineages of closely related MADS box genes. In Arabidopsis these lineages are represented by *AG* and *SHATTERPROOF*. In Antirrhinum, *PLE* is the orthologue of *SHATTERPROOF* while *FARINELLI* is the orthologue of *AG*. It is hypothesized that C function represents the ancestral role of the original, pre-duplication gene, and that this function has been retained by different members of the duplicated gene pair—by the *AG* gene in the Arabidopsis lineage but by the *SHATTERPROOF*-like gene (*PLE*) in the Antirrhinum lineage (Causier *et al.* 2005).

Initial activation of *AG* expression requires the activities of LFY and WUS. Ectopic expression of *LFY* causes ectopic expression of *AG* (Parcy *et al.*, 1998), and the second intron of *AG* has been shown to contain sequences which are bound *in vitro* by LFY (Busch *et al.* 1999). This second intron had previously been shown to be essential for correct regulation of *AG* expression (Sieburth and Meyerowitz 1997). The WUS homeodomain transcription factor is active very early in flower development in whorls 3 and 4, and also binds to the second intron of *AG* to activate its expression (Lohmann *et al.* 2001). AG then downregulates *WUS* expression to ensure the floral meristem remains determinate (Lohmann *et al.* 2001). Gibberellin has also been shown to activate *AG* expression through inactivation of DELLA proteins which inhibit B and C function gene expression (Yu *et al.* 2004).

The expression of *AG* in whorls 3 and 4 of the developing flower is regulated in part by *AP2*, the A function gene as in *ap2* mutants *AG* expression spreads into whorls 1 and 2 (Drews *et al.* 1991). However, *PLE* expression is not regulated by the *LIP1* and *LIP2* genes, as its expression is restricted to whorls 3 and 4 even in a double *lip1/lip2* mutant (Keck *et al.* 2003). Conversely, AG itself represses *AP1* expression, which is initially throughout the floral meristem but is restricted to whorls 1 and 2 following the expression of *AG* (Gustafson-Brown

Figure 9.3 The C function gene, *AGAMOUS*, is repressed from whorls 1 and 2 by one of the A function genes, *APETALA2*, but represses the activity of the other A function gene, *APETALA1*, in whorls 3 and 4.

et al. 1994). It is not yet known whether this repression is direct or indirect. However, it is clear that the postulated mutual A/C repression of the original ABC model operates through both A function genes. AP2 represses *AG* in whorls 1 and 2, but is not itself influenced by AG activity. AG represses *AP1* in whorls 3 and 4, but is not itself influenced by AP1 activity. C function represses one of the A function genes, and is itself repressed by the other one (Fig. 9.3). The expression domain of the C function genes is also determined by a number of other genes, the cadastral genes, which are discussed below.

9.2 The role of D function genes

A new class of homeotic genes, the D function genes, was described in *Petunia* in 1995. The D function genes are required for ovule and placenta development within the carpel, and without their activity the ovules are replaced by carpelloid structures. The D function genes are late acting and work in combination with the C function genes to specify ovule development.

The mature Arabidopsis ovule consists of a haploid embryo sac with two layers of protective integuments, connected to the maternal tissue by a stalk-like structure called the funiculus. The development of the gametes themselves will be discussed in Chapter 11, but the ovule represents an appendage of a floral organ, within the carpels, and so is considered in this section.

The D function genes were defined following the discovery that two MADS box proteins in *Petunia*, FLORAL BINDING PROTEIN 7 (FBP7) and FBP11 were necessary for ovule development. Downregulation of both *FBP7* and *FBP11* by co-suppression resulted in the replacement of

ovules by carpelloid structures (Angenent *et al.* 1995). That the two genes function redundantly was demonstrated later when single loss of function mutants in either gene were shown to have wild type ovule development (Vandenbussche *et al.* 2003). *FBP11* activity is sufficient to induce ovule formation, as ectopic expression of the gene in *Petunia*, using the CaMV 35S promoter, causes the formation of ectopic ovules on the sepals of whorl 1 and the petals of whorl 2 (Colombo *et al.* 1995).

Investigation of *FBP7-* and *FBP11*-like genes in Arabidopsis revealed that the MADS box-encoding gene *SEEDSTICK* (*STK*) is orthologous to both *FBP7* and *FBL11* (Pinyopich *et al.* 2003). *STK* was shown to act redundantly with two other MADS box genes, *SHATTERPROOF 1* (*SHP1*) and *SHP2*, to specify ovule identity in Arabidopsis. Pinyopich *et al.* (2003) showed that the *stk* mutant has defects in funiculus development, but that the ovules are otherwise normal. Similarly, a double mutant between *shp1* and *shp2* has perturbations in seed dehiscence but ovule development is undisturbed. However, a triple *shp1/shp2/stk* mutant has arrested ovule development, indicating that at least one of the three genes must be active to specify the ovule fate. That each of the three genes is also sufficient for ovule development was shown by Favaro *et al.* (2003). Ectopic expression of *STK*, *SHP1*, or *SHP2* was sufficient to induce the formation of ectopic ovules on sepals and petals. However, it was also shown that ectopic expression of *STK* induced expression of *SHP1* and *SHP2* in outer whorls of the flower, implying that the three genes regulate each other's expression and function together to specify ovule identity.

The D function genes are thought to act together with the C function gene *AG* to specify ovule identity. The three Arabidopsis D function genes encode MADS box proteins which are closely similar to AG, and ectopic expression of *STK* induces ectopic *AG* expression (Favaro *et al.* 2003). This new combination of genes is expressed in a limited region of whorl 4 and creates the ovules within the carpels.

9.3 The role of E function genes

The E function genes are a recent addition to the ABC model, and their identification has finally shown how combinations of transcription factors can be sufficient to induce floral organ identity. While the ABC genes are each necessary to specify particular floral organ identities, they are not sufficient. Ectopic expression of the B function genes *AP3* and *PI*, for example, is sufficient to convert all floral organs to petals or stamens, but has no effect on leaves (Krizek and Meyerowitz 1996). The B function genes are sufficient within the flower to modify organ fate, but insufficient to convert leaves into petals. Addition of the E function genes into the equation, however, is sufficient to convert vegetative organs to floral ones. The E function genes are necessary for the development of all floral organs except sepals.

SEPALLATA1 (*SEP1*), *SEP2*, and *SEP3* all encode MADS box transcription factors and they are all expressed in whorls 2, 3, and 4 of the developing flower. *SEP1* and *SEP2* are also expressed in whorl 1 (Flanagan and Ma 1994; Savidge *et al.* 1995; Mandel and Yanofsky 1998). This led Pelaz *et al.* (2000) to develop the hypothesis that the *SEP* genes might be involved in floral organ development. Transposon insertion mutants were identified in each of the three genes, and each single mutant was shown to be phenotypically wild type, as were double mutant combinations. However, in a triple mutant of all 3 genes the petals, stamens, and carpels were all converted to sepals, so the flower develops just like a B and C function double mutant (Fig. 9.4a). The E function genes therefore act redundantly, but are necessary for the B and C function genes to do their job and therefore necessary for petal, stamen, and carpel development (Pelaz *et al.* 2000). The triple mutant also had reduced determinacy of the floral meristem, suggesting that the E function genes act with *AGAMOUS* to maintain the determinate state of the floral meristem.

Transgenic plants ectopically expressing any one of the *SEP* genes using the CaMV 35S promoter are not phenotypically very different to wild type, showing only mild alterations to the petals (Fig. 9.4b). However, combination of the E function genes with the ABC genes is sufficient to convert leaves into floral organs. Specifically, Honma and Goto (2001) showed that ectopic expression of *SEP3*, *SEP2*, *AP3*, *PI*, and *AP1* (i.e., A, B, and E) was

sufficient to convert all leaves to petals. They also showed that ectopic expression of *SEP3*, *AG*, *PI*, and *AP3* (i.e., B, C, and E) was sufficient to convert all cauline leaves (but not rosette leaves) to stamens (Fig. 9.4c). These experiments demonstrated that the ABC genes are sufficient in combination with the E function genes to convert vegetative organs into floral ones.

The interactions between the ABC and E genes are based on physical contact rather than regulation. Pelaz *et al.* (2000) showed that the ABC genes have normal expression patterns in *sep1/sep2/sep3* triple mutant plants, and it had previously been shown that the *SEP* genes are normally expressed in ABC mutants (Flanagan and Ma 1994; Savidge *et al.* 1995). However, yeast two hybrid studies have revealed physical interactions between ABC and E genes. For example, AG has been shown to bind SEP1, SEP2, and SEP3 (Fan *et al.* 1997) and the AP3–PI heterodimer has been shown to associate with SEP3 and AP1 (Honma and Goto 2001). These physical interactions may be the basis of floral organ identity, providing complexes which function together to activate transcription of downstream genes involved in the detailed development of each organ.

The *DEFH84*, *DEFH200*, and *DEFH72* genes are believed to function as the orthologues of *SEP1*, *SEP2*, and *SEP3* in Antirrhinum (reviewed in Jack 2001).

Recently, a fourth E function gene was identified in Arabidopsis. *SEP4* encodes a protein with very strong sequence similarity to the other SEP proteins, and with an initially similar expression pattern in all four whorls of the young flower. This expression later resolves into a strong signal in whorl 1, with weaker expression in the rest of the flower. Ditta *et al.* (2004) identified a mutant line with a T-DNA insertion in *SEP4*, and found the plant to be phenotypically wild type. However, when they constructed a *sep1/sep2/sep3/sep4* quadruple mutant they discovered that the flowers produced four whorls of leaf-like organs. The conversion of floral organs to leaves (rather than sepals, as in the *sep1/sep2/sep3* triple mutant) was apparent from the larger size of the organs, the amoeboid shape of their epidermal cells, and the branched nature of their trichomes (sepal trichomes

Figure 9.4 E function genes. (a) A triple mutant between *sep1/sep2/sep3* has four whorls of sepals and an indeterminate meristem. Image kindly supplied by Gary Ditta and Marty Yanofsky (UCSD). (b) A plant ectopically expressing *SEP3* shows only slight alterations to the sepals. (c) A plant ectopically expressing *AP3*, *PI*, *AG*, and *SEP3* has conversion of leaves to stamens. Images (b) and (c) kindly supplied by Koji Goto (Research Institute for Biological Sciences, Okayama). See also Plate 8.

do not branch). These data indicate that *SEP4* is necessary for sepal development, while the other three *SEP* genes are only necessary for the development of petals, stamens, and carpels.

9.4 The role of cadastral genes

The ABCDE genes rely on precisely defined expression patterns to ensure the production of appropriate floral organs in the correct places. While some of

these expression patterns are determined by mutual inhibition, such as the boundary between the A and C function genes, other expression patterns are determined by additional factors. A gene which delimits whorls and organ primordia territories but has no obvious direct role of its own is called a cadastral gene. Cadastral genes may operate through the spatial regulation of the expression patterns of other genes or through the inhibition of cell proliferation in regions delimiting organogenic territories (Breuil-Boyer *et al.*, 2004). Analysis of cadastral genes determining floral organ positioning has lagged some way behind the analysis of the floral organ identity genes, but information is beginning to emerge on the ways in which the boundaries of ABCDE gene expression are set.

The best studied floral cadastral gene is the *SUPERMAN* (*SUP*) gene of Arabidopsis which negatively regulates B function gene expression in whorl 4, restricting it to whorls 2 and 3. The *sup* mutant produces stamens in whorl 4, as a result of ectopic expression of the B function genes *AP3* and *PI* in those organ primordia (Fig. 9.5a; Bowman *et al.* 1992). *SUP* encodes a complex protein with a zinc-finger domain and a leucine-zipper motif, both reminiscent of transcription factors (Sakai *et al.* 1995). *SUP* is not expressed until after the first detection of the B function genes, when its transcript appears in the stamen primordia, adjacent to the boundary between the stamens and the carpels. This expression pattern has led to the suggestion that *SUP* is initially activated by AP3 and PI, and then acts in a negative feedback loop to inhibit their extension into whorl 4. The precise expression pattern of *SUP* relies on a complex mix of elements, including parts of the *SUP* coding region, suggesting that the gene is under extremely tight regulation (Ito *et al.* 2003).

A number of cadastral genes control the expression of C function genes, ensuring they do not spread into the second whorl and convert the petals into reproductive organs. In Arabidopsis, the repression of *AG* expression from whorls 1 and 2 is under the control of AP2, as discussed above, and several other proteins. The genes encoding LEUNIG (LUG), SEUSS (SEU), and STERILE APETALA (SAP) all interact with AP2 to repress *AG*. In single mutants with lesions in any of these three genes, *AG* transcript is found in whorls 1 and 2 (Liu and Meyerowitz 1995; Byzova *et al.* 1999; Franks *et al.* 2002). *LUG* encodes a protein with several WD repeat domains while SEU encodes a protein with similarity to animal transcriptional regulators (Conner and Liu 2000; Franks *et al.* 2002). It has been hypothesized that the two proteins form a complex in whorls 1 and 2 that is targeted to the second intron of *AG* by the specific DNA-binding activity of AP2 and represses *AG* transcription

(a)

(b)

Figure 9.5 Cadastral genes control the activity patterns of other genes. (a) The *superman* mutant has inappropriate B function in whorl 4, generating extra stamens (sepals and petals removed). (b) The *stylosa/fistulata* double mutant has inappropriate C function in whorls 1 and 2, generating stamens in whorl 2 and carpelloid sepals in whorl 1. Image kindly provided by Zsuzsanna Schwarz-Sommer (MPIZ, Cologne). See also Plate 9.

(Jack 2004). Although *AP2*, *LUG*, and *SEU* are all expressed in all four whorls of the developing flower, their repression of *AG* only occurs in whorls 1 and 2. This is likely to be because the post-translational repression of AP2 in whorls 3 and 4 means that the protein complex cannot be targeted to the *AG* sequence in cells in those whorls (Chen 2004). SEU and LUG may play additional, AG-independent, roles in the development of petals and sepals, including regulation of cell number, vasculature development, and organ polarity (Franks *et al.* 2006).

The RABBIT EARS (RBE) protein plays a similar role to that of SEU and LUG in the repression of AG activity in whorls 1 and 2. The *rbe* mutant has ectopic expression of *AG* in whorl 2, and consequent conversion of petals to stamens. *RBE* encodes a zinc-finger protein with many similarities to SUPERMAN (Takeda *et al.* 2003). Recent reports suggest that the *RBE* gene is required not only to specify the boundary between whorls 2 and 3, but also to specify the boundaries between organs within a whorl, as some *rbe* mutant alleles have fusion of sepals within the first whorl (Krizek *et al.* 2006).

Several cadastral genes also act to control the region of C function activity in Antirrhinum. *STYLOSA* (*STY*) is the orthologue of *LUG* (Navarro *et al.* 2004) and *sty* mutants have ectopic expression of the C function gene *PLE* in whorl 2, and, as a consequence, their petals are staminoid (Motte *et al.* 1998). The *fistulata* (*fis*) mutant also has ectopic *PLE* expression in whorl 2 and develops staminoid petals (McSteen *et al.* 1998). A double mutant between *sty* and *fis* produces carpelloid sepals in whorl 1 and stamens in whorl 2, essentially developing with an A function mutant phenotype (Fig. 9.5b). This confirms that *STY* and *FIS* act as cadastral genes to ensure repression of C function in whorls 1 and 2, allowing A function genes to specify correct sepal and petal development (Motte *et al.* 1998).

9.5 The quartet model of organ identity

The combined data so far obtained on floral organ identity have led to the proposal that tetramers of

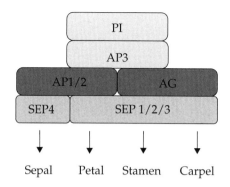

Figure 9.6 The revised ABC model. Tetramers of proteins specify the development of each whorl of organs. In whorls 1 and 4 multiple copies of the SEP and A/C function proteins are believed to make up the tetramers.

MADS box proteins function to specify the identity of each floral organ (Theissen and Saedler 2001). Many MADS box proteins have been shown to associate with other MADS box proteins in yeast two hybrid and *in vitro* assays, and we know that some MADS box proteins (such as the B function genes) operate as a heterodimer. It is therefore hypothesized that two dimers of MADS box proteins bind together and act as a complex to specify the activity of downstream target genes in each whorl. The situation is clearest in whorls 2 and 3, where the tetramers are proposed to consist of AP1, PI, AP3, and a SEP protein, and PI, AP3, AG, and a SEP protein, respectively. In whorl 4 it is postulated that two AG molecules and two SEP proteins interact, while in whorl 1 it is possible that two AP1 proteins interact with two SEP4 molecules (Fig. 9.6). This model remains to be tested, although the molecular data on the binding affinities of the various MADS box proteins are so far supportive. Whether the details of the model are correct, it is clear that the floral meristem identity genes activate the expression of suites of floral organ identity genes, their domains of expression bound by cadastral genes, and that these floral organ identity genes function as transcriptional activators to induce expression of the structural genes required to build the various floral organs.

CHAPTER 10

The ABC Model in Evolution

In Chapter 9 we discussed floral organ development, and the functions of a set of transcription factors in activating the structural genes necessary for correct organ development in the correct position. The majority of these ABC genes encode MADS box transcription factors, a very ancient family of DNA-binding proteins. By tracing the evolutionary history of this gene family we can investigate the molecular processes which lie behind the evolution of flowers. Since flowers are one of the key distinguishing features of the angiosperms, and have been argued to be responsible for their astonishing radiation into the most species-rich plant division by far, the evolution of the MADS box family is of particular interest in reconstructing plant evolutionary history.

10.1 Evolutionary history of MADS box transcription factors

MADS box genes are defined by a highly conserved 180 bp-long DNA sequence, the MADS box, which encodes the DNA-binding domain of the MADS box protein. This MADS domain folds into an antiparallel coiled coil of α-helices that lies flat on the minor groove of DNA (Pellegrini *et al.* 1995). The basic structural similarity of all MADS domains means that all MADS domain proteins bind to similar DNA sequences, with a consensus motif of $CC(A/T)_6GG$, known as a CarG box (Theissen *et al.* 2002). MADS box genes can be divided on the basis of sequence phylogeny and structural properties of the proteins they encode into two types, known as type I and type II. Both types are present in plants, animals, and fungi, suggesting that, not only was a MADS box gene present in the single-celled eukaryote that was the last common ancestor of these three great multicellular kingdoms, but that a gene duplication event had

already occurred in that lineage, giving rise to at least two MADS box genes at least 1 bya (Alvarez-Buylla *et al.* 2000). In contrast to this hypothesis, Tanabe *et al.* (2005) were only able to identify a single MADS box gene in each of three species of green algae, members of the Charophyceae, which are considered to be the sister group to the land plants. This may suggest that type I and type II MADS boxes evolved multiple times within different kingdoms. Further analysis of green algal and basal eukaryotic MADS box genes will be required to resolve this issue. In animals, type I MADS boxes include the human *SERUM RESPONSE FACTOR* (*SRF*) gene family, a highly conserved group of genes with roles in response to growth factors and subsequent development. Fungal type I MADS boxes include *MINICHROMOSOME MAINTE-NANCE 1* (*MCM1*) of yeast, which regulates cell cycle and cell growth. The plant type I MADS boxes, of which there are around 40 in the Arabidopsis genome, are poorly characterized, and none have yet been assigned a clear function (de Bodt *et al.* 2003). The animal type II MADS box genes are involved in muscle-specific gene regulation, while the fungal type II MADS boxes have yet to be fully characterized (reviewed in Becker and Theissen 2003). However, the plant type II MADS box proteins have been extensively studied, and include the ABC genes discussed in Chapter 9.

There are approximately 44 type II MADS box genes in the Arabidopsis genome, and all plant type II MADS boxes share a common domain structure, which has led to them also being known as MIKC MADS boxes (Fig. 10.1; Munster *et al.* 1997). Besides the MADS domain, described above, plant type II MADS boxes contain an I (Intervening) domain, a K (Keratin-like) domain, and a C (C terminal) domain. The MADS domain is responsible for

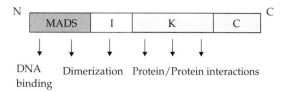

Figure 10.1 The structural domains of plant type II MADS box proteins, with their functions indicated underneath.

DNA binding but also plays a role in dimerization. The I domain is only weakly conserved, consists of around 30 amino acids, and is involved in the selective formation of DNA-binding dimers (Reichmann *et al.* 1996). While dimers can potentially form between any combination of MADS box proteins, only certain dimer combinations bind DNA. Without the I domain, the specificity of the interaction between dimers and DNA is lost. The K domain contains hydrophobic amino acids at regular intervals, generating an amphipathic helix involved in protein/protein interactions (reviewed by Kaufmann *et al.* 2005). The K domain is absent from type II MADS boxes in animals and fungi, and seems to be a plant-specific innovation (Alvarez-Buylla *et al.* 2000). The C terminus of the plant type II MADS boxes is the most variable region of the protein, and can fulfil a number of functions. It may be involved in transcriptional activation, through interaction with components of the cell's core transcriptional machinery, or may be necessary for the formation of multiprotein complexes which themselves are necessary for transcriptional activation (Egea-Cortines *et al.* 1999).

The plant type II MADS box genes can be divided into two types, known as the MIKC^c and MIKC* types (Henschel *et al.* 2002). The differences between these two groups of proteins include much longer I domains in MIKC* MADS boxes and divergent K domains. Both types of gene are present in the moss *Physcomitrella*, as well as in a club-moss (an ancient vascular plant group), suggesting that they are derived from a gene duplication event that occurred in or before the last common ancestor of mosses and vascular plants. This gene duplication can be dated to before the divergence of these groups, which occurred around 450 mya (Henschel *et al.* 2002).

The MIKC^c type II MADS box genes are a monophyletic group which contains all of the functionally characterized plant MADS boxes. This group can be further subdivided into clades, the number of which depends on the method of analysis used. Becker and Theissen (2003) divided the group into 13 clades, each described by the name of the first clade member to be identified. Functional analysis of the genes in these clades has led to hypotheses about the functions of the ancestral gene in each group. Analysis of the groups in different plant species has determined that at least six distinct MIKC^c MADS box genes were present in the last common ancestor of the gymnosperms and the angiosperms, which diverged around 300 mya.

10.2 ABC model in gymnosperms?

Among the clades of MADS box proteins identified in both angiosperms and gymnosperms are the groups containing the B function genes and the C and D function genes. Analysis of the function of these genes in plants that do not produce flowers may help us to understand the evolution of the angiosperm flower we see today. Most molecular work on gymnosperms is conducted on conifers, and a lesser amount on gnetophytes, particularly the tree *Gnetum gnemon* (see Chapter 1 for a reminder of the relationships between the different gymnosperms and the angiosperms). The reproductive structures of conifers differ from those of angiosperms in several main ways. Firstly, conifers do not produce hermaphroditic reproductive structures, but two sorts of cones (Fig. 10.2a). The male cones are usually quite small, and release pollen. The female cones contain the ovules containing the female gametes on individual scales, and are the commonly collected pine cones. Notably, the ovules are not entirely enclosed within maternal tissue, as they are in the angiosperms. Another big difference between gymnosperm and angiosperm reproductive structures is the absence of a perianth on gymnosperms. The perianth is the first and second whorl of floral organs in an angiosperm, and the organs in those whorls are specialized in a number of ways, most commonly for protection of developing floral buds and attraction of animal pollinators to mature flowers. The absence of a perianth in

(a)

(b)

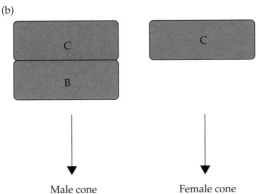

Male cone   Female cone

Figure 10.2 Gymnosperms and the ABC model. (a) Male (left) and female (right) pine cones. (b) B and C function genes are sufficient to specify male cone development, while C function genes alone specify female cone development.

gymnosperms results in much simpler reproductive structures which rarely attract animals to transfer their gametes.

Several B function genes have been isolated from conifers, notably from *Pinus radiata* (Monterey pine) and *Picea abies* (Norway Spruce). Two B function genes have also been isolated from *Gnetum*. The gymnosperm B function genes always resolve as the most ancestral of the B function genes, in phylogenetic analyses (Zahn *et al.* 2005). All of these genes are expressed in the pollen cones, but not in the female cones, and in no other plant tissue at all. This suggests strongly that the gymnosperm B function genes have a role in male reproductive structure development, just as the B function genes

have a role in producing stamens in angiosperm flowers. Interestingly, it appears that the formation of heterodimers which are obligate for functionality is a recent trait in the B function genes. Winter *et al.* (2002) showed that one of the two *Gnetum* B function proteins, GGM2, preferentially formed homodimers *in vitro*, rather than heterodimerizing with AP3, DEF, or GLO. This result was replicated in a yeast two hybrid analysis, and it was further shown that the GGM2 homodimers were able to bind DNA in a perfectly functional manner. Obligate heterodimerization, as shown by the Arabidopsis and Antirrhinum B function proteins, may therefore have evolved later in B function history, following further gene duplication events. In many other respects the gymnosperm B function genes appear very similar to angiosperm ones. Indeed, Sundstrom and Engstrom (2002) showed that ectopic expression of the B function genes *DAL11* and *DAL12* from *P. abies* in Arabidopsis resulted in the same phenotype as that observed when the endogenous B function gene *PI* was ectopically expressed, even though the spruce proteins also form homodimers.

Genes that fall into the C function clade have been isolated from all seed plant groups, including the gymnosperms *Gnetum*, *Cycas* (a Cycad), and *Picea* (conifer). Each of these C function genes has been shown to be expressed in both the seed cones and the pollen cones, indicating that C function is active in all reproductive structures in gymnosperms. This mirrors the situation in angiosperms where C function genes are expressed in both whorls 3 and 4, specifying the reproductive identity of organs in those whorls. In one of the earliest reports of gymnosperm C function, Rutledge *et al.* (1998) showed that the *Picea mariana* (black spruce) *SAG1* gene was expressed in the pollen and seed cones, and that ectopic expression of the gene in Arabidopsis phenocopied ectopic expression of *AGAMOUS* itself, generating first whorl carpels and second whorl stamens. In a very similar paper, published concurrently, Tandre *et al.* (1998) showed that the *P. abies* (Norway spruce) C function gene *DAL2* was also expressed in reproductive organs and could mimic *AGAMOUS* activity when ectopically expressed in Arabidopsis. More recently, Zhang *et al.* (2004) showed that the C function gene from the cycad *Cycas edentate* could complement the Arabidopsis *agamous* mutant when expressed using the Arabidopsis *AGAMOUS* promoter. These data confirm the strong conservation of C function proteins over 300 million years of evolutionary history.

So it appears that the last common ancestor of gymnosperms and angiosperms had a functional set of B and C function genes. Expression of the C function genes, just as in Arabidopsis, confers a reproductive identity on an organ. It seems to be the case in gymnosperms that it is expression of the B function gene that determines which sort of reproductive structure is produced. Expression of B function genes as well as C function genes results in the development of a male reproductive structure, a pollen cone. Expression of just C function genes alone results in the development of a female reproductive structure, a seed cone (Fig. 10.2b). This ability of B function genes to specify male identity to a reproductive structure also fits with the role of B function genes in producing stamens in the third whorl of the angiosperm flower. On the basis of these observations it has been proposed that the development of the classic hermaphroditic reproductive structure of the angiosperms involved a change in B function activity in an axis of a gymnosperm that produced only male or only female cones. Loss of B function gene expression from the upper regions of a male axis would result in the production of female organs within whorls of male organs. Similarly, gain of B function gene expression in the lower regions of a female axis would result in the production of male organs outside whorls of female organs (Theissen *et al*. 2002). The former scenario in particular has many elements in common with the 'mostly male' hypothesis of floral evolution, discussed in Chapter 1 (Frohlich 2002).

However, it appears that the last common ancestor of the angiosperms and gymnosperms did not have A function genes. The clade of MADS box genes containing the A function genes is known as the *SQUA-like* genes, and contains many genes from a diverse range of plants. *SQUA-like* genes have been isolated from monocots (maize, rice, barley, wheat) and basal angiosperms (including Magnolia), as well as from several eudicot species. This clade had therefore been established at the base of the angiosperm phylogenetic tree. However, no gene in this clade has been isolated from any gymnosperm species, suggesting that the clade is younger than the divergence of the angiosperms and the gymnosperms (Becker and Theissen 2003). The absence of A function genes from the gymnosperm lineage may go some way towards explaining the absence of a perianth in gymnosperm reproductive structures. The duplication event which led to the establishment and subsequent radiation of this clade may be responsible for the development of petals and sepals in the angiosperms, and their consequent ability to undergo dramatic speciation in cohort with the many different animals available to serve as pollinators.

10.3 ABC model in monocots?

Within the angiosperms ABC genes have been most extensively studied in the core eudicots (where Arabidopsis and Antirrhinum are placed) and the monocots. Some work has been conducted on the basal-most groups of angiosperms, but functional characterization of the genes identified is currently limited. For example, Stellari *et al*. (2004) isolated B function genes from a range of basal angiosperms, and compared genomic structures of the sequences. Similarly, Kim *et al*. (2005) isolated A, B, and C function genes from a smaller group of basal angiosperms, including *Amborella* itself, the sister to all other flowering plants. They analysed expression patterns of these genes, and concluded that expression of C function genes was conserved between basal angiosperms and the model eudicots, but that both B and A function genes have broader expression patterns in the more basal plants. These analyses will serve as a platform for subsequent analysis of protein function and the evolution of floral organ developmental programmes. In the meantime, since the monocots contain the economically important grasses, work in this group has been quite extensive. Among the clades of type II MADS box genes identified in the monocts are the groups containing the A, B, C/D, and E function genes, suggesting that these clades had been formed by gene duplications by the time the monocots diverged from the eudicots, 160–200 mya.

There are two main floral forms in monocots, each quite distinct from most eudicot flowers. The petaloid, or animal-attracting monocots, produce flowers which look superficially like eudicot flowers. This group includes the tulips, daffodils, lilies, and freesias. However, closer inspection of these flowers reveals that the reproductive organs of some species are surrounded by two whorls of petaloid perianth organs, and that green sepals are not normally present. The two outer whorls of brightly coloured organs are sometimes referred to as tepals. The wind-pollinated monocots, including the grasses such as maize, rice, and wheat, produce very different floral structures again. Most grasses produce separate male and female flowers, and those flowers do not produce petals or sepals. Instead each flower has a lemma and a palea, which function in a similar way to sepals to protect the developing flower. Within these, in most derived grasses, is a pair of small globular structures called the lodicules, nominally in whorl 2 of the flower, which expand on flower opening and force the palea and lemma away from the reproductive structures, allowing access to the wind. Inside the lodicules are the stamens or carpels (Fig. 10.3a).

10.3.1 Grasses

Molecular analysis of ABC genes in grasses has revealed that the lemma and palea are homologous to sepals, and the lodicules are homologous to the petals of eudicots (Fig. 10.3b). E function genes also appear to play roles in grass flower development. In the maize genome at least 8 MADS box genes falling into the E function clade have been identified, and the expression patterns of each are clearly distinct. These expression patterns suggest that the genes play a number of roles in flower development, including in the maintenance of floral meristem determinacy (Theissen *et al*. 2000). Within rice, the E function gene *OsMADS1* has been shown to be necessary for correct development of all floral organs, rather like the E function *SEP* genes of Arabidopsis. The *osmads1* mutant has reduced determinacy of the floral meristem, but also produces flowers with palea- and lemma-like lodicules, a reduced number of stamens, and an increase in carpel number (Jeon *et al*. 2000). Across the

(a)

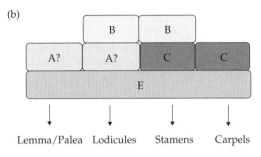

(b)

Lemma/Palea   Lodicules   Stamens   Carpels

Figure 10.3 Grasses and the ABC model. (a) Two flowers of maize, enclosed within a pair of glumes. Each flower has a lemma and a palea in the outer whorl, lodicules in the second whorl (barely visible) and stamens in the third whorl. Maize is monoecious and these are male flowers, so have no fourth whorl carpels. Image kindly provided by Bob Schmidt (U.C.S.D). (b) B, C, E, and possibly A function genes specify organ identity in grass flowers.

angiosperms the diversity of sequence and expression patterns of *SEP*-like genes suggests that they play a range of roles in the development of different types of flowers (Malcomber and Kellogg 2005).

As in the gymnosperms, C function genes have been shown to function in reproductive structure development in the grasses. In maize and rice MADS box genes which fall into the C function clade have been identified and shown to be expressed in the third and fourth whorls of developing flowers (Goto *et al*. 2001). There appears to have been quite extensive duplication of *AGAMOUS*-like genes within the grass family, with at least 4 *AG*-like genes present in the rice genome. Subfunctionalization of these genes has occurred, with no single gene fulfilling the complete role of

AG (that is, specifying stamen and carpel development, inhibiting A function in whorls 3 and 4, and maintaining meristem determinacy). Recent reports suggest that *OsMADS3* is required for stamen development in whorl 3, but that *OsMADS58* is required to maintain floral meristem determinacy (Yamaguchi *et al.* 2006). Such subfunctionalization and subsequent complexity is likely to be found in many species with genomes more complex than that of Arabidopsis.

B function genes have been particularly well characterized in the grasses, and are so functionally equivalent to their eudicot orthologues that there is no doubt of their recruitment to the specification of whorl 2 and 3 development before the monocot/eudicot division. In rice, loss of function of *OsMADS4*, a putative *GLO/PI* orthologue, was shown to result in homeotic transformation of the second whorl lodicules into paleas and lemmas, as normally found in the first whorl. Stamens were also converted into carpels. This phenotype is identical to the sepal/sepal/carpel/carpel phenotype of a eudicot B function mutant, confirming the homology of the genes and of the lodicules with petals (Kang *et al.* 1998). Similarly, loss of function of *OsMADS16*, an *AP3* homologue, also results in a classic B function mutant phenotype (Nagasawa *et al.* 2003). A more detailed study of the maize B function genes revealed that they were able to interact with their eudicot partners. Ambrose *et al.* (2000) described the phenotype of the *silky1* mutant, lacking a functional version of a maize gene orthologous to *AP3* of Arabidopsis. The mutant flowers produce paleas and lemmas in the second whorl and carpels in the third whorl, again identical to a eudicot B function mutant. The *SI1* gene was shown to be expressed in developing organ primordia in whorls 2 and 3 of the flower, confirming its role in the development of those organs. Whipple *et al.* (2004) extended this work by investigating the relationship between SI1 and the maize orthologue of PI, ZMM16. The two proteins were shown to heterodimerize *in vitro* and bind DNA as a heterodimer. Remarkably, ZMM16 was also able to bind DNA as a heterodimer with AP3 (the SI1 orthologue) but not with PI (its own orthologue). Similarly, SI1 was able to bind DNA as a heterodimer with PI but could not interact with AP3. Expression of *SI1* from the *AP3*

promoter in an *ap3* mutant line of Arabidopsis resulted in partial rescue of the mutant phenotype, with both petal and stamen development being restored (although neither organ was completely normal). Similarly, *ZMM16* was partially able to rescue a *pi* mutant (Whipple *et al.* 2004). These data reveal the remarkable degree of conservation between monocot and eudicot B function genes, despite the 200 million years of evolution that separate them and the disparate second whorl structures they induce. That those structures are truly homologous was shown recently by analysis of B function activity in a basal grass without lodicules. *Streptochaeta* does not produce lodicules, but has second whorl bracts, and phylogenetic trees indicate that it is basal to the derived grasses which do produce lodicules. Whipple *et al.* (2007) reasoned that, if B function genes had been recruited separately to determine lodicule development and petal development, then they would not be expressed in the bracts of *Streptochaeta*. *In situ* hybridization confirmed their expression in these second whorl bracts, indicating that the role of B function genes in second whorl organ development is ancient, and that the various morphologies of those organs are derived secondarily.

The A function genes have not yet been well characterized in the grasses, but there are clearly several members of the *SQUA-like* clade which have been isolated from monocots. In maize alone at least five genes which fall into this clade are known, although their functions have yet to be dissected. Analysis of expression patterns suggests that they play a range of roles, including in flower development, floral meristem determination, and vegetative development (Becker and Theissen 2003). Considering the complexities of A function gene activity in Arabidopsis and Antirrhinum, it is not surprising that this aspect of the grass ABC model should be the least clear. However, the presence of these genes in monocots, compared to their absence in the gymnosperms, may provide the basis for the evolution of perianth organs.

10.3.2 Petaloid monocots

ABC genes have been identified from a variety of petaloid monocots, including tulip, lily, *Agapanthus,* and various orchids. In all cases the gene expression patterns have led to the hypothesis that the ABC model applies to these flowers as it does to the eudicots, with one key difference. In those petaloid monocots with two whorls of coloured tepals, B function gene expression is not confined to whorls 2 and 3, but is also found in whorl 1. The two identical whorls of coloured tepals are therefore both developing under the control of petal identity genes (Fig. 10.4). Where mutant or transgenic experiments have been possible with these species, they have confirmed this model. However, unlike the experiments described above which confirm that the B function proteins in maize operate as a heterodimer, there is evidence that the petaloid monocot B function proteins homodimerize. Kanno *et al* (2003)

(a)

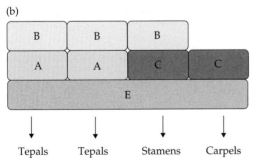

(b)

Figure 10.4 Petaloid monocots and the ABC model. (a) Lilies have two whorls of coloured tepals, an inner whorl of 3 and an outer whorl of 3. (b) A, B, C, and E function genes specify organ identity in tepaloid monocot flowers, with the B function genes being expressed in whorls 1, 2, and 3.

isolated two *AP3*-like genes and one *PI*-like gene from *Tulipa gesneriana*. All three genes are expressed in whorls 1, 2, and 3, confirming their role in the development of both whorls of tepals and the stamens. Each of the AP3-like proteins was shown to heterodimerize with the PI-like protein and bind DNA as a heterodimer. However, the PI-like protein was also able to bind DNA as a homodimer, an ability not shown by the higher eudicot B function proteins. These data suggest that the obligate nature of heterodimer formation in the eudicot B function proteins is a recent evolutionary phenomenon. Similar results have also been reported in *Lilium regale*, where AP3-like proteins heterodimerze with a PI-like protein, but the PI-like protein can also bind DNA as a homodimer (Winter *et al*. 2002). It is likely that the evolution of obligate B function heterodimerization was a two-step process, with the AP3-like descendents of a gene duplication losing the ability to homodimerize first, then the PI-like descendents. In support of this hypothesis, Kramer *et al*. (2006) reviewed both the literature relating to B function proteins and the sequences of B function genes from a large number of angiosperms. They concluded that a duplication event had occurred within the AP3 lineage in the basal eudicots (i.e., after the divergence of monocots) and that a subsequent single nucleotide deletion and frameshift in one of the two duplicates gave rise to the modern AP3 protein and its ability to heterodimerize.

10.4 Variations on the ABC model

The discovery that the petaloid monocots use an AB/AB/BC/C variation on the ABC model to specify their distinctive floral structure illustrates the inherent flexibility of the system. Simply by modifying the expression pattern of one or more of the ABC genes, through mutations to their promoter sequences or to upstream factors, it is possible to produce a flower with a different number or position of each of the floral organs. It is likely that the varying floral structures of many species will be explained through analysis of their ABC gene expression patterns in future years. For example, the monocots are not the only plants to produce brightly coloured first whorl organs, and the expansion of B function gene expression into whorl

1 may turn out to explain the bright blue sepals of *Delphinium*. Similarly, the dogwood *Cornus* attracts pollinators through brightly coloured bracts, which may be the result of ectopic A and B function expression in those organs.

One system that has been shown to develop through a variation on the basic ABC model is the composite inflorescence of *Gerbera hybrida*. *Gerbera* is a member of the Asteraceae and produces a composite inflorescence with female flowers in the outer whorls and hermaphrodite flowers in the inner whorls. The flowers do not have sepals, but structures called pappus hairs form in the first whorl, where they act as parachutes for the seed later in development. MADS box genes have been isolated from *Gerbera* which fall into each of the A, B, C, and E clades, and transgenic and expression analyses indicate that many of these fulfil essentially similar roles to their Arabidopsis counterparts. In particular, analysis of the B function genes has confirmed that the pappus hairs, like the palea and lemma of the grass flower, are simply variations on the theme of a sepal. Loss of function of *Gerbera AP3* or *PI* orthologues through antisense downregulation results in the conversion of the petals to pappus hairs and the stamens to carpels. Similarly, ectopic expression of the *PI* orthologue converts the pappus hairs in whorl 1 to petals and the carpels to stamens (reviewed in Teeri *et al*. 2002). However, it appears that the E function genes operate slightly differently in *Gerbera* from their known mode of action in Arabidopsis. The *Gerbera* gene *GERBERA REGULATOR OF CAPITULUM DEVELOPMENT 1* (*GRCD1*) is very

similar in sequence to the E function *SEP* genes of Arabidopsis, and, like the *SEP* genes, is expressed in all floral organs. However, downregulation of *GRCD1* function through an antisense approach resulted in the conversion of the sterile stamens (found in female flowers) into a second whorl of petals. Analysis of C function activity in these transgenic plants confirmed that GRCD1 was not operating through any transcriptional control of C function gene activity. These data suggest that this particular *Gerbera* E function gene has non-redundant whorl-specific activity, operating with the B and C function genes to specify stamen development (Kotilainen *et al*. 2000). This contrasts markedly with the position in Arabidopsis where the E function genes are redundant, only resulting in a phenotypic change when all three are inactivated together, and influence the development of all floral organs, rather than just one whorl.

The production of unisexual flowers is not unique to the Asteraceae, occurring also in monoecious systems such as maize and dioecious systems such as *Silene dioica* (red campion). These will be discussed in more detail in Chapter 12. In some species producing unisexual flowers the abortion of one type of reproductive organ has shown to be correlated with the downregulation of the appropriate B or C function genes, but it is not yet clear how the relationship between floral gender and ABC gene function is controlled. These and many other variations on the basic ABC model remain to be analysed, but the current depth of knowledge on the control of floral organ identity in Arabidopsis serves as an excellent starting point for such studies.

CHAPTER 11

Function and Development of Gametophytes

In previous chapters we have considered the induction and development of flowers, without paying much attention to the reason for their existence. The role of flowers is to ensure the transfer of a small but precious package, the pollen grain, from one plant to another. That pollen grain contains the sperm, which fertilizes an egg cell retained within the tissues of another flower, resulting in the formation of a zygote and, in due course, a new generation of plants which will produce flowers in their turn. While the details of seed and fruit development are beyond the scope of this book, it is necessary to provide a brief summary of the events that occur within the male and female reproductive organs of the flower. Here, meiosis and then haploid mitosis produce an independent generation, the gametophyte, including the gametes. In this chapter we will discuss the development of male and female gametophytes, and briefly consider the events that occur when they are brought together.

11.1 Alternation of generations in multicellular organisms

The alternation of generations in plants is a topic often considered deeply confusing. In fact, this confusion stems from a failure to understand the life cycles of all multicellular organisms, including animals. Put simply, in multicellular eukaryotes there are three types of life cycle, shown in Fig. 11.1. This is most likely a simple consequence of multicellularity evolving multiple times independently. There are several possible ways of developing a multicellular sexual life cycle from a unicellular one, and it seems that animals, plants, and fungi have each adopted different systems. These different life

cycles have different genetic effects and so have different evolutionary consequences. The different life cycles are defined on the basis of where ordinary mitotic cell divisions occur relative to the process of meiosis.

The animal (and fucoid) life cycle is based on gametic meiosis (Fig. 11.1a). In organisms that have gametic meiosis most of the life cycle is spent in the diploid stage, and accordingly animal bodies are composed of diploid cells. When reproduction occurs, those diploid cells undergo meiosis to produce gametes directly, and those gametes immediately fuse to make a diploid zygote. So animal cells only undergo mitosis in the diploid state, and the haploid cells of animals have one job and one job alone—to fuse together to make a new diploid cell.

The fungi, and some of the algae, have a life cycle which operates in the completely opposite way (Fig. 11.1b). The zygote is diploid, but instead of undergoing mitosis to produce a body, it first undergoes meiosis and becomes haploid. These haploid cells then undergo mitosis to produce the fungal body, which is still haploid. Gametes are produced by this haploid body through more mitotic divisions. This system is known as zygotic meiosis.

The third sort of life cycle is the one that the green algae and all land plants use, and is known as sporic meiosis (Fig. 11.1c). In an organism with sporic meiosis, mitosis happens in both the haploid and diploid stages of the life cycle. The diploid zygote undergoes mitosis to produce a diploid body, just like an animal. This diploid body undergoes meiosis to produce haploid cells, which are not gametes but are called spores. The spores undergo mitosis to produce a haploid body. And

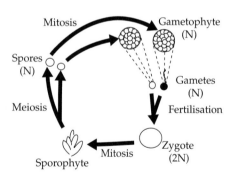

Figure 11.1 Different multicellular life cycles. (a) Gametic meiosis, as seen in animals. (b) Zygotic meiosis, as found in fungi. (c) Sporic meiosis, found in plants.

the haploid body then undergoes further mitosis to produce the haploid gametes, which fuse to make a diploid zygote again. So there are essentially two types of body—a haploid type and a diploid type. The haploid body is known as a gametophyte, because it produces gametes, and the diploid body

is called the sporophyte, because it produces spores. All land plants have both diploid and haploid phases in their life cycles. Although all terrestrial plants show sporic meiosis they vary greatly in the length of time they spend in the haploid and diploid states, and also in the independence or interdependence of the two phases. Over the course of their evolution, terrestrial plants have gradually changed from spending most of their time in the haploid state to spending most of their time in the diploid state; that is, they have moved from haploid or gametophyte dominance to diploid or sporophyte dominance.

Different life cycles carry different genetic consequences. Undergoing mitosis in the haploid state, as fungi and the plant gametophyte do, places enormous selective pressure on expressed genes. A gene expressed in the haploid stage cannot be masked by a corresponding dominant allele, so all alleles of genes expressed in the haploid stage are directly exposed to selection. Since the haploid gametophyte undergoes the complex processes of cell growth, cell division, and cell differentiation, many genes are expressed during this stage. In animals, the haploid phase is limited to a single cell, the gamete, which does not have to do much—no mitosis, no growth, no development of complex organs. So lethal recessive alleles can pass unnoticed in the animal haploid stage and then be masked by dominant alleles in the diploid stage. But for a plant, those lethal alleles will never get through the gametophyte stage. Accordingly, plants cannot carry lethal recessive alleles and so have a much lower genetic load than animals do. Plant populations have generally healthier genetic make-ups than animal populations and are less subject to inbreeding depression.

It is possible that the increasing dominance of the plant life cycle by the diploid sporophyte through evolutionary time reflects the high mortality of the gametophyte stage. By reducing the gametophyte stage to only a few cell divisions, angiosperms may have succeeded in minimizing this mortality while retaining the ability to lose catastrophic mutations before they enter a diploid phase. In more basal plants, such as mosses, the gametophyte stage of the life cycle is dominant. Thus, the familiar tufted green moss plant is the haploid gametophyte, while the diploid sporophyte is the stalked head which grows on top of the moss body at certain times of year. However, within the angiosperms the gametophyte generation has been reduced to only a few cell divisions, which are retained within the sporophyte body, the familiar flowering plant. The male gametophyte is simply the pollen grain, while the female gametophyte is the egg sac.

11.2 The angiosperm female gametophyte

The female gametophyte is also known as the megagametophyte. It is found within the ovule, within the carpel of the flower (Fig. 11.2a). Several female gametophyte forms are known in flowering plants, but the most common form (and the one found in Arabidopsis), called the *Polygonum* form, consists of seven haploid cells, one of which contains two haploid nuclei. These seven cells are of four types. At the top of the embryo sac, furthest from the point at which sperm will later enter (the micropyle), are three antipodal cells. Below these is the central cell, which contains two nuclei, and will later be fertilized by a single sperm to generate the triploid nutritive endosperm. Below this is a single egg cell (the gamete) flanked by two synergids. The female gametophyte is retained at all times within the tissues of the sporophyte parent, but from there it plays a crucial role in directing the growth of the incoming pollen tube. It also expresses genes which control the initiation of seed development after fertilization, and later control the development of both the embryo and the endosperm (Yadegari and Drews 2004).

The development of the megagametophyte is a two-stage process. Firstly a cell within the sporophyte tissue (within the ovule), known as the megaspore mother cell, undergoes meiosis and gives rise to four haploid nuclei. This process is called megasporogenesis. Usually three of these nuclei degrade, and the fourth, the megaspore, then undergoes mitosis to produce the eight haploid nuclei of the mature gametophyte. This process is called megagametogenesis (Yadegari and Drews 2004).

11.2.1 Megasporogenesis

Megasporogenesis is a relatively simple process, during which a single meiotic event gives rise to

(a)

(b)

(c)

Figure 11.2 The angiosperm female gametophyte. (a) Position of ovules within the Arabidopsis carpel. (b) Megasporogenesis in Arabidopsis. (c) Megagametogenesis in Arabidopsis.

four haploid nuclei. Ma (2005) provides an excellent summary of chromosome behaviour during meiosis in plants. In different plant species the four hapoid nuclei may be packaged into one, two, or four cells. Most commonly, in plants which produce a *Polygonum*-type megagametophyte, the megaspore mother cell within the ovule produces four haploid cells, as cell division occurs after both meiosis I and meiosis II. Three of these haploid cells undergo cell death, leaving a single haploid megaspore (Fig. 11.2b). However, in other species cell division occurs after only one, or even neither, of the meiotic divisions. In species with bisporic megasporogenesis, cell division occurs only after meiosis I, so meiosis II results in the production of two cells, each containing two haploid nuclei. One of these survives and enters megagametogenesis to produce a final total of six haploid nuclei. In species with tetrasporic megasporogenesis, cell division does not occur after either meiotic division. The final megaspore has four haploid nuclei, which undergo mitosis to produce around 16 nuclei in the mature megagametophyte.

Since the development of the megaspore is a phase of sporophyte development, it is genes expressed in the sporophyte that control megasporogenesis. Mutations in such genes often manifest themselves as female infertility, due to failure of the megaspore mother cell to develop, or of meiosis to occur, or due to incorrect positioning or polarity of the embryo sac within the ovule (reviewed by Yadegari and Drews 2004).

11.2.2 Megagametogenesis

Megagametogenesis consists of a number of mitotic cell divisions, resulting in the mature megagametophyte. In the most common system this simply involves the single haploid spore undergoing three rounds of mitosis to produce eight haploid nuclei (Fig. 11.2c). In Arabidopsis these mitotic divisions require the activity of CHROMATIN REMODELLING PROTEIN 11 (CHR11), as transgenic lines with *CHR11* silenced are unable to proliferate nuclei and so fail to undergo megagametogenesis (Huanca-Mamani *et al.* 2005). It is likely that a number of such 'housekeeping' genes are essential for these mitotic divisions, and it is the inability of the gametophyte to form gametes in their absence that protects the plant gene pool from accumulating deleterious mutations. Cell walls do not usually form until all divisions have occurred, allowing migration of nuclei during development. Specifically, one nucleus from the top of the embryo sac, and one from the bottom (the micropylar end, where the pollen tube will arrive) migrate to the centre of the structure and fuse, either immediately or at fertilization, to generate the diploid central cell which will be fertilized to give rise to the triploid endosperm. This fusion requires the RPL21M protein, a mitochondrial 50S ribosomal subunit, and in a mutant defective in the production of this protein the nuclear membranes of the two central cell nuclei fail to merge (Portereiko *et al.* 2006). In different species other gametophyte cells may also undergo different differentiation events. For example, in Arabidopsis the three antipodal cells, those furthest from the micropyle, undergo cell death shortly before fertilization occurs.

Little is known about the evolutionary history of this simple eight-celled megagametophyte. More basal lineages of land plants, including gymnosperms, inevitably produce megagametophytes containing many more cells. Recent reports of a nine-celled megagametophyte in *Amborella trichopoda*, the sister to all other angiosperms, suggest that the modern angiosperm situation arose by reduction in the cell divisions which occur during gametogenesis (Freidman 2006; Rudall 2006).

Megagametogenesis is the whole of the gametophyte's life, and it is genes expressed by the gametophyte that determine correct gametogenesis. Mutations which perturb cell division or any essential metabolic function will be lethal at this stage, and can usually be identified because of perturbed segregation ratios in the progeny of a plant. Classically, a megagametophytic mutation will result in an absence of 50% of the seeds in a pod or silique. For example, Ebel *et al.* (2004) showed that the plant retinoblastoma protein, RBR1, is essential for correct megagametogenesis in Arabidopsis. The *rbr1* mutants could only be maintained as heterozygotes, as the megagametophyte failed to develop normally if it carried the *rbr1* mutant allele. The RBR1 protein is involved in cell cycle control in plants and animals, and the *rbr1* mutant megagametophytes underwent excessive mitotic divisions, resulting in a non-viable embryo sac. This mutant phenotype also supports the suggestion that the ancestral angiosperm megagametophyte contained more cells, but that genes such as *RBR1* have been recruited to reduce cell division and minimize gametophyte lifespan. In theory, mutations in genes expressed in the sporophyte may also affect megagametogenesis, by altering the environment surrounding the developing gametophyte or inhibiting its access to water or other essential compounds.

Yadegari and Drews (2004) reported the identification of over 150 mutants with perturbed megagametogenesis, and described their classification into mutations affecting different stages of gametophyte formation. Detailed characterization of the lesions occurring in these mutants is complex, as visualization of gametogenesis is not simple, given that it occurs within living tissues of the sporophyte. However, molecular characterization of these mutated genes will greatly enhance our understanding of how the angiosperm female gametophyte develops and functions. Similarly, Pagnussat *et al.* (2005) described 130 gametophytic mutants in Arabidopsis, generated using transposon insertions perturbing segregation ratios in the next generation. These mutants affect all processes from the first mitotic division of megagametogenesis, through megagametophyte development, to later embryo sac functions including post-fertilization processes such as embryo development. The authors were able to amplify sequence flanking the inserted transposons, and catalogue the genes perturbed in the mutants. The proportions of proteins of different types involved at different stages varied quite significantly. During megagametogenesis itself, genes with roles in secondary metabolism seemed to be particularly important, along with genes involved in protein degradation. During the later fertilization process genes related to energy metabolism and stress responses seemed more important. This initial categorization of genes important at different stages has provided surprising insights into female gametophyte development and function, and it is clear that many further insights will follow from detailed characterization of the genes involved.

11.3 The angiosperm male gametophyte

The male gametophyte, also known as the microgametophyte, begins life within the stamens of the flower, and completes its development by germinating to produce a pollen tube (usually within the style of (hopefully a different) flower). It is even more reduced than the female gametophyte, and less variable in structure between different plants. Essentially, the microgametophytes are pollen grains, and the locules of the stamens of most plants are filled with many hundreds or even thousands of them. The mature microgametophyte consists of three haploid cells. A larger vegetative cell contains two sperm cells, which go on to fertilize the egg cell and the diploid central cell of the female gametophyte. The vegetative nucleus itself also travels down the developing pollen tube, alongside the sperm nuclei. The whole microgametophyte is coated with a specialized outer layer of sporopollenin, which protects the contents from dehydration and mechanical damage as they are transferred between plants (Fig. 11.3b). This pollen coat also serves to mediate interactions between the microgametophyte and the female sporophyte tissue that it lands on, and so plays an important

(a)

(b)

Figure 11.3 The angiosperm male gametophyte. (a) Microsporogensis and microgametogenesis in Arabidopsis. The tetrad of microspores present at stage 2 disassociate before undergoing microgametogenesis. (b) Mature microgametophytes are coated with elaborate sporopollenin.

role in mating success. The pollen coat is mainly derived from sporophytic tissue on the inside of the stamen locules.

The development of the microgametophyte is a two-stage process (Fig. 11.3a). Firstly a cell within the sporophyte tissue (within the stamen), known as the microspore mother cell, undergoes meiosis and gives rise to four haploid cells, known as a tetrad (see Ma (2005) for a summary of the events that occur during meiosis). This process is called microsporogenesis. The tetrads are initially enclosed in a thick callose wall, which is broken down by an enzyme secreted from the stamen wall, releasing the microspores. These microspores then enlarge and divide asymmetrically to produce a large vegetative cell and a small generative cell, which is retained within the vegetative cell. This process is called microgametogenesis. The generative cell divides again to produce two sperm cells, although this division may not occur until after pollination has been achieved (McCormick 2004).

11.3.1 Microsporogenesis

Microsporogenesis occurs within the stamen. Specifically, on top of the stamen filament is formed an anther, which consists of four cell layers enclosing the microspore mother cells. The innermost of these four layers, the tapetum, has a key role in pollen grain development. When the microspore mother cells initiate meiosis, callose is deposited on their outer surfaces. Thus the tetrads of microspores are surrounded by callose. These tetrads then disassociate from the tapetum, generating a space, called the locule. Within the locule callase released by the tapetum degrades the callose outer walls and frees the four microspores from each tetrad (Ma 2005). They are then free to enlarge and begin mitotic divisions to produce the microgametophyte.

As with megasporogenesis, microsporogenesis is controlled by genes expressed by the sporophyte itself (that is, the diploid parental tissue). Such genes must ensure the correct formation of the cell layers in the anther, along with regulating the onset of meiosis. For example, the *EXCESS MALE SPOROCYTES 1* (*EMS1*) gene is necessary for the normal development of tapetum cells in the innermost layer of the anther and microspore mother cells attached to them (Zhao *et al*. 2002). In the *ems1* mutant excess microspore mother cells develop at the expense of the tapetal layer. However, since the tapetum is necessary for pollen development, this mutant is male sterile. It seems that the EMS1 protein is essential for the division of cells following meiosis, as the microspore mother cells in the *ems1* mutant undergo meiotic separation of chromosomes and chromatids but do not form new cell membranes around the new nuclei. These incomplete microspores are then incapable of undergoing mitosis. *EMS1* encodes a leucine-rich receptor-like protein that localizes to cell membranes and is hypothesized to be present in the membranes of tapetal precursor cells (Zhao *et al*. 2002). It may be bound by the TAPETUM DETERMINANT 1 (TPD1) protein, which is a very small, possibly secreted, protein, encoded by a gene expressed in the developing microspore mother cells (C Yang *et al*. 2003). Mutation of *TPD1* results in a very similar phenotype to that of the *ems1* mutant, namely an increase in microspore mother cells at the expense of tapetum, suggesting that these two proteins act together to regulate the respective numbers of tapetal cells and microspore mother cells. Other leucine-rich repeat receptor kinases are also involved in distinguishing tapetal and microspore mother cells. Colcombet *et al*. (2005) reported that a

double mutant between the *SOMATIC EMBRYO-GENESIS RECEPTOR KINASE 1* and *2* genes (*SERK1* and *SERK2*) showed a phenotype similar to that of the *ems1* mutant, with an absence of tapetum and an excess of microspore mother cells. It is likely that a number of receptors and ligands are involved in this feedback loop, to provide multiple checkpoints in this essential process. Following this establishment of cell fate and cell number, a collection of downstream genes, including several encoding transcriptional regulators, act together to ensure correct differentiation of microspore mother cells (reviewed by Ma 2005). Mutation of any of these genes results in a failure of microsporogenesis and subsequent male sterility.

The microspores themselves also develop in response to sporophyte-expressed genes. These genes control the meiotic divisions which produce the microspores. Many essential genes are required for correct pairing and separation of chromosomes during meiosis. Later, the *TETRASPORE* (*TES*) gene is essential for the formation of cell plates between the meiotically produced haploid nuclei. The *tes* mutants produce large microspores containing four haploid nuclei in a single cell. The TES protein is a kinesin which may determine microtubule positioning within the cell, a necessary component of cell division. The absence of the TES protein is correlated with an absence of the radial arrays of microtubules usually seen when the microspores undergo cell division (S. Yang *et al*. 2003).

Detailed analysis of microsporogenesis has been used in the dissection of evolutionary relationships between flowering plants. One key character is the timing of the cell divisions with relation to the nuclear divisions, while another is the orientation of formation of successive cell plates during meiosis. Analysis of these traits in basal angiosperms has revealed a great degree of variability between taxa and suggests that microsporogenesis is an evolutionarily very labile stage of development (Furness *et al*. 2002).

11.3.2 Microgametogenesis

Following microsporogenesis, the anther locules contain free microspores. These become heavily vacuolized, pushing the nucleus to the edge of the cell. The microspores then undergo a stereotyped set of cell divisions (one asymmetric, one symmetric), to produce the three nuclei of the microgametophyte. The first mitosis is asymmetric because the nucleus is at the periphery of the cell. The large daughter cell is the vegetative cell, and contains most of the cytoplasm from the microspore. The smaller cell, the generative cell, is entirely engulfed by the vegetative cell. It then undergoes a further, symmetric division to produce two sperm cells. This final division occurs within the developing pollen grain in some species (those with trinucleate pollen), but in other species (those with binucleate pollen) it occurs during pollen tube growth after pollination (Ma 2005). The developing pollen grain is coated with two protective layers, the exine and the intine. The exine is made of highly elaborate sporopollenin, which mediates interactions with stigmatic cells, while the intine is made from cellulose, pectin, and proteins. Both layers appear to be mainly derived from the anther tapetum. The mature pollen grain is then dehydrated ready for release into the outside world.

As with megagametogenesis, microgametogenesis is regulated by genes expressed by the gametophyte nuclei themselves. However, because the tapetum contributes to pollen coat development, there is also a clear role for sporophytically expressed genes in microgametogenesis. For example, the genes described earlier which are necessary to correctly specify tapetal and microspore fate also influence gametophyte development. Those microspores which survive in mutants defective in these genes fail to undergo correct gametogenesis due to the abnormal tapetal layer surrounding them. Similarly, the sporophyte expresses many genes involved in building the pollen coat. Plants carrying mutations in these genes are usually male sterile, as a result of perturbed exine and intine development. For example, the *male sterile 2* mutant of Arabidopsis carries a lesion in a gene encoding a protein that promotes the synthesis of long-chain molecules, such as sporopollenin (cross-linked carotenoids). The mutant produces non-viable pollen grains with thin cell walls lacking an exine (Aarts *et al.* 1997). A number of such mutants are reviewed by Ma (2005).

Male gametophytic mutants influence segregation ratios in the progeny of plants, usually resulting in abortion of 50% of pollen grains. However, unlike the situation with female gametophytic mutants, where this results in 50% reduction in seed set, the vast excess of pollen relative to egg cells means that male gametophytic mutants can be difficult to identify using conventional screens. Instead, perturbations in the segregation ratios of other markers are used to pick out male gametophyte mutations. Since many thousands of genes are likely to be expressed in the male gametophyte, it is not surprising that many mutants have been described. Indeed, recent microarray analysis of genes expressed in the pollen grain has revealed that 992 genes of the 7792 on the array were expressed, suggesting a likely total of around 4000 pollen-expressed genes when the whole genome is considered. The expressed genes included a high proportion of essential genes with roles in signal transduction and cell wall biogenesis (Honys and Twell 2003). Many of the mutants that have been described so far underline the importance of the asymmetry of the first mitotic division. For example, the *gemini pollen 1* mutant produces pollen grains with apparently randomly oriented first divisions, which then arrest after that first division (Park *et al.* 1998). The cells produced by this first division express vegetative cell markers, indicating that the vegetative cell is the default state in the absence of asymmetry. The GEM1 protein binds to microtubules and appears to be involved in microtubule positioning, apparently essential for the appropriate asymmetry of the first mitotic division (Twell *et al.* 2002). Other mutants are affected in the number of mitotic divisions that occur, the positioning of the vegetative and generative cells with respect to one another, and the positioning of the different cells within the pollen grain itself. It is clear that many genes are involved in microgametogenesis, and it should therefore be noted that the reduced size and duration of the male gametophyte stage of the angiosperm life cycle is still quite sufficient to ensure that few lethal mutations are retained within the genome.

11.4 Events following pollination

Mature pollen arrives at the stigmatic surface at the distil end of a carpel through any number of routes, although the important role of animal pollinators will be discussed in detail in Section III of this book. Once at the stigma of a suitable flower of the same species, the pollen grain germinates. The developing pollen tube grows through the carpel tissue, in a special tissue called the transmitting tract. Genes expressed in the stigma and style (that is, sporophytic genes) play important roles in guiding the germinating pollen tube towards the embryo sac. The vegetative nucleus and the sperm nuclei migrate with the pollen tube, staying close to the growing tip. The pollen tube is attracted to the micropylar end of an ovule, by a number of gametophytic chemical messengers, and there bursts within a synergid, releasing the two sperm cells into the entrance to the embryo sac. The synergids die, allowing the sperm cells to fuse with the embryo sac proper. One sperm nucleus enters mitosis with the egg cell, to produce a diploid embryo, while the other enters mitosis with the diploid central cell to produce a triploid tissue, the endosperm. Analysis of a number of Arabidopsis mutants has shown that the female gametophyte guides the pollen tube to the micropyle, from there into the synergids, arrests pollen tube growth, causes pollen tube bursting, and prevents further pollen tubes entering the embryo sac (reviewed by Dresselhaus 2006). Mechanisms are also in place to ensure that the male and female gametes are at the same stage of the cell cycle when fertilization occurs. The appropriate cell cycle stage varies between plant species, but is consistent within a single species (reviewed by Weterings and Russell 2004). The endosperm grows and divides to provide the nutritive tissue in the seed, while the embryo undergoes cell division and development to produce the next sporophyte generation. The outer layers of the ovule, the integuments, become the seed coat which protects the developing embryo. Development of this new generation is beyond the scope of this book.

CHAPTER 12

Preventing Self-fertilization

This section of the book has focused on the development of the angiosperm flower. The flower that has been discussed so far is a 'perfect' (hermaphrodite) flower, containing both male and female reproductive structures, and producing, through sporogenesis and gametogenesis, both male and female gametes. Hermaphroditism carries both advantages and disadvantages. Perfect flowers can, assuming no other constraints, self-pollinate and fertilize their own ovules. This sexual reproduction without the need for another individual of the same species gives the plant a certain set of advantages. The most important of these is guaranteed reproduction, irrespective of the availability of likely mates, and without relying solely on asexual reproduction (such as the production of runners or tubers). Since self-fertilization is a sexual process, involving independent assortment of chromosomes at meiosis and the possibility of recombination between chromosomes, it does not result in the same lack of genetic variability that is the consequence of asexual reproduction. This guaranteed sexual reproduction gives self-pollinating plants the ability to colonize new habitats, and it is a common trait in weedy species. A second advantage of self-fertilization is the saving in energy and nutrients that would otherwise be spent producing costly rewards for pollinators. It is perhaps not surprising, then, that around 20% of angiosperm species are predominantly self-fertilizing (Barrett 2002). However, self-fertilization does carry a disadvantage relative to outcrossing, which is that the genetic variability produced, although greater than in an asexual population, is considerably less than that seen in an outbreeding population. The reduction in viability of inbred progeny compared to outcrossed progeny is known as inbreeding depression, and is recognized as the primary

selective pressure resulting in strategies to avoid self-fertilization (Barrett 2002). The balance between the relative importance of assured reproduction and genetic variability differs in different species, largely as a result of their habitats, life cycles, and the niches that they occupy (Charlesworth 2006). In this chapter we will consider the developmental and biochemical ways in which self-fertilization can be reduced or prevented.

12.1 Reducing self-pollination in a hermaphroditic flower

Self-pollination can occur within a flower, or between flowers on the same plant. Pollination within a single flower is called autogamy, and is an obvious consequence of producing hermaphrodite flowers. Autogamy may occur simply as a result of the positioning of stamens and anthers, or it may be facilitated by visiting animals. Weedy plants that self-pollinate (such as Arabidopsis) do not usually require an animal pollinator, as pollen is transferred directly from the stamens onto the stigma, which are in very close proximity. Autogamy can be reduced by separating the stamens and stigma physically within the flower (herkogamy) or by separating the timing of their development (dichogamy).

12.1.1 Herkogamy

Herkogamous flowers have their anthers and stigma positioned such that pollen cannot reach the stigma without help (Fig. 12.1a). This trait should act to prevent pollen reaching the stigma, unless transferred by an animal. However, a pollinating animal is likely to carry pollen from other individuals, meaning that a herkogamous flower should

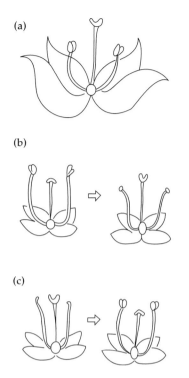

Figure 12.1 Reducing self pollination in a hermaphrodite flower. (a) Herkogamy. (b) Dichogamy—protandry. (c) Dichogamy—protogyny.

receive a mixed load of self and non-self pollen. Takebayashi *et al.* (2006) investigated within-population variability in herkogamy in *Gilia achilleifolia*, a small blue-flowered Californian phlox which has the stigma extended some way out of the flower, beyond the stamens. They found that a greater degree of herkogamy (measured as distance between the stigma and the top of the stamen) was correlated with a greater degree of outcrossing, confirming the hypothesis that herkogamy reduces self-pollination. However, not all studies have found correlations between herkogamy and outcrossing. Medrano *et al.* (2005) found no increase in outcrossing associated with increased herkogamy in six populations of *Narcissus longispathus*. This may reflect a number of problems with this method of reducing outcrossing. Herkogamy is dependent on pollinator foraging behaviour to ensure the non-autogamous pollen deposited on the stigma is collected from a different individual, not from a different flower on the same plant. It is also reliant on the pollinator not brushing by the stigma when

it leaves the flower, a problem in the *N. longispathus* system as the pollinator often grooms on the edge of the flower, near the stigma, after feeding.

12.1.2 Dichogamy

Dichogamy is a slightly more reliable method of preventing autogamous pollination than herkogamy. Dichogamous flowers have the timing of anther dehiscence separated from the timing of stigma receptivity. Dichogamy is extremely common, with as many as 87% of angiosperm species showing some degree of temporal separation of male and female function (Bertin and Newman 1993). The great advantage of dichogamy over herkogamy is that it is not possible for an animal pollinator to accidentally transfer pollen from the flower onto the receptive stigma. However, the same difficulty occurs with foraging behaviour—if pollinators forage primarily within an inflorescence then the pollen they transfer to receptive stigmas will still be self pollen, just from different flowers on the same plant.

There are two variations of dichogamy, according to which reproductive structure matures first. In protandrous flowers the anthers dehisce and the pollen is shed several days before the stigma is receptive (Fig. 12.1b). In protogynous flowers, which are less common, the stigma is receptive several days before the anthers dehisce and then ceases to be receptive before the pollen is actually released (Fig. 12.1c). A possible explanation for the relative infrequency of protogyny is that it requires a reversal of the usual order of development of the reproductive organs. As we have seen in previous chapters, the fourth whorl organs, the carpels, usually develop last in floral meristems.

Protogyny is predicted to be a more efficient inhibitor of autogamy than protandry. In protandrous flowers pollen may persist until after the stigma becomes receptive, causing self-pollination. Another advantage of protogynous flowers is that they have the ability to 'top up' any ovules which have not received outcross pollen with their own pollen when it is released. Since protogyny is more effective than protandry, but protandry is more common, it has been proposed that one of the main benefits of protandry is in preventing interference between stamens and stigmas, by separating them, rather than in reducing inbreeding depression. Such interference results in the wastage of pollen on self stigmas, when it could contribute to male fitness if properly dispersed. Sargent *et al.* (2006) modelled the evolution of dichogamy, and found that both avoiding inbreeding depression and avoiding interference between stigma and stamens played important roles in the appearance of dichogamy under different conditions. This may go some way towards explaining the surprising prevalence of dichogamy (particularly protandry) in biochemically self-incompatible species (which should not need additional mechanisms to prevent self-fertilization; Bertin 1993).

Species-level comparative studies have also demonstrated a correlation between pollinator type and dichogamy. Protandrous species are more often pollinated by bees or flies, while protogynous species are often wind- or beetle-pollinated (Sargent and Otto 2004). These correlations suggest that, as we saw with herkogamy, pollinator foraging behaviour can have significant consequences for the effectiveness of strategies to minimize self-pollination.

12.2 Monoecy

The only developmental way to ensure that autogamy does not occur is to separate male and female functions into different flowers. Monoecious plants are hermaphroditic, as they produce both male and female gametophytes (and thus gametes), but the individual flowers on a monoecious plant are of a single sex. This strategy still allows self-pollination between different flowers of the same individual (geitonogamy) to occur, and so is not a very effective method of preventing self-pollination. It is, however, quite common, with around 5% of angiosperm species being entirely monoecious, and another 4.5% producing both monoecious and hermpahroditic flowers.

Monoecy has often been associated with inefficient pollination systems. It is common in grasses and wind-pollinated trees, and has recently been found to be particularly frequent in tropical Australian tree species (Gross 2005). This may reflect the infrequency or absence of specialized pollinators such as birds, bats, and large bees in the

Australian forests, and the prevalence of generalist pollinators such as beetles, flies, small bees, and thrips. When pollination is unreliable or inefficient, monoecy may be selectively advantageous over dioecy (see below) because pollen does not have to travel between individuals, just between flowers, to achieve fertilization. It may also be advantageous to advertise strongly to such inefficient pollinators by providing 'spare' male flowers with plenty of rewarding pollen.

A second major advantage of monoecy is that it allows the plant to vary the number of male flowers and the number of female flowers it produces according to its age, health, or environment. For instance, in many monoecious trees young plants have almost entirely male flowers, whereas older trees have mostly female flowers. Older trees have usually built up more resources and can afford to support fruit, whereas younger trees benefit from the ability to contribute to the next generation without the investment in costly fruit and seed. Similarly, drought-stressed monoecious plants in dry habitats tend to produce mostly male flowers while ones in more moderate habitats tend to produce more female flowers (Freeman *et al.* 1981). This ability to selectively allocate resources to different reproductive functions is often considered a much bigger advantage of monoecy than any effect on outcrossing rate, which is seen as a secondary consequence of responding to selective pressure to maximize reproductive output under different conditions.

Monoecy is believed to be a derived condition (Richards 1997). A number of plant species produce both unisexual and hermaphroditic flowers, and these are believed to represent evolutionary intermediates. Andromonoecious plants, such as *Cucumis melo*, the melon, produce both male and hermaphroditic flowers, while gynomonoecious plants, such as various *Poa* species, produce both female and hermaphroditic flowers. It has been observed that andromonoecious plants with few resources produce a greater proportion of male flowers than their better resourced contemporaries. These data suggest that the ability to vary male and female output, rather than avoidance of autogamy, is the driving force behind the evolution of monoecy.

Figure 12.2 Monoecy. Flowers of each sex are often developmentally separated in the plant.

The development of monoecy from hermaphroditic flowers requires male sterility and female sterility to be expressed developmentally and spatially within the same individual plant. It is likely that this involves specific loss of B or C function gene activity in whorl 3 to make female flowers and specific loss of C function gene activity in whorl 4 to make male flowers (see Chapter 9). It is usually the case that there is a distinct pattern of arrangement of male and female flowers within a monoecious plant. For instance, most grasses have female flowers at the bottom of the inflorescence and male flowers at the top (Fig. 12.2). This architectural separation of reproductive function is probably easier to produce developmentally than a more integrated arrangement, as promoter activity of B and C function genes can be tied to developmental stage. However, it should be remembered that the B and C function genes are only necessary to differentiate male and female function, but that other genes may actually specify that function. Perl-Treves *et al.* (1998) examined three *AGAMOUS*-like (C function) genes in *Cucumis sativus*, cucumber, and found no changes in expression correlated with hormone treatments that alter floral sex. The sex of maize flowers is determined by a number of genes that inhibit stamen development or abort carpel development at an early stage. Gibberellin has also been shown to influence floral sex, and many of the maize mutants with disruptions in moneocy have been shown to be perturbed in gibberellin synthesis or perception (Dellaporta and Calderon-Urrea 1994). A considerable amount of further work is required to provide a full picture of the development of monoecious flowers.

12.3 Dioecy

Dioecy is a relatively common solution to the problem of self-fertilization, with around 6% of angiosperm species showing complete separation of the sexes (Barrett 2002). Dioecy is scattered throughout the angiosperm phylogenetic tree, suggesting that it has evolved independently many times. As many as 160 plant families include dioecious species, and it has been predicted that the trait has evolved around 100 times independently. However, there are a few plant families where dioecy is particularly prevalent. These include the Euphorbiaceae, the Cucurbitaceae, and the Urticaceae (Ainsworth, 2000).

Dioecy is often viewed as the logical endpoint of mechanisms to separate sexual function in space or time. However, although it does effectively eliminate self-fertilization, it also carries a great risk of reproduction failing entirely. Dioecy would clearly be a big disadvantage to weedy species and those that regularly colonize new habitats. Attempts to model the evolution of dioecy have concluded that avoiding inbreeding is one selective pressure that may be responsible for its appearance, but sexual specialization may be equally important (Freeman *et al.* 1997). Sexual specialiation involves the development of characters which enhance male or female reproductive success. For example, the enlarged feathery stigmas of female wind-pollinated flowers would interfere with activity of stamens, and so retaining the two organs in separate flowers or on separate plants is advantageous. Similarly, female flowers must be borne on branches strong enough to support developing fruit, while male flowers may benefit from thinner, more flexible branches which encourage pollen dispersal. This differential development to maximize a single sexual function may, like the differential allocation of resources to sexual functions in monoecious plants, actually represent the driving force behind the evolution of dioecy, with the elimination of self-pollination simply a fortuitous consequence.

Dioecy can evolve through monoecy or through one of the intermediate states of androdioecy or gynodioecy. Androdioecy (male individuals and hermaphrodite individuals) is extremely rare and thought to occur when selection favours the reappearance of hermaphroditism in a dioecious population (Pannell 2002). Gynodioecy is much more common, and the more likely route from hermaphroditism to dioecy. The common European weed *Plantago coronopus* is a familiar example of a gynodioecious species, and has been well characterized at the level of population dynamics (Koelewijn and van Damme 1996).

Sex determination in dioecious plants is relatively poorly understood, largely because the multiple origins of the trait seem to be associated with multiple developmental mechanisms. In some plants sex is determined by the segregation of a single locus or a few loci, present on ordinary chromosomes (Vyskot and Hobza 2004). In other species plant growth regulators (hormones) play key roles. In six plant families it has been shown that sex chromosomes control the development of dioecy (Ainsworth 2000). However, even these six families have clearly distinct sex determining mechanisms. *Silene* (campion) contains sex chromosomes which operate under the same active Y system as seen in mammals, with XY individuals developing as males and XX individuals developing as females (Grant *et al.* 1994). In contrast, *Rumex* (sorrel) species use a dosage-dependent system similar to that of *Drosophila melanogaster*, with an X:autosome ration of more than 1.0 resulting in female development and a ratio of less than 0.5 resulting in male development (Ainsworth 2000). Plant sex chromosomes are often considerably larger than autosomes, although the reasons why are not well understood.

The differentiation of male or female flowers on a dioecious individual requires the same developmental modifications as already discussed for monoecious plants. The timing of abortion of male or female organs varies greatly across different species, with some plants failing to initiate the redundant organs at all, others aborting them some way through development and others still producing organs which appear normal but fail to function. The prediction that ABC function genes are involved in sexual differentiation in dioecious plants has been proved true for *Rumex acetosa*. B function genes in this species are expressed only in male flowers, and only in the stamen whorl. They are not expressed in the sepaloid perianth organs of either male or female flowers. C function genes are expressed in both sexes, initially in both the stamen and carpel whorl. However, as the 'inappropriate' set of organs aborts, transcript of the C function gene becomes undetectable in those organs. Thus, in the female flowers C function expression is lost from the developing stamens at the time when their development ceases, and in the male flowers C function expression is lost from the carpel at the time when its development ceases (Ainsworth *et al.* 1995). It is not yet known whether this loss of activity is caused by abortion of organ development or is itself the cause of the cessation of development. It is also currently unclear how sex determination through sex chromosomes is linked to sexual differentiation at the level of floral organ development. It has been suggested that the relatively basal angiosperm *Thalictrum dioicum*, a member of the Ranunculaceae, might provide a system in which to link these two stages of dioecious development. *T. dioicum* flowers do not even initiate the inappropriate organs—no carpel primordia are found in male plants and no stamen primordia in females. B and C function gene expression has been analysed in this species, and, although the presence of multiple copies of both types of genes makes the analysis complex, there is clear segregation of expression patterns with plant gender. This suggests that sex determination methods in this species operate through complete repression of the transcription of organ identity genes (Di Stilio *et al.* 2005).

12.4 Self-incompatibility (SI)

It is clear that developmental separation of male and female reproductive functions occurs for complex reasons, only some of which may be related to preventing self-fertilization. Such separation can vary greatly in its effectiveness, and all mechanisms except dioecy do allow some self-fertilization. However, in around 90 angiosperm families, and up to 50% of angiosperm species, it has been reported that some form of biochemical self-incompatibility exists (McClure and Franklin-Tong 2006). Biochemical self-incompatibility results in the failure of self pollen to fertilize an ovule of the same individual, whether the pollen grain was produced on the same flower or on a different flower. In those families where self-incompatibility has been studied, the control of this block to self-fertilization can be attributed to a single locus, the S locus. The S locus does not necessarily represent one single gene, but is a closely linked region of the chromosome containing several genes which together control self-incompatibility. The S locus exists in multiple forms, or haplotypes (rather than alleles, since multiple genes are present), providing different 'mating types'. An incompatible pollination occurs when one or both of the S haplotypes in the male parent is the same as one or both of the S haplotypes in the female parent.

There are two main forms of self-incompatibility. In gametophytic self-incompatibility it is the genotype of the pollen grain, the gametophyte, that is key. Since pollen grains are haploid, they carry only one S haplotype (Fig. 12.4). In sporophytic self-incompatibility it is the genotype of the parent plant from which the pollen grain came, the diploid sporophyte, that matters. Since sporophytes are

diploid the pollen grain will carry the imprint of 2 S haplotypes (Fig. 12.3). Differences in the mechanism of preventing fertilization also separate gametophytic and sporophytic systems. In gametophytic self-incompatibility the growth of the pollen tube is inhibited within the tissue of the style. In sporophytic self-incompatibility the germination of the pollen grain is inhibited on the surface of the stigma and no pollen tube enters the style. Of the 90 families of angiosperms that have self-incompatibility, no family consists entirely of self-incompatible species. It is also the case that sporophytic and gametophytic self-incompatibility systems are not found within a single family. Gametophytic self-incompatibility is thought to be the more common of the two types, but sporophytic self-incompatibility has been more thoroughly characterized.

12.5 Sporophytic self-incompatibility (SSI)

The best studied sporophytic self-incompatibility system is that found in the Brassicaceae, because of the importance of this family in crop production. However, sporophytic self-incompatibility is also found in other large families, including the Asteraceae, the Convolvulaceae, and the Caryophyllaceae. Sporophytic self-incompatibility can be genetically complex, because different S locus haplotypes may be dominant, recessive, or co-dominant, and these relationships may differ between the male components of the system and the female components. Thus, a particular S locus haplotype may be dominant in the stigma but co-dominant in the pollen, making the analysis of relationships between haplotypes rather complex (Hiscock and McInnis 2003). The *Brassica* self-incompatibility reaction can be divided into two stages. Firstly, self pollen, or at least pollen with the appropriate genotype to trigger the response, is recognized by the presence of the same S haplotype in the male and female parents. Secondly, this recognition stimulates a signal transduction cascade that leads to pollen rejection. The SI response in *Brassica* is controlled by a single S locus, at which up to 80 different S haplotypes have been identified. The locus itself contains as many as 14 expressed genes (Watanabe *et al.* 2000) and the genes determining

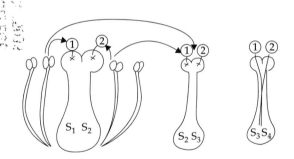

Figure 12.3 Sporophytic self-incompatibility. Pollen grains carry the imprint of the parent plant's diploid genotype.

Figure 12.4 Gametophytic self-incompatibility. Pollen grains carry only a single mating type, according to their own haploid genotype.

the male and female components of the SI response are distinct, although closely linked. This ensures that male and female SI alleles are not brought together in inappropriate combinations through meiotic recombination. In contrast, the genes involved in the signal transduction cascade causing pollen rejection appear to be entirely unlinked and are scattered throughout the genome.

12.5.1 The female determinant of SSI

The S locus contains multiple genes, so identifying the female and male components of self-incompatibility involves analysis of their expression patterns and functions. Two *Brassica* S locus genes encode proteins only detected in the stigma. These two highly polymorphic genes are called the *S Locus Glycoprotein* (*SLG*) and the *S locus Receptor Kinase* (*SRK*). *SLG* encodes an abundant glycoprotein, which is secreted into the cell wall matrix of the stigmatic papillae (Nasrallah *et al.* 1985; Kandasamy *et al.* 1989), while *SRK* encodes a receptor serine-threonine kinase protein, with an extracellular domain, a transmembrane domain, and a cytoplasmic domain with serine/threonine kinase activity (Stein *et al.* 1991). The extracellular domain is structurally very similar to the SLG protein. Because of this similarity it has been hypothesized that *SLG* was produced by a partial duplication of *SRK*. Stein *et al.* (1996) confirmed that the SRK protein is localized to the plasma membrane of stigmatic papillae.

Analysis of mutant lines of *Brassica* in which self-compatibility has been lost has confirmed the importance of the *SRK* gene in determining female SI. Several such lines have been shown to contain mutations within *SRK*. Transgenic work also supports this conclusion. Introduction of the *SRK* gene from an S locus haplotype called S28 into a *Brassica rapa* line containing the S60 haplotype resulted in the production of plants which rejected pollen carrying the S28 haplotype. These data confirmed that the *SRK* gene is sufficient to determine which S type the female reproductive organs function as. The transgenics continued to reject S60 pollen, so the *SRK28* transgene had simply added to their female SI type. No other pollen was rejected by these plants, confirming that the response was specific to the allele of *SRK* that was added (Takasaki *et al.* 2000).

Similar experiments have been conducted with *SLG*, and have concluded that it is not sufficient alone to determine female S locus type, but that it acts with *SRK* and enhances its activity. Introduction of the *SLG28* gene into a heterozygous line of *B. rapa* carrying S52 and S60 failed to confer recognition or rejection of S28 pollen, suggesting that *SLG* is not sufficient alone to confer the female phenotype. However, crossing these plants to the transgenic plants described above, which express *SRK28*, resulted in a significant enhancement of rejection of S28 pollen, with much lower seed set than shown by the plants containing *SRK28* alone. From these data, it is apparent that *SRK* is the female specificity determinant in *Brassica*, and that *SLG* enhances the recognition process in the stigma (Takasaki *et al.* 2000). One possibility is that *SLG* determines the strength of the incompatible reaction, with strong incompatibility resulting from particular *SLG* alleles but weak incompatibility resulting from other, or even absent, *SLG* alleles (Hiscock and McInnis 2003).

12.5.2 The male determinant of SSI

Similar approaches have been taken to establish which genes at the S locus determine the male mating type of a plant. The *S locus Cysteine-Rich* (*SCR*) gene (also sometimes known an *SP11, S-locus pollen protein 11*) encodes a small cysteine-rich protein which is present only in the anthers and has a signal peptide for secretion (Schopfer *et al.* 1999; Takayama *et al.* 2000). Schopfer *et al.* (1999) demonstrated the role of the *SCR* gene in male mating type using both loss-of-function and gain-of-function approaches. First, they showed that a self-compatible mutant line of *Brassica oleracea* lacked the *SCR* transcript. Then they introduced the *SCR6* gene transgenically into a line homozygous for the S2 haplotype. Pollen from all 12 resulting transformants shown to express the transgene was rejected by the stigmas of plants homozygous for the S6 haplotype, confirming the importance of the *SCR* gene in the male phenotype.

Further analysis of SCR protein function has revealed how haploid pollen grains can carry the

PRESTON RIDGE LIBRARY
COLLIN COLLEGE
FRISCO, TX 75035

imprint of two S locus haplotypes, resulting in sporophytic self-incompatibility. When the *SCR9* gene was expressed in bacterial cells, and the SCR9 protein purified, the protein alone resulted in an SI response when applied to stigmatic cells of a plant carrying the S9 haplotype, but not of a plant carrying the S8 haplotype (Takayama *et al.* 2000). Within the plant, the *SCR* genes are expressed early in anther development in the tapetal cells which line the inside of the anther, as well as in the developing microspores (Takayama *et al.* 2000). As pollen grains develop within the anther, SCR protein is rubbed from the tapetum onto the surface of the pollen grain. As a result, pollen grains carry SCR proteins encoded by both *SCR* genes present in the parent plant, even though the gametes within them only carry a single allele at the *SCR* locus.

The dominance of *SCR* alleles is a reflection of their expression patterns. Dominant alleles are expressed tapetally and gametophytically, while recessive alleles are only expressed in the tapetum. That is sufficient to provide a coating of SCR protein to the outside of the pollen grain, maintaining self-incompatibility in a homozygous recessive individual. However, surprisingly, recessive alleles are not expressed at all or are expressed at extremely low levels in the presence of a dominant allele, a result which holds true for *B. rapa*, *B. oleracea*, and *Arabidopsis lyrata* (Kusaba *et al.* 2002; Shiba *et al.* 2002). This transcriptional control of allele expression may explain how S locus alleles can differ in their dominance relationships between male and female functions (Charlesworth 2002).

12.5.3 The SSI response

When a pollen grain lands on a stigma, the pollen coat SCR diffuses into the cell wall of the stigmatic papillae, where it binds to the receptor domain of the SRK. Prior to SCR binding, the SRK is present as dimers associated with two thioredoxin-H-like proteins (Mazzurco *et al.* 2001). The thioredoxin-H-like proteins are thought to prevent autophosphorylation of the SRK, maintaining it in an inactive state in the absence of the SCR ligand (Cabrillac *et al.* 2001). If the SCR and SRK are of the same type, then the SCR binds to the SRK and residues in the SRK kinase domain are phosphorylated, generating the active form of the protein

(Takayama *et al.* 2001). This leads to an intracellular signal transduction cascade that inhibits pollen germination. SLG appears to bind to the SRK, stabilizing it and increasing its affinity for SCR (Takayama *et al.* 2001). The signal transduction cascade downstream of SRK is thought to act through degradation of factors essential for pollen tube growth. ARC1, a stigma-specific protein containing an Armadillo motif, has been shown to bind to the SRK (Gu *et al.* 1998). Further experiments indicate that transgenic plants expressing an antisense version of *ARC1* show loss of their SI responses (Stone *et al.* 1999). ARC1 has been shown to have ubiquitin ligase activity and to promote protein ubiquitination during the rejection of pollen by the stigma (Stone *et al.* 2003). Yeast two hybrid studies also suggest that a kinase-associated protein phosphatase, calmodulin, and a nexin all bind to the SRK (Vanoosthuyse *et al.* 2003). It is likely that a number of other proteins are also involved in the rejection response, being necessary for pollen tube growth.

12.5.4 SSI in other plant families

Sporophytic self-incompatibility has been most extensively studied in the Brassicaceae, but recent reports of its control in the Convolvulaceae and the Asteraceae suggest that it has evolved multiple times through different molecular mechanisms. Within the Convolvulaceae research has focused on *Ipomoea trifida*. Little is yet known about what does constitute the S locus in this species, but it has been shown that genes with sequence similarity to *SRK* and *SLG* do not segregate with the S locus, indicating that they are not involved in self-incompatibility, and are expressed in all tissues rather than in a stigma-specific manner (Kowyama *et al.* 1996; reviewed by Hiscock and Tabah 2005). In the Asteraceae, similar conclusions have been drawn from work with *Senecio squalidus*. Hiscock *et al.* (2003) reported that *SRK*-like sequences did not segregate with the S locus and were not specifically expressed in the stigma. More recently, work in *Senecio* has focused on genes encoding glycoproteins which are expressed in the stigma and do segregate with the S locus, although their molecular function is as yet unknown (Hiscock and Tabah 2005).

12.6 Gametophytic self-incompatibility (GSI)

Gametophytic SI has been most widely studied in the Solanaceae, although it is found in 60–90 families, including the Rosaceae, Leguminosae, Plantaginaceae, and Papaveraceae (Franklin-Tong and Franklin 2003). As with sporophytic self-incompatibility, GSI is controlled by a single S locus which is highly polymorphic and contains the determinants of both male and female function tightly linked. However, since male S type is determined by the genotype of the haploid pollen grain, the gametophyte, there is no problem of dominance with respect to the male function. All male S haplotypes are dominant, because they interact independently with the stigma. Gametophytic self-incompatibility differs from sporophytic self-incompatibility in that pollen grains do germinate but are inhibited within the transmitting tract of the style, rather than on the stigmatic surface.

12.6.1 The female determinant of GSI

A small glycoprotein, similar to the sporophytic SLG and present in mature styles, is the female determinant in the gametophytic SI system. The protein was initially identified on the basis of its abundance in the style and its segregation with the S locus in *Nicotiana alata* (Anderson *et al.* 1986). The protein has significant structural and sequence similarity with a group of fungal ribonucleases, or RNases, and has been shown to have ribonuclease activity, and so was named S-RNase. Gray *et al.* (1991) confirmed that S-RNases prevented protein production, both within a pollen tube and in an *in vitro* system, and showed that S-RNases degraded all mRNAs, not specific substrates. Transformation experiments similar to those used to test the female determinant of sporophytic self-incompatibility have shown conclusively that S-RNase is sufficient to determine female mating type in *N. alata*. Introduction of the gene encoding S-RNase from the SA2 haplotype into a different genetic background (a hybrid between *N. alata* and *Nicotiana langsdorffii*) resulted in the production of plants which rejected SA2 pollen but not pollen grains carrying other S haplotypes (Murfett *et al.*, 1994). To date, it appears that S-RNase is the only protein involved in specifying female mating type in gametophytic self-incompatibility systems.

12.6.2 The male determinant of GSI

F box proteins, which target other proteins for degradation, have been shown to be the male determinants of gametophytic SI in both Petunia (Solanaceae) and Antirrhinum (Plantaginaceae). F box proteins in many systems bind specific proteins to the SCF E3 ubiquitin ligase complex, targeting them for degradation by the 26S proteasome. In gametophytic SI it is thought that F box proteins target S-RNases for degradation, allowing pollen tubes to grow. Genes encoding F box proteins were first shown to be linked to the S locus by sequencing of the genomic region surrounding S-RNase in *Antirrhinum hispanicum* (Lai *et al.* 2002). These genes were named *SLF* genes (*S Locus F box*) and were then shown to influence compatibility when introduced transgenically into different backgrounds (Qiao *et al.* 2004; Sijacic *et al.* 2004). F box proteins were initially thought to act within the pollen tube to degrade all the S-RNases entering from the style. In this model, when a matching S-RNase enters the pollen tube it is not degraded by the F box protein and is able to inhibit pollen tube growth. Thus, Sijacic *et al.* (2004) introduced the *SLF2* transgene into a line of *Petunia inflata* carrying both S2 and S3 haplotypes at the S locus. Under normal conditions, both S2 and S3 pollen grains would be rejected, as the matching S-RNase escapes degradation and inhibits pollen tube growth. However, self-incompatibility broke down in the transgenic plants. S2 pollen grains carrying the *SLF2* transgene behaved exactly as normal S2 pollen grains, being rejected because the S2-RNase escaped degradation. However, the S3 pollen grains carrying the *SLF2* transgene were not inhibited. In these pollen tubes, the S3-RNase was degraded by the transgenic SLF2 protein, while the S2-RNase was degraded by the endogenous SLF3 protein. With both RNases inactivated the pollen tube was free to grow normally.

12.6.3 The GSI response

Before the male determinant of gametophytic SI was identified, debate centred on its likely role in prohibiting RNase activity. The observation that compatible S-RNases do enter the pollen tube (Luu *et al.* 2000) indicated that the male determinant was likely to inhibit RNase activity, rather than act as a gatekeeper, guarding the pollen tube. The identification of the male determinant as an F box protein seemed to reinforce this model, and SLF has been shown to bind S-RNase, using yeast two hybrid analysis (Qiao *et al.* 2004). However, recent reports have suggested that this model might be too simple. Goldraij *et al.* (2006) reported that non-self S-RNases are not degraded in *N. alata*, where compatible and incompatible pollen tubes contain similar quantities of RNase. Instead, Goldraij *et al.* (2006) showed that S-RNases were sequestered in a vacuolar compartment in the pollen tube, which broke down after pollination with an incompatible pollen grain. They further showed that a stylar protein called HT-B was necessary for the release of S-RNases from the vacuole. HT-B was significantly degraded following pollination with an incompatible pollen grain, but less so after a compatible pollination. In transgenic plants expressing an antisense *HT-B* construct the pollen tube vacuole did not break down after an incompatible pollination. An alternative to the RNase degradation model was proposed by Goldraij *et al.* (2006) and reviewed by McClure and Franklin-Tong (2006). These authors suggested that S-RNases might be compartmentalized into pollen tube vacuoles in compatible reactions, and only released to inhibit pollen tube growth in incompatible reactions. The SLF protein might therefore be involved in targeting the HT-B protein for degradation, preventing vacuolar breakdown, rather than in directly degrading the S-RNases themselves. A different problem with the degradation model was revealed by the analysis of two mutant lines of *Prunus avium*, sweet cherry, both of which contain a deletion of the *SLF* gene. Pollen from these plants is perfectly viable on all stigmas, including self ones, indicating that targeted degradation of S-RNases is not necessary to allow pollen tube growth, at least in this system (Sonneveld *et al.* 2005). Sonneveld *et al.* (2005) proposed that the pollen tube itself may contain a system for degradation of S-RNases, with the SLF protein simply introducing specificity into the reaction. The recent nature of the discovery of the male

determinant of gametophytic SI means that our understanding of this system lags considerably behind our understanding of sporophytic SI. Much exciting work remains to be done, to uncover the details of the incompatibility reaction and any downstream responses.

12.6.4 GSI in other plant families

Gametophytic SI has been studied in several plant families, and extremely similar mechanisms seem to be at work in the Solanaceae (*Petunia* and *Nicotiana*), Plantaginaceae (Antirrhinum) and the Rosacaeae (*Prunus*). However, it seems that an entirely different system of gametophytic SI operates in the Papaveraceae. In *Papaver rhoeas* small secreted proteins known as S proteins determine the female mating type (Foote *et al.* 1994), while the male determinant is as yet unknown. The function of the S proteins is also unclear, although it is likely that they bind to receptors on the outer face of the pollen tube. Inhibition of pollen tube growth in this system operates through changes to calcium signalling (reviewed by McClure and Franklin-Tong 2006).

12.7 Heteromorphic self-incompatibility

In around 20 different families of angiosperms there are species which have a heteromorphic self-incompatibility system, where the biochemical self-incompatibility type is associated with certain morphological traits (Barrett 2002). The common primrose, *Primula vulgaris*, is a good example of this type of system. It has sporophytic SI with only 2 S haplotypes, which also control the position of the anthers and stigma within the flower. In 'pin' flowers the style extends above the anthers, and in 'thrum' flowers the style sits below the anthers (Fig. 12.5). Plants producing pin flowers are genetically homozygous at the S locus and plants producing thrum flowers are heterozygous. Crosses between the two therefore produce 50% heterozygous, and therefore thrum, flowers and 50% homozygous, and therefore pin, flowers. This system places pollen on specific parts of the pollinator's body, from where it is deposited only on a stigma of the opposite flower type. This prevents

Figure 12.5 Heteromorphic self-incompatibility, a form of reciprocal herkogamy.

half of the pollen being wasted on the wrong stigma type, which would reject it biochemically, and thus saves energy and resources as well as reducing blockage of the stigma by incompatible pollen. Other species have three different floral forms, with stigmas at short, medium, and tall heights (known as tristyly, as opposed to the distyly seen in *P. vulgaris*).

The self-incompatibility present in such systems may work in different ways in different floral morphs. For example, Massinga *et al.* (2005) showed that incompatible pollen tubes were inhibited on the stigmatic surface of the long styled form of *Pentanisia prunelloides* (wild Verbena) but within the style of the short styled form. The style length

polymorphism in this species is associated with an anther length polymorphism and these influence pollination by butterflies, which carry pollen on their heads or proboscises, according to the anther length of the flower from which it came.

The existence of reciprocal floral types with differences in style and anther height is not always associated with biochemical self incompatibility. When such reciprocal herkogamy occurs alone it is known as heterostyly, which may be more common than full heteromorphic self-incompatibility. Both heterostyly and heteromorphic SI require the linkage of genes controlling a number of morphological traits, plus biochemical SI (if appropriate), at a single locus. Little is currently known about the molecular nature of this locus, as research to date has focused on the population dynamics and evolution of heterostyly (discussed by Mast and Conti 2006).

If plants have evolved mechanisms to reduce self-fertilization, then they must also ensure that they achieve outcrossing. The interactions of plants with their pollinators, and the adaptations which facilitate those interactions, are the subject of the third section of this book.

POLLINATION SUCCESS: MOLECULAR AND ECOLOGICAL INTERACTIONS

How and Why Does Floral Form Vary?

CHAPTER 13

Why Are Flowers Different? Pollination Syndromes—The Theory

Until this point, we have largely ignored the concept that the flowers of different species may be very different from one another. However, it is clear from a merely cursory glance around any garden in the summer months that flowers come in an enormous variety of sizes, shapes, colours, and scents. In Section III of this book we focus on the differences between flowers, as opposed to the molecular similarities that unite them. In this chapter we begin by considering the different ways that flowers can be pollinated. It is a basic premise underlying much of floral biology that differences in pollination system explain many of the differences in floral form. The evidence to support this premise is not as compelling as we might like to think, as will be discussed in later chapters. However, to set the stage for those discussions, in this chapter we look at the historical concept of the pollination syndrome and the predictions it makes about floral morphology.

13.1 Cross pollination

If cross pollination is to occur within a population of flowering plants, it is necessary for pollen to be transported from the anthers in a flower on one individual to the stigma in a flower of a second individual. The alternative is self-pollination, when pollen is transferred from the anther of a single flower onto the stigma of the same flower (autogamy), or, confusingly, is transferred from the anther of one flower to the stigma of a second flower on the same plant (geitonogamy). Self-pollination is thought of as an adaptation to unreliable pollen vectors, particularly seen in short-lived annual plants and those species which regularly

invade new habitats. However, self-pollination is not optimal for genetic recombination and so in the long term may compromise the genetic diversity and evolutionary lability of a population (see discussion in Chapter 12). Therefore a great many flowering plants have adaptations which maximize their potential for cross pollination. Since plants are immobile, this transport of pollen from plant to plant must be accomplished with the aid of a pollen vector. The pollen vector may be abiotic or biotic.

13.2 Abiotic pollen vectors

13.2.1 Wind pollination

Abiotic pollen vectors consist primarily of the wind and water. Of these, wind pollination (anemophily) is by far the more common, being found in 18% of angiosperm families (including the ecologically dominant Poaceae, the grasses) as well as in the conifers (Ackerman 2000). Wind pollination is a secondarily derived state in angiosperm flowers, as the earliest flowers are believed to have been pollinated by beetles (Thien *et al.* 2000). Anemophily is thought to evolve in response to changes in the environment which decrease the efficiency of biotic pollination while enhancing the success of wind pollination. Examples of environmental changes which might cause a shift towards wind pollination include variations in pollinator abundance, migration into an area with a very dry climate, or the arrival of another plant species which might compete for pollinator attention (Culley *et al.* 2002). Many plant species have been shown to be both wind- and animal-pollinated, either simultaneously or at different times within the same season (Cox 1991). This combination of biotic and abiotic pollination systems is known as ambophily, and may be the transitional route through which full wind pollination most commonly evolves.

Angiosperm species that are wind-pollinated typically show a particular set of characteristics. Most of these characteristics have been shown to enhance the success of wind pollination, so can be thought of as adaptations to a wind-pollinated reproductive system (Culley *et al.* 2002). As wind pollination is a passive process, many of these adaptations are to do with the environments and habitats in which anemophilous plants live. To begin with, wind-pollinated species typically inhabit environments with stronger winds but lower humidity than those of biotically pollinated species. Wind-pollinated plants are found in habitats with low rainfall, ensuring pollen is not washed away. Wind-pollinated species are usually found in moderately high densities within an area, but with low densities of other species surrounding them. This combination of features maximizes the chances of released pollen landing on another individual of the same species.

Besides common features of environment and habitat, wind-pollinated species also share a typical set of floral characteristics. Flowers and inflorescences are commonly held above or away from vegetation, which maximizes access by the wind to reproductive organs. Very often the inflorescences are pendulous, like those of willow trees. The flowers within these inflorescences are usually unisexual. It would be hard to prevent self pollen clogging the stigmas of flowers exposed to the wind, if both male and female reproductive organs were present in a small area. The outer two whorls of floral organs, the petals and sepals, are usually much reduced or absent in wind-pollinated flowers, and, where present, do not usually produce much scent or pigmentation. The loss of a brightly coloured and strongly scented corolla is unsurprising in view of the unconscious nature of the wind as a pollen vector. The stamen filaments of wind-pollinated flowers are usually long, exposing the locules to the wind, and the pollen grains they produce are smooth-surfaced, small in size, and enormous in quantity. Small size and a smooth surface have been shown to enhance the aerodynamic properties of pollen travelling in wind (Niklas, 1985), while large quantities are necessary to increase the chances of wind-blown pollen reaching a stigma. The stigmas themselves are usually long and feathery, and protrude a considerable way

Figure 13.1 The long tasselled flowers of a wind-pollinated grass hang far from the main body of the plant.

beyond the outer edges of the flower, maximizing surface area available for pollen capture (Culley *et al*. 2002). Wind-pollinated grass flowers are shown in Fig. 13.1.

13.2.2 Water pollination

Water pollination (hydrophily) is a relatively rare system of gamete transfer, but it is employed by a few species of grasses and waterweeds, occurring in only 31 genera of 11 different families, 9 of the families in the Monocots (Cox 1988). In most hydrophilous species, pollen is released below the water surface and carried passively by currents to female reproductive structures. This mechanism of water pollination is used by many marine plants, such as the Caribbean turtle grass, *Thalassia testudinum*, which releases pollen grains underwater, bound up in strands of mucilage. These are carried by currents below the water surface, and the mucilage encourages them to stick to the female flowers, which are also underwater.

However, in some species pollen is released onto the surface of the water and is carried on the surface to female reproductive structures. The pollen of such plants, for example *Elodea canadensis* (the Canadian waterweed), is ornamented with many tiny spikes, which trap pockets of air and ensure the pollen floats. Anther dehiscence is often explosive, scattering the pollen grains widely across the water surface. The female reproductive structures may always be held at surface level in some hydrophilous angiosperms, but in others the female flower closes around the pollen and is then withdrawn below the water surface after pollination. A variation on the theme of releasing pollen onto the water surface is shown by species of tapegrass, *Vallisneria*, which release entire male flowers onto the water, with the pollen never leaving the anthers until it arrives at a female flower. The female flowers of many hydrophilous species create depressions in the surface tension of the water, encouraging passing pollen grains to slide down towards them. While hydrophilous flowers do produce relatively large quantities of pollen, this use of water surface tension dynamics to facilitate pollination does mean that less pollen is required than might be predicted (Cox 1988).

13.3 Biotic pollen vectors

Biotic pollen vectors provide some or all of the pollination service for the majority of flowering plants. The frequency with which we observe this interaction leads to a common assumption that plants and animals are cooperating in a mutualistic association. However, animals rarely deliberately pollinate flowers, but instead view flowers as food sources. The reward offered is usually a mixture of nectar, a sugar solution (which may contain various concentrations of several different sugars, along with other nutrients), and pollen itself, which is very rich in amino acids. In collecting this food, animals inadvertently brush against the reproductive organs of the flower, causing pollen to be transferred from stamens to their bodies, and from their bodies onto stigmatic surfaces. Many flowers show structural features which enhance this inadvertent contact between pollinator body and floral reproductive organs, such as narrow corolla tubes or dorsally positioned anthers which deposit pollen on the back or head of the pollinator.

13.4 Principles underlying the pollination syndrome concept

Biotic pollen vectors consist of a range of animals—most commonly insects, as well as birds, bats, and a small number of other vertebrates. In a seminal book published in 1966, Faegri and van der Pijl set out to 'formulate the general principles of pollination ecology, applicable anywhere'. A major device used by Faegri and van der Pijl was the pollination syndrome, a suite of floral traits associated with the attraction of a particular group of pollinators. The assumption behind the pollination syndrome concept is that coevolution between plant and animal has led to the acquisition by plants of floral features which maximize their chances of attracting and being pollinated by a particular animal or group of animals.

13.4.1 Do flowers act as specialist or general advertisements?

The pollination syndrome concept explains the differences between different flowers as consequences of the types of animals that they attract. This concept relies on the principle that characteristics such as particular colours, scents, shapes, and markings are specific to flowers pollinated by specific types of animals. The different floral characteristics can be viewed as animal-specific advertisements, adapted to maximize the chances of attracting a particular type of honeybee or butterfly or beetle.

However, there is an alternative way of explaining the many differences between different flowers. It was Darwin himself who first articulated the reason flowers are brightly coloured:

Flowers rank amongst the most beautiful productions of nature; but they have been rendered conspicuous on contrast with the green leaves, and in consequence at the same time beautiful, so that they may be easily observed by insects. (Darwin 1859)

On first sight this might be read as consistent with the pollination syndrome concept, in saying that different colours attract different animals. In fact all that this quote says is that colour (and by extension shape and scent) simply provides contrast between flowers and the green vegetation around them. In this case any feature that makes a flower stand out even more against the green backdrop is likely to enhance the success of that flower in attracting pollinators. In this argument, all the different floral characteristics are simply a number of different ways that plants have hit upon to make their flowers more attractive to a broad spectrum of pollinators. Instead of providing species-specific advertisements, the different floral characteristics might simply represent an almost infinite number of different advertisements, all of which serve to attract animals in general. While it is true that each possibility is likely to be correct in the case of certain plant species, it should be borne in mind that the pollination syndrome concept rests on the idea of specialization, and may not be compatible with a view of flowers as non-specific advertisers of rewards to generalist pollinators.

13.4.2 Plants and pollinators want different things, making species-specific interactions unlikely

Plants and their pollinators have different requirements of the interaction, and these differences may reduce the likelihood of coevolution occurring between individual species. A plant will experience maximum reproductive fitness if it maximizes pollen dispersal between individuals of the same species, without its stigmas becoming clogged with pollen of other species and without wasting much energy on nectar production. Therefore its optimal pollen vector will be an animal which alights only briefly on each flower, moves rapidly between individuals, is faithful to only one plant species, and does not eat much.

The animal will conserve energy if there is no need to forage, and so should prefer flowers with a large reward, where it can stay a long time. If the reward is insufficient in one species it should forage on a range of plant species.

However, very specific interactions between plant and pollinator carry their own risks. A one-on-one relationship between flower and pollinator should mean that little pollen is wasted, and so its production can be less prolific, saving energy. Female fertility should also be enhanced as foreign pollen will not be applied to the stigma, where it can block acceptable pollen from germinating. However, such an extreme specialist can expect to set no seed at all if the pollinator is absent in any one season, through disease or climatic changes, a situation which would be catastrophic for an annual plant and may also be very disadvantageous for a perennial.

It is generally argued that for these reasons the evolution of plants and their pollinators has followed a middle road, with groups of pollinators coevolving with groups of flowers, rather than one-on-one species-specific interactions. It is the broad association between a group of flowers and the group of animals that pollinates them that has come to be called a pollination syndrome (Faegri and van der Pijl 1966).

13.5 The pollination syndromes

A pollination syndrome classically describes a suite of adaptations shown by a plant to a taxonomic order of animals, and by those animals to a particular group of flowers, which may not be phylogenetically related to each other. The adaptations shown by the animals may be behavioural or morphological, while the plants can only show morphological adaptations, which may include flower size, structure, colour, reward, and timing of floral induction and opening. A table summarizing the key floral traits traditionally associated with different pollinators is shown in Fig. 13.2.

13.5.1 Beetle pollination (Cantherophily)

Beetle pollination is widely believed to have been the first pollination syndrome, the one used by the first angiosperms, as the Coleoptera, the beetles, constitute one of the oldest groups of insects and were already numerous at the time that the angiosperms came into existence, having themselves arisen around 260–280 mya (Ponomarenko 1995). Fossilized pollen found in the digestive tracts of beetles has shown that they had already acquired the habit of grazing on the pollen of cycads, conifers, and other gymnosperms, before the angiosperms appeared (Labandeira 1997; Thien *et al.* 2000). When considering a pollination syndrome, it is useful to begin with the features of the animal's biology relevant to its role as an agent of pollen transfer. Beetles have mouth parts positioned parallel to the axis of the body, which limits

their ability to manipulate food sources, particularly those with depth. They are also quite big animals, with a moderately high demand for protein as well as carbohydrate. Beetles do not have good colour vision, but do have a strong sense of smell and a particular attraction to fruity smells. Flowers that have become adapted for pollination by beetles might be expected to have characteristics which make them both attractive to beetles and easy to obtain food from once the animal has landed. These features would include the provision of a floral reward consisting of both nectar (for sugar) and pollen (for protein) in a flat structure from which the reward can be lapped. Beetle-pollinated flowers are usually saucer or bowl-shaped, with nectar secreted into the shallow bowl, and often have excess anthers to produce extra pollen as a reward. Since colour vision is unimportant to beetles, beetle-pollinated flowers might be expected to waste little energy on the production and modification of pigments, but instead to produce a strong scent. Many beetle-pollinated flowers do produce a fruity fragrance, and they are often greenish or off-white in colour. The classic example of a beetle-pollinated flower is magnolia, essentially unchanged for

	COLOUR	SHAPE	SCENT	REWARD
BEETLE	White/cream	Dish/bowl	Fruity, quite strong	Nectar to lap, some excess pollen
FLY	White/cream/ pale yellow	Dish-like or more complex	Minimal	Small amount of nectar
BEE	Blue/yellow/ ultraviolet	Deep tube, bilateral symmetry	Minimal or some	More nectar, some excess pollen
BUTTERFLY	Yellow/red/ orange	Deep tubes with landing platforms	Minimal	Only nectar
MOTH	White	Bilateral symmetry, deep tubes	Strong, sweet, at night	Plentiful nectar
BIRD	Red/orange	Pendant or bilateral and upright, tube	Minimal	Much nectar
BAT	White	Large, saucer shaped	Strong, butyric acid	Copious nectar and much excess pollen

Figure 13.2 Table to show the general features of flowers involved in different pollination syndromes.

100 million years and still pollinated by the same sorts of animals (Fig. 13.3a). Lilies, wild roses, and some poppies are also often beetle-pollinated.

13.5.2 Fly pollination (Myiophily)

The Diptera, the fly order, show the greatest variation in methods and habits of pollination of any

group of insects. This makes it hard to define the morphological traits that we would expect to see in a fly-pollinated flower. Indeed, some flies behave very like beetles, and pollinate flowers morphologically very similar to beetle-pollinated flowers (Thien *et al.* 2000). Other flies may have more in common with wasps, and pollinate flowers with very different structural features. One feature of flies which has been utilized by flowering plants is their persistence throughout all seasons. Flies are one of the few groups of insects that are not strictly periodic, and, as such, many plants flowering under adverse conditions or at odd times of the year can be entirely dependent on flies for their pollination. Another feature common to all flies is that they do not feed their offspring, and are usually lighter-bodied than many other insects. In consequence they do not need much food, and fly-pollinated flowers usually supply only a small quantity of nectar. Although there is much variation within the order, in general flies are more visually acute animals than beetles, and have a positive preference for pale and yellow colours, and for nectar guides. Fly-pollinated flowers are often coloured cream or yellow, and do not usually have much scent. Classic examples of fly-pollinated flowers include species such as carrot and other members of the Umbelliferae, as well as species such as groundsel and other Asteraceae. A fly-pollinated *Fatsia japonica* flower is shown in Fig. 13.3b.

13.5.3 Bee pollination (Melittophily)

Of the Hymenoptera, the bees have specialized most towards a diet of nectar and pollen. Wasps will take nectar to meet their sugar requirements, but do not actively collect either nectar or pollen. There are a few specialized instances of ants feeding on nectar, often from extra-floral nectaries, but for the most part it is the bees that are the great pollinators of the Hymenopteran order. Bees are large animals, and have a substantial energy requirement. They usually forage for nectar for themselves, and pollen to feed to larvae back in their hive or brood chamber. The pollen is carried on their bodies, in specialized structures that range from simply having hairy feet to having pollen baskets on the hind legs, and is then groomed off back at the hive. In response to these

Figure 13.3 Insect-pollinated flowers. (a) Magnolia flowers are beetle-pollinated. (b) The fly-pollinated flowers of *Fatsia japonica*. (c) Bumblebee (*Bombus terrestris*) entering a *Hebe* flower. (d) Many daisies are butterfly-pollinated. (e) The flowers of *Angraecum sesquipedale* have very long nectar spurs and are pollinated by extremely long-tongued moths. Photographs (a), (d), and (e) kindly supplied by Cambridge University Botanic Garden and H. Rice. See also Plate 10.

features, bee-pollinated flowers are usually quite large to bear the weight of the animal, and often have a clear landing platform. They are also often closed, with petals or other structures which must be pushed aside by the bee to access the nectar. This sort of mechanism prevents nectar robbing by other animals, and helps to keep bees constant, as the supply of nectar is likely to be good. Inside these large flowers is usually a reasonable volume of nectar, and a small amount of excess pollen. Bees have the ability to perceive depth and many species have long tongues. The nectar is often secreted at the base of deep corolla tubes or nectar spurs, again denying access to species without the size and tongue length to access it. Since pollen found on the bee's body is groomed off and fed to the larvae, many bee-pollinated flowers have dorsal anthers which deposit the pollen on the back of the bee's neck, a position from which it cannot easily be groomed. Bees have good colour vision and can see in ultraviolet, blue, and yellow, but do not have receptors to perceive red which they see only as a weak green signal. Bee-pollinated flowers are usually brightly coloured, with yellow and blue being considered classic bee colours. However, many bee-pollinated flowers are red or pink. It has been shown that in many of these cases the presence of UV absorbing pigments modifies the red to a colour more highly visible to the bee (see Chapter 18; Chittka and Waser 1997). As quite intelligent animals, bees are sensitive to nectar guides, which enable them to handle the flowers more quickly by directing them straight to the nectar. Many bee-pollinated flowers have nectar guides visible either in the range of the spectrum that we can see or else in the ultraviolet part. Classic examples of bee-pollinated flowers include the garden snapdragon (*Antirrhinum majus*), nettles, and foxgloves. A bumblebee visiting a *Hebe* inflorescence is shown in Fig. 13.3c.

13.5.4 Butterfly pollination (Psychophily)

In contrast to bees, the other most prominent group of pollinating invertebrates in temperate climates does not feed its offspring and has relatively low energy requirements. Butterflies are relatively light in weight, and usually alight on flowers, which

reduces their energy expenditure. They also have long tongues, often 1 to 2 cm in depth. Flowers pollinated by butterflies usually combine a flat structure to alight on with deep tubes in which the nectar is secreted. This can be achieved either by presenting a tube-shaped flower with a landing rim around it, like a buddleia, or else by clustering a number of small tubes together to produce a larger flat surface, the system used by members of the Asteraceae, such as Michaelmas daisies (Fig. 13.3d). Butterflies are not known to have much sense of smell at all, and butterfly-pollinated flowers do not usually produce much scent. However, butterflies do have good colour vision, and can see red. Butterfly-pollinated flowers are usually brightly coloured, with reds and yellows often predominant.

13.5.5 Moth pollination (Phalaenophily or Sphingophily)

Although moths and butterflies are members of the same insect order, the Lepidoptera, their methods of pollination and the structures of the flowers they visit are completely different. The major reason for this is the difference in their behaviour—most moths are nocturnal while butterflies are diurnal. Moths also prefer not to land on flowers, collecting their nectar while hovering. Many moths are also much heavier in the body than butterflies. The combination of a heavy body and hovering flight means that moths have very high energy requirements indeed, and moth-pollinated flowers usually produce more nectar than butterfly- or bee-pollinated flowers. To facilitate nectar collection while hovering, moth-pollinated flowers are usually bilaterally symmetrical, and presented to the pollinator with the petal lobes bent backwards and the corolla tube open for easy access. To prevent theft of this large supply of easily accessible nectar by other animals, it is common for moth-pollinated flowers to close up or fold over during the day, and only open at night. Alternatively, the nectar may be presented in a tube so long that it is inaccessible to other animals. Moth tongues are longer than those of any other insect, and moth-pollinated flowers often have much longer nectar spurs than any other flower type. The extreme example of this relationship is the story of the Madagascan orchid, *Angraecum sesquipedale*, which has a nectar spur up to 30 cm in length (Fig. 13.3e). In 1862, on seeing one of these plants in flower at the Royal Botanic Gardens, Kew, Darwin predicted that there must exist a moth with a proboscis long enough to reach the nectar at the bottom of this spur. He was proved correct in 1903 with the identification of the hawk-moth *Xanthopan morganii* ssp. *praedicta*, which has a tongue of a similar length to the nectar spur, and has been shown to pollinate the flower in a caged environment. Colour vision is irrelevant to moths, which fly at night, and moth-pollinated flowers are often white or cream. This allows them to stand out against the dark vegetation at night. However, moths do have a good sense of smell, and it is usual for moth-pollinated flowers to release a strong scent, often in the evening. Classic examples of moth-pollinated flowers include gardenia and some honeysuckles.

13.5.6 Bird pollination (Ornithophily)

Since the first pollinating animals were insects, the later emergence of pollinating vertebrates meant that these animals had access to flowers already adapted for pollination by another animal. There are sufficient similarities between butterfly- and bird-pollinated flowers to suggest that the early pollinating birds fed from flowers which had coevolved with butterflies. There is some speculation as to how bird pollination evolved, with some authors suggesting that birds hunting nectar-gathering insects in flowers accidentally discovered nectar. It is certainly true that even hummingbirds, which take almost all of their energy from nectar, still eat the occasional insect to meet their protein requirements. The classic pollinating bird is the hummingbird, which is found only in the Americas, but a variety of other groups also feed on nectar, including the African sunbirds, Australian lorikeets, and American honey-creepers. Different types of birds have different feeding strategies. For example, hummingbirds hover beneath or in front of flowers while feeding, so hummingbird-pollinated flowers tend to be either pendant (like a fuchsia, Fig. 13.4a) or stand out with free space in front. Sunbirds, on the other hand, perch while

feeding, so sunbird-pollinated flowers have a perch with the nectary facing towards it (Fig. 13.4b). Bird-pollinated flowers are usually either brush or tube-shaped, and the nectar is secreted into spurs, which are usually shorter and wider than those on butterfly flowers. They also must be quite tough, as a beak is both stronger and harder than a butterfly's tongue. One of the key features of a bird-pollinated flower is the quantity of nectar secreted. Birds are larger animals than the invertebrates discussed previously, and their energy requirements can be very high indeed. They feed from flowers which produce very large quantities of very concentrated nectar, so much so that it will actually drip from the flowers at certain times of the year. Birds have good colour vision (see Chapter 18) and the flowers they pollinate are usually red, often with contrasting yellow marks to act as nectar guides. However, scent is not important in bird pollination. Classic examples of bird-pollinated flowers include red columbine, poinsettia, eucalyptus, hibiscus, and passion flower.

13.5.7 Bat pollination (Chiropterophily)

A quarter of all bat species use flowers for food to some extent, and a few species rely on flowers for all of their nutritional requirements. Bats are large, heavy animals which sometimes land on the flowers they feed from. Those flowers tend to be large and robust, and usually saucer-shaped for ease of lapping nectar. Because bats must find the flowers at night they are not usually within the foliage, but either hang below it for easy access (Fig. 13.4c) or actually develop on the trunk of the plant itself. Bat–pollinated flowers produce the most nectar of any flower type, with as much as 15 ml sometimes recorded from single flowers. Those bat species which have become entirely dependent on flowers eat pollen as their only protein source. The flowers they visit usually have greatly enlarged anthers, and sometimes a very great number of them. The flowers of some bat-pollinated species commonly develop over 2000 anthers per flower. Those anthers may only open at night, to prevent pollen robbing by beetles. Since bats are nocturnal, the flowers themselves may sometimes only open at night and may only last one night before senescing.

Figure 13.4 Vertebrate-pollinated flowers. (a) The pendant form of *Fuchsia* flowers is ideal for hovering hummingbirds. (b) Bird of Paradise (*Strelitzia regina*) flowers provide a sturdy landing platform for non-hovering birds. Photograph kindly supplied by Cambridge University Botanic Garden. (c) The flowers of *Strongylodon macrobotrys*, the jade vine, hang far below the foliage, making them readily accessible to bats. See also Plate 11.

Bats are colour-blind, so flower colour is irrelevant in attracting them. Bat-pollinated flowers are usually white to cream or sometimes a greenish pink colour. The main attractant used by flowers to attract bats is scent. Bat-pollinated flowers generate a very strong scent, often containing butyric acid. Bats which are specialized to feed on nectar often have a nasal cavity larger than that of insectivorous bats, suggesting that scent is more important to flower bats than to insectivorous bats. In contrast the sonar apparatus is sometimes reduced, as little time is spent hunting insects. Flower bats have longer tongues and narrower snouts than insectivores, and the tongue often has papillae on the end for lapping up the nectar. Classic examples of bat-pollinated flowers include many cacti, as well as members of the Bignoniaceae (trumpet creepers) and the Bombaceae.

13.5.8 Deceit pollination

The term deceit pollination covers such an enormous range of relationships that it would be impossible to describe them all in this section. Instead we will consider three examples of deceit pollination, each involving increasingly specialized flowers which appear to have coevolved with particular pollinating animals. Although the relationships considered in this section are more complex than the basic pollination syndromes described earlier, they still represent recognizable syndromes with clearly apparent matches between flower morphology and animal behaviour.

The simplest form of deceit pollination is a version of Batesian mimicry, a phenomenon which occurs when an organism mimics another and has particular features of the model attributed to it. For example, the yellow and black banding of hoverflies is a form of Batesian mimicry, causing predatory birds to avoid the hoverfly in case it has the same poisonous sting as the wasps and bees it mimics. A similar situation exists where species of plants mimic the flowers of other species, but do not provide a reward. This sort of pollination system requires the mimic to develop the appropriate colour, shape, and scent to match the model, but to fail to produce nectar. The match does not have to be perfect, as the visual acuity of many pollinators is insufficient to distinguish between generally similar flowers. For instance, the rewardless orchid *Orchis israelitica* has been shown to function as a mimic of the nectar-supplying *Bellevalia flexuosa*, even though the similarity between them is only at the very general level of inflorescence architecture, approximate floral shape, and white colour (Galizia *et al.* 2005). Another orchid species, *Dactylorhiza sambucina*, produces no reward and mimics a range of rewarding species. Within Europe there are two colour morphs of this species, yellow and magenta, and recent reports indicate that negative frequency-dependent selection maintains this colour polymorphism, as naïve bees preferentially visit the rarer morph, having not yet learned to associate it with a lack of reward (Gigord *et al.* 2001). This form of mimicry occurs where a large number of flowering plants of different species grow in a defined area and flower at similar times. The

flowers mimicked may be pollinated by any of the common pollinating animals, so the mimic may develop all the features (except the reward) of any of the standard pollination systems. This sort of pollination by deceit saves the plant energy, but it can only be successful when the mimic is at low frequencies in the community; otherwise the pollinators will abandon both the mimic and the genuine flowers.

A less diffuse system of deceit pollination is called *sapromyiophily*. Sapromyiophilous flowers attract carrion and dung flies and carrion and dung beetles, by mimicking the appearance and scent of a piece of rotting flesh or of carnivore or herbivore faeces. The insects are attracted to the flower to feed or lay their eggs, and transfer pollen in the process. The flower provides no reward to the insect, and may in fact reduce its fitness by causing eggs to be laid on a structure which will not provide sustenance to any maggots that hatch. A range of different colours and scents are used by sapromyiophilous flowers, but they are commonly dark red, brown, or blue-tinged. They may have a surface covering of fine hairs, believed to enhance the visual mimicry of dead flesh. Most flowers that attract egg-laying insects also release a strong scent. A recent report of the components of the scents of a range of sapromyiophilous flowers in the family Apocynaceae observed that different species released different mixtures of hexanoic acid, carboxylic acids, pyrazines, heptanal, octanal, dimethyl oligosulphides, indole, and cresol, and that the mix of compounds released gave specific corpse, urine, or dung odors (Juergens *et al.* 2006). To facilitate the dispersal of scent, some flowers are thermogenic, generating heat through a salicylic acid-activated signal transduction cascade. There are many examples of sapromyiophilous flowers, particularly in the orchid family, but the most famous example is the world's largest inflorescence, the titan arum (Fig. 13.5a).

The most specialized pollination system by deceit is that used by flowers which mimic the female of a species of insect, and invite the male to attempt to mate with them. The colours and scents produced by these mimicking flowers are highly species specific, and may include the release of compounds which mimic the insect's pheromones.

(a)

(b)

(c)

Figure 13.5 Floral mimicry. (a) The Titan arum (*Amorphophallus titanium*) attracts pollinators by releasing a strong scent reminiscent of rotting flesh. Image kindly provided by Cambridge University Botanic Garden. (b) *Ophrys episcopalis*, which mimics female insects to achieve pollination through pseudocopulation. Image kindly provided by Richard Bateman (Royal Botanic Garden, Kew). (c) The composite inflorescence of *Gorteria diffusa* mimics its pollinating flies. See also Plate 12.

Colours usually involve yellows, purples, and browns, and may be distributed on the flower to mimic the appearance of the wing cases of the female. No reward is provided to the pollinating animal, but, because this kind of mimicry releases instinctive behaviour, the animal cannot learn to avoid these flowers, and so they are not as frequency-limited as food deceivers. A classic example of a flower pollinated in this way is the fly orchid,

Ophrys insectifera, which mimics the female of the scoliid wasp *Campsoscolia ciliate*. Although the visual similarity between the flower and the wasp is very striking, it is the scents produced by the flower which are most astonishing. The orchid releases a range of very specific volatiles, which closely mimic the sex pheremones of the female wasp. Some of these, such as (omega-1)-hydroxy acids and (omega-1)-oxo acids, are not found elsewhere in the plant kingdom (Ayasse *et al.* 2003). A number of related species produce similarly impressive mimics (Fig. 13.5b). A less extreme, but nonetheless remarkable, example is the South African 'beetle daisy', *Gorteria diffusa*. The composite inflorescence of these daisies contains bright orange ray florets, several of which develop large black spots (Fig. 13.5c). The spots are composed of both intense pigmentation and specialized cells which add three-dimensionality and sheen. They attract male bee flies, *Megapalpus nitidus*, by mimicking the female flies which often spend the night in the closed inflorescences (Johnson and Midgely 1997). These are perhaps some of the most extreme examples of coevolution between plant and pollinator, and certainly give much credit to the idea that floral features have evolved in response to pollinator preferences.

The different pollination syndromes discussed in this chapter vary from the extremely diffuse (fly pollination) to the very specialized (some of the deceit pollination examples). While there is little doubt that some specialized flowers have evolved traits which make them more attractive to particular animals, there is a great deal of debate in the literature as to whether the majority of 'ordinary' flowers can really be said to show pollination syndromes. In the next few chapters we will discuss the mechanisms by which the great diversity of shapes and colours is generated, before returning to consider the adaptive significance of these features in Section IIIB.

CHAPTER 14

Changing Floral Shape and Structure

This section of the book is focused on the changes that can occur to floral form, and on the effects of those changes on pollinator behaviour. In Chapter 9 we discussed the ABC model of flower development, and the molecular changes that occur to generate the correct numbers and arrangements of the four whorls of floral organs—sepals, petals, stamens, and carpels. In Chapter 13 we considered the traditional idea that different pollinators might have preferences for different shapes, structures, and sizes of flowers. In later chapters we will investigate the evidence in support of this hypothesis, but before that we must consider the molecular and developmental changes which can lead to variation in floral form. In this chapter, then, we start with a basic flower, such as the flower of Arabidopsis, and consider the ways in which its form can be altered by known developmental programmes. We will focus on the petals, or corolla, as the whorl of organs of primary importance in attracting potential pollinators, and consider changes to its size, its symmetry, the shape of its component petals, and its position within an inflorescence.

14.1 Changing corolla size

As we saw in Chapter 9, a petal develops from a primordium of dividing cells, which arises on the floral meristem in response to the activity of organ identity genes. The final size of the petals, and thus the corolla they form, is a factor of the combined amount of cell division and cell expansion that occurs throughout the development of each organ. Only these processes can influence plant organ size. In animals cell death and cell migration may also play roles, but plant cells do not migrate, and

programmed cell death in plants results only in dead structures (such as xylem vessels), as there is no mechanism to remove dead cells (Meyerowitz 1996). Cell division is the process by which a single cell replicates its DNA and then divides into two daughter cells, partitioning the DNA equally between the two daughters. Cell expansion is the process by which post-mitotic cells increase in size, largely through vacuolation. Earlier cell growth is driven by an increase in cytoplasmic mass, but the vast majority of increase in cell size is due to post-mitotic expansion through vacuolar expansion (Sugimoto-Shirasu and Roberts 2003). In theory, then, it would seem relatively easy to change final petal size. An increase in cell division rate should result in more cells, increasing petal size, while a decrease in cell division rate should have the opposite effect. Similarly, an increase in the extent to which cells expand should increase overall organ size, while a decrease should result in smaller cells and a smaller organ.

An increasing body of evidence suggests that the control of organ size is a rather more complex process than these basic mechanisms might suggest. In particular, it is becoming apparent that cell division and cell expansion are necessary to organ growth, but do not themselves control the final organ size. Instead, there appear to be mechanisms present that monitor organ growth and balance the number and size of cells to achieve the correct end point (Shpak *et al.* 2003). A number of studies have now shown that altering cell division or expansion through the use of transgenes does not necessarily lead to a change in organ size, because the other process adapts to compensate. For example, ectopic expression of the auxin-binding protein (ABP1) from Arabidopsis in tobacco causes an increase in cell expansion. However, cell division is reduced and the organs therefore end up the same size as wild type, but are composed of fewer, larger cells (Jones *et al.* 1998). Similarly, ectopic expression of *CycD3;1*, a cyclin gene which induces progress through the cell cycle, causes an increase in cell number in Arabidopsis but no overall change in organ size. Instead, organs are found to be composed of numerous but very small and poorly differentiated cells (Dewitte *et al.* 2003). Studies such as these suggest that an intrinsic mechanism coordinates cell division and cell expansion, to control overall organ growth.

Recent analysis of the *AINTEGUMENTA* (*ANT*) gene of Arabidopsis suggests that it might function as a checkpoint through which such control is exerted. Unlike the examples described above, changes in ANT function do result in changes in overall organ size. Loss of ANT function results in leaves and petals smaller than wild type, while increased *ANT* expression in transgenic plants increases the sizes of all organs (Mizukami and Fischer 2000). These changes result from changes in total cell number, which are not compensated for by changes in cell size. ANT appears to control cell number not by affecting the rate of cell division, but by regulating the duration of cell division in a developing organ. ANT sustains the expression of D-type cyclins, which are necessary for progress through the cell cycle, increasing the window of time during which cell division can occur. The ANT protein is a transcription factor with similar domains to APETALA2, the A function protein we met in Chapters 8 and 9, and may act by directly maintaining transcription of these cyclin genes (Mizukami 2001). It has recently been shown that an auxin-inducible gene, *ARGOS*, prolongs the expression of *ANT*, and therefore acts upstream of it to maintain this competence for division (sometimes called meristematic competence) of the cells within developing organs (Hu *et al.* 2003). The apparent involvement of auxin in this process suggests that larger scale forces are at work, and that overall organ size is likely to represent a balance between a variety of environmental signals integrated at the point of controlling cell division and expansion. Similarly, the *ARGOS-LIKE* (*ARL*) gene

has been shown to respond to brassinosteroid application, and maintains cell expansion in developing tissues. Transgenic plants expressing reduced levels of *ARL* through RNAi inactivation have smaller lateral organs than wild type, including petals (Hu *et al.* 2006). *ARL* has significant sequence similarity to *ARGOS*, both encoding proteins with putative transmembrane domains, and it is likely that this protein family is essential to the control of organ size.

An unrelated protein, BIG BROTHER (BB), has recently been shown to control maximum organ size by targeting the proteins necessary for cell division for degradation. Plants mutants at the *BB* locus have much larger floral organs than wild type, and transgenic plants ectopically expressing *BB* have much smaller organs (Fig. 14.1). The protein regulates the duration of the growth phase, not the intensity of growth, and affects final cell number, not cell size. *BB* encodes an E3 ubiquitin ligase, a protein which binds other proteins to ubiquitin conjugating enzymes, preparing them for degradation. These data strongly suggest that maximum organ size is regulated by preventing further cell division through degradation of key proteins (Disch *et al.* 2006).

The extent to which the control of organ size is both robust and delicate can be seen from the astonishing degree to which petal shape and size is uniform within a species, but has the potential to be extraordinarily variable between species. Analysis of patterns of cell division and expansion within a single petal has shown that the processes are controlled in a coordinate way. Using markers for cell cycle and electron microscopical techniques to

BB ox   WT   *bb*-1

Figure 14.1 The *bb* mutant (right) has larger floral organs than wild type, while plants overexpressing *BB* (left) have smaller floral organs. Images kindly provided by Michael Lenhard (University of Freiburg).

follow cell expansion, Reale *et al.* (2002) described the patterns of cell behaviour in a developing petal of *Petunia hybrida*. They found that cell division was initially distributed uniformly throughout the tissue. However, the decline of cell division was not uniform, beginning first at the base of the petal and then spreading to form a gradient from base to tip. Cell expansion was coordinated with decline of cell division, beginning first in the basal part of the petal and then spreading throughout the whole tissue. The complexity of this developmental pattern confirms that control mechanisms of the sort discussed earlier must be at work to ensure appropriate final size of the organ.

Changing petal size may not be as simple a matter as we initially thought, but it has clearly happened many thousands of times throughout the evolutionary history of the angiosperms, to create the enormous range of petal sizes we see today. Understanding the control mechanism that coordinates cell division and cell expansion will be necessary for us to understand how petal size can be adapted. The current work on *ANT* and *ARGOS* in Arabidopsis should pave the way for future analysis of the regulation of organ, and particularly petal, size in more complex types of flower.

14.2 Changing corolla symmetry

The Arabidopsis flower is radially symmetrical, as are the flowers of many common species such as buttercup, rose, lily, and tulip (Fig. 14.2a). It is possible to draw many lines of symmetry on the face of these flowers. A radially symmetrical flower (also called an actinomorphic flower or polysymmetric flower) is the default form that develops through the activities of the ABC genes, as they direct identical development in all regions of each concentric whorl on the floral meristem. However, many plants produce flowers that are bilaterally symmetrical (also called zygomorphic or monosymmetric). It is only possible to draw a single line of symmetry on the face of these flowers, as not all organs within a whorl are identical (Fig. 14.2b). The changes may be to any of the four types of floral organ, but the most commonly seen zygomorphic flowers have alterations to either the petals or the stamens, or both. Classic examples of zygomorphic flowers

(a)

(b)

Figure 14.2 Zygomorphy and actinomorphy. (a) Many flowers are radially symmetrical, or actinomorphic. (b) The flowers of *Antirrhinum* species are bilaterally symmetrical, or zygomorphic. See also Plate 13.

include Antirrhinum, sweet peas, and many orchid species. A consequence of zygomorphy is that flowers come to have a clear top and bottom, and so are also said to have dorsoventral symmetry, or dorsoventrality. One advantage of dorsoventrality is that it confers position on each organ within the flower, allowing organs in different positions to adopt different morphologies. Organs can be identified and described as the dorsal (top) petals, lateral (side) sepals, ventral (bottom) stamens, and so on. When each developing organ has a specific position then it is possible to modify the development of individual organs so that the flower produced has a clear structure, such as the ventral keel on pea flowers.

The key to producing a zygomorphic flower is to supply it with a signal which marks a region of the developing floral meristem as top or bottom. It is not possible to study this sort of process in a species like Arabidopsis, with actinomorphic flowers, but a considerable amount of work has been done on this problem in Antirrhinum. The Antirrhinum flower is zygomorphic with respect to both petals and stamens. The flowers have five petals; two large dorsal ones, two smaller lateral ones, and a single ventral petal which has a distinctive bend in it, creating the lip or hinge which acts as the landing platform for pollinating bees. There are also five stamen primordia, but the single dorsal stamen aborts, leaving two lateral stamens and two longer, hairier ventral stamens.

Analysis of the generation and maintenance of zygomorphy in Antirrhinum has been primarily concerned with four mutant lines, *cycloidea* (*cyc*), *dichotoma* (*dich*), *radialis* (*rad*), and *divaricata* (*div*). The *cyc* mutant has an incomplete conversion from zygomorphy to actinomorphy. The lateral petals of the mutant are converted into ventral petals, and the dorsal petals are morphologically somewhere between dorsal and lateral petals. The ventral petal is unaffected, giving a flower with three ventral petals and two dorsal/lateral petals. This incomplete loss of zygomorphy results in a flower described as semi-peloric, based on Linnaeus's description of fully actinomorphic mutants as peloric (or monstrous). The *dich* mutant produces flowers with a similar phenotype to those of *cyc*, but less dramatic in their conversion. However, if the two mutant lines are crossed together to create a *cyc/dich* double mutant then the flowers produced are fully peloric or completely radially symmetrical, with all the petals converted to the ventral form (Luo *et al.* 1996).

The key to understanding these mutants is the ventralization of organs in the mutant form. This led to the conclusion that *CYC* and *DICH* provide dorsalizing signals to the developing floral meristem. In fact, both the *CYC* and *DICH* genes are expressed in overlapping domains in the dorsal portion of the floral bud, and remain so expressed throughout flower development. *CYC* is expressed in the dorsal petals and dorsal stamen primordium, while *DICH* is expressed in the dorsal stamen primordium and the dorsal part of the dorsal petals. *CYC* and *DICH* both encode TCP family transcription factors. Both proteins seem to act early in development to limit growth in the dorsal region of the meristem, ensuring that the dorsal organ primordia stay within the region they define. The mutant floral meristem starts to differ from wild type at the time that sepal primordia become visible, and this restriction of dorsal primordium growth culminates in the abortion of the dorsal stamen. Later in flower development, CYC and DICH promote petal lobe growth so that the dorsal petals end up larger than the others. The result of this expression pattern is that every organ primordium in a developing Antirrhinum flower has a unique combination of expression of *CYC*, *DICH*, and the organ identity (ABC) genes. The only primordia that have the same patterns are those which are mirror images of each other, and this provides the bilateral symmetry to the flower (Luo *et al.* 1996).

DIVARICATA, by contrast, acts to provide a ventralizing signal to the floral meristem, and the *div* mutant has an abnormal ventral petal with little or no hinge and lip region. *DIV* encodes a transcription factor from the MYB family, and is expressed throughout the developing flower (Galego and Almeida 2002). The explanation for the ability of *DIV* to act only in the ventral and lateral petals while expressed throughout the flower comes from an analysis of the role of *RADIALIS* (*RAD*). *RAD* also encodes a MYB transcription factor, but one from a different subfamily to DIV. Most plant MYB proteins contain two repeats of the MYB domain, and are classed as R2R3 MYBs. The RAD protein contains a single MYB domain, and is therefore classed as a one-repeat MYB. There is some evidence to suggest that the MYB domain in one-repeat MYBs is not able to bind DNA. *RAD* is expressed in the dorsal regions of the developing flower, in a spatial and temporal pattern which suggest that it is activated by CYC and DICH (Corley *et al.* 2005). This activation is confirmed by the reduction of *RAD* expression in a *cyc* mutant and its complete absence from the *cyc/dich* double mutant. Corley *et al.* (2005) suggested that RAD acts as a repressor of transcription, and proposed that it competes with DIV in the dorsal part of the flower, inhibiting DIV binding to promoters of target genes and thus preventing any ventralizing activity. Zygomorphy in Antirrhinum can therefore be seen as a result of the activities of two TCP proteins and two MYB proteins, providing dorsalizing and ventralizing signals to the floral meristem.

Within the same order as Antirrhinum, many species show zygomorphy which is not identical to that of Antirrhinum. For example, the flower of *Mohavea confertiflora* aborts the top three stamens (i.e., the dorsal and both lateral stamens), not just one, while its corolla shows little deviation from radial symmetry. These differences have been correlated with differences in the expression patterns of the *CYC* and *DICH* orthologues. Unlike in Antirrhinum, both genes in *Mohavea* are expressed in the lateral stamen primordia as well as the dorsal one, resulting in the abortion of all three. However, in the petals the expression of the *CYC* and *DICH* orthologues terminates much earlier than it does in Antirrhinum, possibly contributing to the reduced zygomorphy of the corolla (Hileman *et al.* 2003).

Zygomorphy has evolved multiple times from actinomorphy, and there are also examples of reversion from a zygomorphic form to an actinomorphic form (Endress 2001). Current work in the field is focused on assessing whether evolution of zygomorphy involves repeated recruitment of the same molecular components to direct similar developmental programmes, or whether different developmental pathways apply in different taxa (Cubas 2003). Within the Leguminosae, recent reports suggest that *CYC*-like genes have been recruited to the development of zygomorphy in *Lotus japonicus* and *Lupinus nanus*. The *LjCYC2* gene was isolated from *L. japonicus* by Feng *et al.* (2006) and shown to be expressed in the dorsal part of developing flowers, just as *CYC* itself is in Antirrhinum. Loss of *LjCYC2* expression resulted in ventralization of the dorsal petal, while ectopic expression of *LjCYC2* resulted in dorsalization of the lateral and ventral petals. Similarly, Citerne *et al.* (2006) showed that a *CYC*-like gene was expressed in the dorsal part of the developing flower of *L. nanus*. Surprisingly, however, the radially symmetrical phenotype of a close relative, *Cadia purpurea*, was not caused by loss of expression of this *CYC*-like gene, but by its expression throughout the developing flower.

Therefore *CYC*-like genes have not only been recruited to specify floral zygomorphy in the Leguminosae (a separate evolutionary development from that of Antirrhinum) but their ectopic expression has also resulted in dorsalization and therefore actinomorphy of whole flowers within this group.

A rarer form of symmetry seen in some species is left–right asymmetry, most commonly seen in the reproductive organs. Asymmetrical female reproductive organs result in a condition known as enantiostyly, which occurs in at least 10 families of angiosperms (Fig. 14.3, see also Plate 14; Endress 2001). This curving of the stigma and style to one side of the flower occurs mainly in flowers that are buzz-pollinated (pollinated through vibratile pollen release by pollen-gathering bees), and it may be that it reduces damage to the style when large animals land on and grasp the stamens. Jesson and Barrett (2002) showed that enantiostyly increased outcrossing rate, even where both right- and left-styled flowers are present on the same individual plant, presumably by decreasing the frequency with which self pollen is passed to the stigma. In most described cases both left- and right-handed flowers are found on the same individual, suggesting an essentially random process, although in three families left- and right-handedness is separated on different individual plants (Barrett *et al.* 2000). In one of these cases, *Heteranthera multiflora*

Figure 14.3 The enantiostylous flower of *Solanum heterodoxum*. The style is the filamentous pink organ. Image kindly provided by Sandy Knapp (Natural History Museum, London).

(an American monocot), it has been shown that style direction segregates in a Mendelian fashion, with the right-styled form dominant over the left (Jesson and Barrett 2002). Little is yet known about the developmental control of this form of asymmetry.

A little understood form of floral symmetry is the development of petals contorted to the left or the right. Such contortions are most prominent in the floral bud and are similar in appearance to a spiral phyllotactic arrangement of the petals. It is possible for contort flowers to show fixed behaviour, where only the right- or left-handed form is present in a species. Alternatively, an unfixed pattern where left- and right-handedness is shown by an individual plant is also possible. Within the Rosids (one of the two major subclasses of the Eudicots) most contort species show an unfixed pattern. However, in the other subclass, the Asterids, a fixed pattern is shown at the species or higher taxonomic level. There is currently nothing known about the development of this form of asymmetry, and no explanation as to why the Rosids and the Asterids should show such different behaviour (Endress, 2001).

14.3 Changing petal shape

The shape of a petal is a result of the distribution of cell divisions and the distribution and direction of cell expansion which occur during its development. As such, the development of petal shape is an integral part of the development of petal size, controlled by the coordinating forces described earlier in this chapter. The development of petal shape has been most successfully studied using clonal analysis. Rolland-Lagan *et al.* (2003) used a line of Antirrhinum with a temperature-sensitive transposon inserted in the *PALLIDA* gene. Flowers that are *pal* mutant are white. Transposon excision was induced by exposure of plants to a 15°C growth regime, resulting in clones of red epidermal cells descended from the single cell in which the transposon had excised from the *PAL* locus. Transposon excision was induced at a number of time points in petal development, and the size and shape of the resulting sectors were mapped onto models of petal development. This study revealed that petal growth rate was relatively constant, and that the asymmetric shape of the Antirrhinum petal results from

changes in the direction of growth rather than from local differences in the rate of growth. However, petal growth was shown to be anisotropic (stronger in one direction than another), with an average 15% more cell divisions along the principal axis of growth than along the axis perpendicular to it. The main growth direction was shown to rotate as the petal grew, giving the final asymmetric shape. Models such as this will be useful in describing the development of other petal shapes, and comparisons between species should allow us to dissect the factors responsible for the many different petal shapes seen in modern angiosperms.

Final petal shape is also dependent upon the generation of surface curvature. Curvature of the petal surface is produced by differential growth of the margins relative to the central region of the tissue. In fact, many petals are essentially flat, showing no such curvature. However, the chances of this occurring by chance are very low, implying that genetic mechanisms which regulate growth rate to ensure no curvature is produced exist (Nath *et al.* 2003). In some species petal curvature is an important part of corolla appearance, and in these cases it is likely that similar genetic mechanisms regulate relative growth differently to produce the final petal shape. In Antirrhinum, organ curvature is inhibited by the action of the *CINCINNATA* gene. Plants that are *cin* mutants have curved leaves as a result of excessive growth of the margins relative to the central region of the tissue. Further analysis indicated that this was a result of delayed cell expansion in the centre of the leaf relative to the margins. The *CIN* gene encodes a transcription factor from the TCP family, to which *CYC* and *DICH* also belong, and may promote flatness by making cells in marginal regions of the leaf more sensitive to a signal causing arrest of cell division and expansion (Nath *et al.* 2003). Similar processes may be involved in the control of petal curvature, although CIN itself does not seem to operate in the same way in the petal. Plants that are *cin* mutants have smaller petals than wild type flowers, and fewer cells. These data indicate that CIN promotes petal growth through extending the cell division window in the petal lobes (Crawford *et al.* 2004).

The petals of some species show distinct architectural features. The ventral petal of Antirrhinum, for example, has a distinctive hinge or lip region, which forms the landing platform for pollinating bees. While all Antirrhinum petals show differences between the basal portions (which fuse to form the corolla tube) and the distal regions, which remain separate as lobes, the division between the two is most marked in the ventral petal. Elaboration of landing platforms is a common developmental process in many animal-pollinated flowers, and one which involves a very clear fold in a petal. A number of genes have been shown to play roles in landing platform development in Antirrhinum. The *div* mutant has a very disturbed ventral petal, with no hinge formation. This may reflect the loss of the ventralizing signal and the lack of activity of downstream genes which contribute to hinge development. One such gene is the *Antirrhinum majus MYB MIXTA LIKE 1* (*AmMYBML1*) gene, expression of which is significantly reduced in the div mutant flower. *AmMYBML1* encodes a MYB transcription factor expressed in the ventral petal, in the epidermis of both lobe and tube, but also in the adaxial-most mesophyll in the hinge region. Examination of this region indicated that the adaxial-most mesophyll cells expand more than the abaxial ones, causing the bend of the petal to make the hinge. Ectopic expression of *AmMYBML1* in tobacco petals causes the excess expansion of all petal mesophyll cells, supporting the idea that the gene is important in hinge development (Perez-Rodriguez *et al.* 2005). The *CIN* gene is also expressed in the cells of this region and may act to reinforce other signals causing tissue bending (Crawford *et al.* 2004).

Overall petal shape may be modified by the elaboration of detailed structures, either as part of the whole petal or by modification of epidermal cells. For example, the flowers of many species, including *Linaria vulgaris* (toadflax) and many species of *Aquilegia* and orchids, produce a nectar spur, either on the ventral petal or on all petals (Fig. 14.4, see also Plate 15). The spur contains secreted nectar, only accessible to particular pollinators. *Linaria* itself is primarily pollinated by long-tongued bumblebees, although shorter-tongued species may rob the flower of nectar by chewing through the nectar spur from the outside (Newman and Thomson 2005). Petal spurs require elaborate outgrowth of a small region of the petal, and their development is

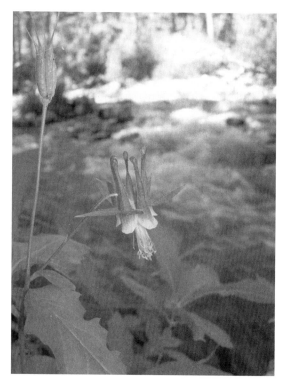

Figure 14.4 The petals of *Aquilegia formosa* are heavily modified to produce nectar spurs. Image kindly provided by Scott Hodges (U.C.S.B.).

not well understood. An actinomorphic mutant of *Linaria* with all petals converted to the ventral form and making spurs was described by Linnaeus and has been reported many times since. This mutant has been shown to be defective in the function of a *CYC*-like gene, demonstrating that spur development acts downstream of genes controlling floral symmetry (Cubas *et al.* 1999). More recently, two mutants of *Antirrhinum majus* which produced ectopic nectar spurs on the ventral petal were identified. Antirrhinum, unusually among its close relatives, does not normally produce a nectar spur. Both mutants were shown to contain transposon insertions in genes encoding homeodomain transcription factors, similar to SHOOT MERIS-TEMLESS (STM) of Arabidopsis. STM is required for apical meristem maintenance, and causes cells to undergo division. Ectopic expression of *STM*-like genes in the developing petals of these Antirrhinum mutants resulted in nectar spur development (Golz *et al*. 2002). These data suggest that normal nectar spur development involves the activity of homeodomain transcription factors to cause plentiful cell division in appropriate petal tissue.

Elaboration of epidermal surfaces may also contribute to final petal appearance. The ventral petal of *Antirrhinum* contains a number of specialized epidermal cell types, including conical-papillate cells in the lobes, long trichomes on the lip, and twin lines of shorter trichomes leading down into the throat. These specialized cells may act as nectar guides or to trap pollen from visiting bees, and they may also be involved in regulating petal temperature and scent production (Perez-Rodriguez *et al*. 2005).

14.4 Generating a composite inflorescence

While modifications to corolla size and shape can have dramatic effects on the appearance of the whole flower, an even more dramatic effect can be achieved by modification of the entire inflorescence, so that it mimics the appearance of a single, large, flower. This form of floral development is characteristic of the family Asteraceae, where the entire inflorescence is condensed into a head-like structure (the capitulum), containing hundreds, or sometimes thousands, of flowers (often referred to as florets). The common daisy is a good example of this growth form, but other familiar plants with this type of composite inflorescence include dandelions, the cultivated Chrysanthemums and Gerbera (Fig. 14.5a), and the groundsels. A composite inflorescence may be composed of a single type of flower (such as the inflorescence of *Senecio vulgaris*, groundsel) or multiple different morphological forms. Where different floral forms are present they may vary in colour (for example, the yellow and white flowers of the daisy inflorescence), symmetry (daisies have both zygomorphic flowers (the outer white ones) and actinomorphic flowers (the central yellow ones)), and sexuality. It is relatively common for the different flowers in a composite inflorescence to vary in sexuality. *Gerbera hybrida*, for example, has hermaphrodite central flowers but female outer flowers.

(a)

(b)

(c)

Figure 14.5 Composite inflorescences. (a) The capitulum of *Gerbera hybrida*, with zygomorphic outer florets (each with one large petal) and actinomorphic inner florets. (b) The capitulum of *Senecio vulgaris* usually contains only actinomorphic disc florets. (c) Scanning electron micrograph of developing capitulum of *Chrysanthemum*. The individual developing flowers can be seen, each with five petals, with developing pappus bristles between the flowers. See also Plate 16.

The development of a composite inflorescence involves a reduction in internode length relative to an ordinary inflorescence. This reduction in internode length causes all the flowers to develop closely together, and is associated with a widening of the inflorescence meristem to form the typical head-shaped structure. Flowers are usually produced in a spiral phyllotaxis in members of the Asteraceae, and the developing floral meristems produce some of the clearest examples of spiral patterning seen in plants. It was traditionally thought that the inflorescence meristem of the Asteraceae was indeterminate, with the potential to produce an infinite number of flowers. Most species in fact produce a relatively consistent number of flowers, but this was explained as a function of spatial constraint on the densely packed head. However, recent work has demonstrated that the inflorescence meristem of *G. hybrida* is indeterminate, and requires the activity of a gene called *GRCD2*, encoding a MADS box transcription factor, to confer determinacy and ensure formation of the correct number of flowers. Reduction of *GRCD2* expression levels, using antisense technology, resulted in the continued development of the centre of the inflorescence, producing new flowers long after a wild type inflorescence had stopped and formed a single seed head (Uimara *et al*. 2004).

The flowers of members of the Asteraceae typically lack a whorl of sepals, which would be cumbersome when producing a composite inflorescence that mimics a single flower. Instead, most species have a whorl of fine hairs, known as pappus bristles, around the base of each corolla (seen developing in Fig. 14.5c). These bristles are involved in dispersal of the resulting seed, using the wind as the agent of dispersal. Through isolation and analysis of the ABC genes from *G. hybrida*, Yu *et al*. (1999) demonstrated that the pappus bristles are highly modified sepals, produced through the activities of genes orthologous to those that produce sepals in Arabidopsis.

Analysis of the differentiation of the flowers of a composite inflorescence into multiple types has focused on floral symmetry and the activities of *CYC*-like genes. In particular, extensive use has been made of *S. vulgaris*. In the UK *S. vulgaris* exists in two forms, a radiate morph, which has both zygomorphic outer flowers and actinomorphic central flowers, and a non-radiate morph, which has only actinomorphic flowers (shown in Fig. 14.5b). The non-radiate form is believed to be ancestral, while the radiate form is thought to have arisen by introgression of genetic material from *Senecio squalidus* (the Oxford ragwort, a non-native but naturalized species with both actinomorphic and zygomorphic flowers; Abbott *et al*. 1992). The ray floret trait segregates as a single locus, and it has been hypothesized that a *CYC*-like gene may be responsible for the development of ray florets (Gillies *et al*. 2002). Further work in this system should allow characterization of floral symmetry changes and floral sexuality changes in composite inflorescences.

CHAPTER 15

Colouring the Flower

One of the most obvious ways in which flowers differ from one another, more striking even than the changes in size, shape, and symmetry discussed in the previous chapter, is in their colour. This colour is usually the result of pigment deposition in the petals, but in some species brightly coloured bracts, sepals, or stamens can provide a similarly striking display. The colour of a flower has traditionally been viewed as one of the ways in which plants attract pollinating animals, and there is considerable literature on the preferences of different pollinators for different colours. In recent years these data have often been reinterpreted, with the emphasis more on contrast and search images, rather than on particular colours as 'favourites' of particular animals. In this chapter we consider the different biochemical pathways through which flowers can become coloured, paving the way for a discussion of the regulation, patterning, and enhancement of colour in the next chapter.

15.1 Colour as a signal

Colour is a signal used by plants for a wide variety of purposes. Colour is used to signal ripening of fruit, and to attract animals to eat that fruit and disperse the seeds it contains. Colour can also be used to signify the unpalatibility or toxicity of certain tissues, again often including fruits, particularly unripe fruits. Colour can also be an accident—the orange colour of autumn leaves is usually attributed to the breakdown and reabsorption of green chlorophyll by the plant, leaving behind the orange and brown waste products of the breakdown reactions. Similarly, visible red colouration of leaves and stems is often a side effect of the production of UV-absorbing anthocyanins and flavonoids in

response to a range of stresses, particularly high light levels.

The bright colours of flowers are primarily a signal to attract pollinating insects by making the floral tissue stand out against a background of vegetation. As we noted in Chapter 13, Darwin himself observed the importance of pigmentation in flowers, and argued that colour is simply a means of providing flowers with contrast against vegetation. This argument is supported by modern analysis of insect visual acuity, which indicates that vegetation is visually very similar to bark, soil, and stone from an insect's point of view (Kevan *et al.* 1996). All these materials weakly reflect light across the whole range of an insect's visual spectrum. Leaves differ from the rest only in absorbing red light, but since red is at the very periphery of the visual spectrum for most insects, this makes little difference. All flowers, then, whatever their colour, stand out against this generally dull background (Kevan *et al.* 1996). However, it has also been argued that particular colours may act as signals to particular groups of animals. It is certainly the case that the visual ranges of different animals vary somewhat, with the insects able to see into the ultraviolet (unlike some pollinating vertebrates) but usually, with the exception of the Lepidoptera, unable to easily detect light at the red end of the spectrum (see Chapter 18 for a full description of bee and bird visual capability). In these circumstances there is little point in a bat-pollinated flower using UV reflecting pigments, nor in a beetle-pollinated flower investing heavily in red-reflecting anthocyanins, and the colours of flowers (or at least a biochemical analysis of the pigments they contain) may indeed provide clues as to the identity of their pollinators. It is also the case that some rather dramatic changes in floral colour are associated with pollination and changing levels of nectar availability. For example, it has been shown that pollination of *Viola cornuta* is necessary to cause a change in pigmentation from white to purple, a change which presumably signals a change in nectar availability and floral receptivity to pollinating insects (Farzad *et al.* 2002). There are many reports of similar changes in floral colour, often associated with anthesis or pollination. As another example, Darwin forwarded a letter to *Nature* in 1877 describing the change in colour of a Brazilian species of *Lantana* which progresses from yellow, through orange to purple, following pollination. Recent reports have confirmed that this change in flower colour serves to direct pollinating butterflies to flowers still receptive, while retaining a large coloured inflorescence which is more visible from a distance than a smaller inflorescence would be (Weiss 1991). It is clearly the case that flower colour can act as a more specific signal than simply contrast with vegetation. However, analysis of the attractiveness of particular colours to particular animals has resulted in less convincing evidence for the significance of colours as unique signals for individual animals. This issue will be considered in more detail in Chapters 18 and 19.

Biological colours can be produced in two different ways. Some animals produce 'structural' colours, caused by the refraction of light from complex physical surfaces. Plants usually produce colour by synthesizing pigments which absorb subsets of the visible spectrum, reflecting back only what they do not absorb and causing the tissue to be perceived as the reflected colours. Chlorophyll absorbs light in both the red and blue parts of the spectrum, reflecting only green light, and causes leaves to appear green to us. Similarly, a flower that we perceive as red contains pigments which absorb yellow, green, and blue light, leaving red light as the only wavelength visible to us which is reflected. White tissues reflect all visible wavelengths equally, while black tissues absorb all visible wavelengths (light reflection by white, black, and red flowers is shown in Fig. 15.1). Where an animal has a different set of photoreceptors to us, then similar black and white effects can be obtained by absorbing or reflecting light across all the wavelengths visible to that animal.

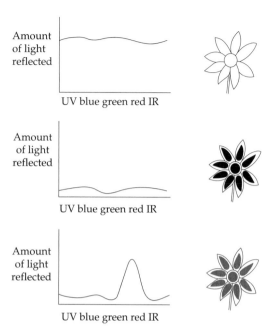

Figure 15.1 Flower colour depends on absorption and reflection of light by pigments. If a flower contains no pigment, and absorbs no light, the reflected light of all colours makes it appear white (top). If the flower absorbs light of all visible wavelengths it appears black. If the flower absorbs all light except red, and reflects the red light, it will appear red (bottom).

15.2 Plant pigments

Whichever organs they are found in, plant pigments can be divided into three chemical classes: the flavonoids, the betalains, and the carotenoids. The flavonoids are the major floral pigments in higher plants, and give rise to ivory and cream colours (flavonols and flavones), yellow and orange colours (aurones and chalcones), and the red-pink-purple-blue range (the anthocyanins). They are water soluble and accumulate in cell vacuoles. Flavonoids also play a range of other important roles in plants, including defence against pathogens and predators, protection against damaging UV light, and participation in pollen development and germination. The betalains are a group of pigments found exclusively in the angiosperm order Caryophyllales, and nowhere else in the plant kingdom. They give the red colour to beetroot and also to some flowers. The carotenoids are much more widespread, although less significant as floral pigments than the anthocyanins. They are lipid-soluble and are found in plastids throughout the plant. Carotenoids give yellow and orange colour to flowers, although their more important role is as an essential component of the photosynthetic apparatus.

15.3 Carotenoid synthesis

Carotenoids are amongst the most widespread pigments in the natural world, and have the most varied functions of any group of pigments. In plants they play important roles in photosynthesis, where they act as accessory light harvesting pigments and as photo-protectants (Hirschberg 1999), as well as acting as floral and fruit pigments. In mammals they are the precursors for vitamin A synthesis, and in fish they are essential for phototropic responses (Goodwin 1980). Animals cannot synthesize carotenoids directly, but modify plant-derived carotenoids present in their diets for diverse purposes. For example, the red plumage of flamingos is produced using modified carotenoids (Grotewold 2006). The carotenoids are a family of isoprenoid derivatives. Isoprenoids themselves are lipid molecules, with an estimated 22,000 different types known. They have essential roles as membrane sterols, components of chlorophyll, cytokinins, abscisic acid, and a variety of roles in plant secondary metabolism. They are usually classified by the number of carbon atoms they contain. The carotenoid family contains conjugated polyene molecules, composed of 40 carbon atoms. The molecule is produced by the joining of eight 5-carbon isoprenoid units such that the arrangement of the units is reversed at the centre. Carotenoid hydrocarbons are known as carotenes, and include pigments such as zeaxanthin. The addition of oxygen to these molecules creates the oxygenated carotenoids, or xanthophylls, which are also important pigments.

The absorption spectrum of a particular carotenoid molecule is determined by the conjugated polyene system and additional structural features. Each conjugated double bond increases the wavelength of maximum absorption by 7 to 35 nm (Goodwin 1980). Thus deep orange flowers may contain lycopene, orange flowers contain β-carotene, and yellow flowers often contain highly oxidized xanthophylls (Fig. 15.2).

(a)

(b)

(c)

Figure 15.2 Carotenoids give the yellow and orange colours to (a) *Gerbera hybrida*, (b) *Freesia*, and (c) Lilies. See also Plate 17.

Carotenoids are synthesized within plastids, although the genes encoding the enzymes necessary for their synthesis are located within the nucleus, implying that the enzymes are transported after translation across the plastid membrane. All isoprenoids, including carotenoids, are produced from a 5-carbon isoprenoid called isopentenyl pyrophosphate (IPP). The major source of plastidic IPP is from pyruvate and glyceraldehyde-3-phosphate (Hirschberg 1999). IPP is converted into dimethylallyl pyrophosphate (DMAPP) by an isomerization reaction catalysed by isopentenyl pyrophosphate isomerase (IPPI). This enzyme has been shown to be the rate-limiting step in carotenoid synthesis in bacteria and yeast (Kajiwara *et al.* 1997).

Assuming IPP and DMAPP are both present, the synthesis of carotenoids proceeds along a regular pathway, which can be followed in Fig. 15.3. Prenyl transferases with different substrate specificities build up a 20-carbon molecule in a succession of steps. One IPP molecule is condensed with one DMAPP molecule to produce a 10-carbon compound known as geranyl pyrophosphate (GPP). The addition of another IPP molecule produces the 15-carbon farnesyl pyrophosphate (FPP) and a final addition creates the 20-carbon molecule geranylgeranyl pyrophosphate (GGPP). The prenyl transferases involved in these different steps have not yet been well characterized. One reason for this is that carotenoid synthesis occurs in all organs of all

Figure 15.3 Carotenoid synthesis leads to the production of lycopene, from which all other carotenoids can be derived.

plants, but different genes may encode the enzymes active in different locations and at different times. It is therefore quite difficult to define the precise enzymes involved in the production of a particular carotenoid in the flowers of a particular species. However, it is believed that different prenyl transferases interact with the different substrates to produce each of GPP, FPP, and GGPP. In Arabidopsis at least five genes encode GGPP synthase. Of the proteins encoded by these genes, two are located in the plastids, two in the endoplasmic reticulum, and one in the mitochondria. Two of the genes are expressed throughout the plant, one is specific to flowers and root tissue, one is entirely flower-specific, and one is entirely root-specific (Okada *et al.* 2000). When a simple plant with a small genome, such as Arabidopsis, shows such complexity in the activity of only one of these enzymes, it is not surprising that we cannot yet say precisely which genes are active in the petal carotenoid pathway of many species.

Up to this point in the pathway, intermediates may be used to make cytokinins, monoterpenes, sterols, triterpenes and sesquiterpenes, phytol, and gibberellins as well as carotenoids. The first committed step of carotenoid synthesis is the condensation of two molecules of GGPP to form the 40-carbon molecule phytoene, which is a colourless molecule formed by the head-to-head condensation of the two GGPP molecules. This generates the reversal of the molecular arrangement at the centre characteristic of carotenoids. This coupling reaction, an electrophilic attack on a double bond, is a two-step process and is catalysed by the enzyme phytoene synthase (Chappell 1995). Genes encoding phytoene synthase have been isolated from a number of species, and share domains encoding prenyl transferase activity with other enzymes, notably squalene synthases (Chappell 1995).

The conversion of phytoene to lycopene involves two steps, although in bacteria and fungi the conversion uses only one bifunctional enzyme (Moise *et al.* 2005). In plants, phytoene is converted first into a ζ-carotene by phytoene desaturase. Phytoene desaturase uses plastoquinone and then oxygen as electron acceptors during this reaction, and is encoded by genes expressed at constant levels throughout flower development (Al-Babili *et al.*

1996). The enzyme ζ-carotene desaturase (ZDS) converts the ζ-carotene to 7,9,9′,7′-tetra-*cis*-lycopene, and then lycopene itself is produced using the enzyme carotenoid isomerase. Mutation in genes encoding carotenoid isomerase results in the accumulation of tetra-*cis*-lycopene and a subsequent shift in the absorption spectrum of the tissue. The pathway to this point is shown in Fig. 15.3.

Lycopene itself is a carotenoid, and all other carotenoids can be derived from it. The different pathways that give rise to the different carotenoid end products are complex; however all carotenoids can be formally derived from lycopene by reactions involving hydrogenation, dehydrogenation, cyclization, insertion of oxygen, double bond migration, methyl migration, chain elongation, and chain shortening (Goodwin 1980).

Final flower colour can be attributed as much to the amounts of carotenoid present as to the types of carotenoid involved. Moehs *et al.* (2001) showed that variation in marigold flower colour was correlated with amount of xanthophylls present, and also with the expression levels of all the carotenoid biosynthetic genes. These data suggest that regulation of carotenoid synthesis occurs primarily through the transcriptional control of the genes encoding the biosynthetic enzymes, perhaps in ways similar to the regulation of anthocyanin synthesis (discussed in Chapter 16).

15.4 Flavonoid synthesis

The flavonoids are a group of phenolic compounds in which two six-carbon rings are linked by a three-carbon unit. All flavonoids absorb in the wavelength ranges 250 to 270 nm and 330 to 350 nm, and some in the range 520 to 550 nm, so that they appear as visible pigmentation to all animals, although the colour they appear will vary with the animal's ability to perceive reflected ultraviolet light. Flavonoids are water soluble and accumulate in the vacuoles of higher plant cells. Flavonoids play a number of roles in plant physiology and development. They may be involved in defence against pathogens and predators, are a component of the legume-*Rhizobium* signalling pathway, are required for correct pollen development and pollen tube growth, protect sensitive tissues from ultraviolet radiation, and act as antioxidants and metal chelators. Within the flower their main role is as pigments and co-pigments, with different classes of flavonoids contributing different colours.

The first committed step of flavonoid synthesis is the condensation of three molecules of malonyl CoA (derived from acetyl CoA) with one of p-coumaroyl CoA (derived from the skimate pathway and general phenylpropanoid metabolism) to produce chalcone, the key intermediate in the synthesis of all flavonoids (Martin and Gerats 1993). Chalcone itself is usually yellow or orange, and may accumulate as a pigment in its own right. In a few species it is converted into a yellow aurone, the pigment which provides the yellow colour to some Asteraceae flowers, such as Dahlia (Nakayama 2002). Aurones are derived from chalcones by hydroxylation of the B-ring moiety and/or oxidative cyclization. In Antirrhinum, both of these reactions are catalysed by aureusidin synthase, which is encoded by a gene with strong similarity to more general plant polyphenol oxidases (Nakayama et al. 2000; Sato et al. 2001).

Usually, however, chalcone is modified to a colourless flavanone, and flavanone then feeds into one of three further pathways. Flavanone may be directly converted into flavones by enzymes called flavone synthases. Flavones vary in colour from very pale to bright yellow, depending on their degree of hydroxylation. Flavones may provide some yellow colouration to flowers, but are more usually thought of as co-pigments (see Chapter 16). Alternatively flavanone can be converted into dihydroflavonol by the enzyme flavanone 3-hydroxylase, a dioxygenase. Dihydroflavonols can be modified by another dioxygenase, flavonol synthase, to form various flavonols. The flavonols are usually colourless, but act as co-pigments, stabilizing and modifying the colour of other pigment molecules (see Chapter 16). They also absorb light strongly in the UV and so may appear coloured to insects that can see in that range. Alternatively, dihydroflavonols may be converted to anthocyanidins by the activity of two enzymes, dihydroflavonol 4 reductase and anthocyanidin synthase. The unstable products of anthocyanidin synthase may be further modified by glycosylation, methylation, and acetylation to form anthocyanins. Anthocyanins provide a number of different colours,

depending on a number of factors (Fig. 15.4). One factor is the degree of hydroxylation of the B ring. There are at least 17 known anthocyanins, which may range from orange/brick red (pelargonidins; one hydroxyl on the B ring), red/magenta (cyanidins; two hydroxyls on the B ring), to purple/blue (delphinidins; 3 hydroxyls on the B ring), with increased blueness determined by an increase in hydroxyl groups. Methylation of an anthocyanin tends to shift its colour towards red, compared to bluer unmethylated molecules. Acylation with aromatic groups shifts the pigments towards the blue range, whereas aliphatic acylation increases the stability of the anthocyanins in solution.

The amount of pigment present in a tissue can also determine the perceived colour. Flavonoid concentrations vary dramatically both between and within species. For example, wild cornflower petals are reported to comprise 0.70% dry weight anthocyanin, while more intensely coloured garden varieties accumulate up to 15% dry weight anthocyanin. Similarly, the concentration of anthocyanin in deep purple pansy varieties may be as high as 30% dry weight.

Anthocyanin synthesis has been characterized both from a biochemical perspective and a molecular one, and the pathway is summarized in Fig. 15.5. The most detailed studies have used the petals of Antirrhinum and Petunia as models. Antirrhinum flowers normally produce a magenta anthocyanin, called cyanidin. Petunia flowers produce a purple anthocyanin, delphinidin, as well as petunidin. Since the blueness of the anthocyanin is determined by the degree of hydroxylation of the B ring, in theory a plant can always hydroxylate the molecule less, and thus make less blue anthocyanins, but it will not necessarily have the enzymes to hydroxylate it more, and so cannot make bluer anthocyanins. In fact, investigation of this hypothesis in Petunia revealed that the enzymes of the biosynthetic pathway have a greater efficiency for conversion of their usual substrates, and cannot therefore easily produce orange pelargonidins (Mol et al. 1998).

The p-coumaroyl CoA for flavonoid biosynthesis comes from general phenylpropanoid metabolism, and is shown in Fig. 15.5. The first enzyme of phenylpropanoid metabolism is phenylalanine ammonia lyase (PAL), which catalyses the

(a)

(b)

(c)

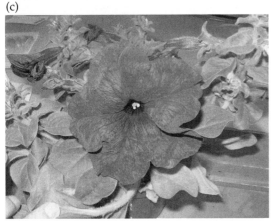

Figure 15.4 Anthocyanins give the pink, magenta, and purple colours to (a) *Pelargonium* (pelargonidin), (b) *Antirrhinum majus* (cyanidin), and (c) *Petunia hybrida* (delphinidin and petunidin). See also Plate 18.

conversion of phenylalanine to cinnamate (Martin and Gerats 1993). Cinnamate is converted to a hydroxycinnamic CoA ester (4-coumaroyl CoA) through two steps, and this intermediate feeds into a number of pathways, including those responsible for lignin production and the synthesis of cyanogenic compounds (Martin and Gerats 1993).

The first committed enzyme of flavonoid synthesis, which condenses malonyl CoA with this 4-coumaroyl CoA to produce a yellow chalcone, is chalcone synthase (CHS). The *NIVEA (NIV)* gene encodes CHS in Antirrhinum and loss of gene function through mutation results in pure white flowers in which no flavonoids are produced at all. A large number of *nivea* mutant alleles have been described, and it is clear that *NIVEA* function is determined partly by accurate regulation of its expression. Although *NIV* appears to be the only locus encoding chalcone synthase in Antirrhinum, multiple CHS encoding genes are present in many species. Eight CHS loci have been described in Petunia (Koes *et al.* 1986), and the potential for redundancy between so many genes may explain the absence of a CHS mutant in this species. Maize has also been shown to contain two loci encoding CHS (Coe *et al.* 1981; Dooner 1983), although the expression patterns of the two transcripts do not appear to overlap.

The next step in the pathway, the isomerization of chalcone into a colourless flavanone by the closure of the C ring, occurs spontaneously at a low rate but is accelerated by the enzyme chalcone isomerase (CHI) (Mol *et al.* 1985). CHI mutants usually accumulate chalcone as well as reduced quantities of anthocyanin, indicating that, while CHI may not be strictly necessary for anthocyanin synthesis, it is required to prevent this step of the pathway limiting the rate of pigment production. Petunia has been shown to contain two transcripts of *CHI*, with different expression patterns (van Tunen *et al.* 1988).

The flavanone produced in step two is one of the molecules which can be hydroxylated on the B ring to generate bluer end products. The dihydroflavonol produced in the next step (below) is the other intermediate that can act as substrate for extra hydroxylations. In Antirrhinum, flavonoid 3'-hydroxylase (F3'H) activity adds a single hydroxyl group to the B ring, leading to the final

Figure 15.5 The anthocyanin synthetic pathway. The different products depend on the groups found at positions R_1 and R_2.

accumulation of cyanidin (magenta) as the pathway's end product, instead of pelargonidin (orange/brick red). The *EOSINA (EOS)* gene of Antirrhinum encodes F3'H, and *eos* mutants accumulate pelargonidin as a consequence (Martin and Gerats 1993). Two loci encode F3'H in Petunia, each active in different parts of the flower (Brugliera *et al.* 1999). In some plants, including Petunia, a related enzyme, flavonoid 3'5'-hydroxylase (F3'5'H), is able to add two hydroxyl groups to the B ring, leading to the final production of the blue anthocyanin, delphinidin. Two genes encode F3'5'H in Petunia, and their expression patterns differ, ensuring both the petal tube and limbs contain the enzyme (Stotz *et al.* 1985; Holton *et al.* 1993a). Interestingly, sequence analysis of F3'H and F3'5'H from various species indicates that F3'5'H was derived from F3'H before the divergence of angiosperms and gymnosperms. However, within the Asteraceae a separate, and quite recent, origin of F3'5'H has occurred (Seitz *et al.* 2006).

In the third step of the anthocyanin pathway flavanone-3-hydroxylase (F3H) catalyses the hydroxylation of flavanone to produce very palely-pigmented dihydroflavonols. These dihydroflavonols may be used as substrates for flavonol production by flavonol synthase or stay within the anthocyanin synthetic pathway. Within the anthocyanin pathway they may also be subject to one or more extra hydroxylations by F3'H and F3'5'H. The *INCOLORATA (INC)* locus of Antirrhinum encodes F3H and mutations at this locus result in an ivory phenotype. Unusually for a locus encoding an enzyme, the wild type *INC* allele is only semi-dominant. Therefore the heterozygote *INC/inc* produces a reduced amount of anthocyanin relative to the wild type homozygote. This observation suggests that F3H may be a rate-limiting step in anthocyanin production in Antirrhinum (Martin and Gerats 1993). In Petunia the *AN3 (ANTHO-CYANIN3)* locus encodes F3H.

The ivory dihydroflavonol produced in step three is reduced by dihydroflavonol 4-reductase (DFR) to give an ivory-coloured leucoanthocyanidin. The *PALLIDA (PAL)* gene of Antirrhinum is responsible for DFR production and mutation results in ivory-coloured flowers (Martin *et al.* 1985; Coen *et al.* 1986), while the *ANTHOCYANIN6 (AN6)*

locus encodes DFR in Petunia (Beld *et al.* 1989). It is DFR in Petunia which exhibits substrate specificity and which is therefore responsible for the failure of the petal to produce pelargonidin if F3'H and F3'5'H activities are not present. Petunia DFR shows strong substrate specificity for dihydroquercetin (one extra hydroxylation) and dihydromyricetin (two hydroxylations), as opposed to dihydrokaempferol (no extra hydroxylations) (Martin and Gerats 1993).

Leucoanthocyanidin is converted into a coloured anthocyanidin by the action of a leucoanthocyanidin dioxygenase or anthocyanidin synthase. The Antirrhinum locus *CANDICA (CANDI)* is required for these conversions, and encodes a protein with strong sequence similarity to dioxygenases, including INC (Martin *et al.* 1991). Mutations at this locus again result in ivory-coloured flowers. The Petunia orthologue of *CANDI, ANT17*, is highly expressed in petals and its transcript is absent from regulatory mutants which do not produce anthocyanin (Weiss *et al.* 1993).

Anthocyanidin is glycosylated to form an anthocyanin by UDP glucose-flavonoid 3-o-glucosyltransferase (UF3GT). The addition of this glucose molecule stabilizes the pigment and enables it to be transported across the vacuolar membrane from the cytosol, where all previous steps in the pathway have occurred. In maize the *BRONZE1 (BZ1)* locus encodes UF3GT, and in *bz1* mutants anthocyanidin accumulates in the cytosol and polymerizes with proteins to form a bronze-coloured complex (Larson and Coe, 1977).

Anthocyanin is transported across the vacuolar membrane by the glutathione pump (Marrs *et al.* 1995). A glutathione S-transferase (GST) is required for transport of the anthocyanin, acting as a molecular chaperone for the transporter. The *BRONZE2 (BZ2)* locus of maize encodes GST, and *bz2* mutants accumulate anthocyanin in the cytoplasm, where it is oxidized to a bronze colour. It has also been shown that inhibition of the transporter by vanadate prevents anthocyanin accumulation in protoplast vacuoles (Marrs *et al.* 1995). In Petunia the *ANTHOCYANINLESS 9 (AN9)* locus also encodes a GST responsible for the final step of floral anthocyanin synthesis. However, plants contain many GSTs and phylogenetic analysis suggests that *BZ2* and *AN9* evolved independently from one another and from different types of GST. This conclusion is particularly striking in light of the ability of *BZ2* to rescue an *an9* mutant and vice versa. However, since both genes are transcriptionally regulated by the same proteins, which control the anthocyanin biosynthetic pathway in both species, it is likely that their recent evolutionary history has followed very similar patterns (Alfenito *et al.* 1998).

In Antirrhinum, Petunia, and many other species the anthocyanin undergoes a second glycosylation, adding an additional sugar. In both these cases the sugar is rhamnose, and the glycosylation is catalysed by the enzyme rhamnosyl transferase (RT). This second glycosylation may serve to stabilize and enhance colour—Petunia mutants lacking RT

activity are slightly less blue than wild type (Martin and Gerats 1993). The activities of acyl transferases and methyl transferases can also influence anthocyanin colour at this stage.

Within the vacuole, anthocyanin is sometimes stored in membraneless matrixes, known as anthocyanic vacuolar inclusions, which act as traps for anthocyanin (Markham *et al.* 2000). The packing and storage of anthocyanins may also influence final petal colour (Grotewold 2006).

15.5 Betalains

The betalains have been less well studied than the carotenoids and anthocyanins. They replace anthocyanins as the colouration in flowers and fruit of

Figure 15.6 The final product of betalain synthesis depends on what is conjugated to betalamic acid.

the order Caryophyllales. However, within that order the families Caryophyllaceae (the pinks) and Molluginaceae (the carpetweeds) do not make betalains and do make anthocyanins. It has been proposed that an ancestor of the Caryophyllales made both betalains and anthocyanins, but that within each family one pathway was lost (Strack *et al.* 2003). It is not unusual for a plant species to produce both carotenoids and betalains, or both flavonoids and betalains, but there is no known species producing both anthocyanins and betalains. This mutual exclusivity may reflect the overlapping absorption spectra of anthocyanins and betalains, and therefore the lack of advantage in producing two similarly coloured pigments (Grotewold 2006). Common betalain-pigmented plants include beetroot, the petals of the Christmas cactus, and the brightly coloured bracts of *Bougainvillea*.

The betalains are water-soluble nitrogen-containing pigments synthesized from tyrosine by the condensation of betalamic acid, and come in two main colour groups. The betacyanins, formed when betalamic acid interacts with a derivative of dihydroxyphenylalanne, are red to purple, while the betaxanthins, formed from the condensation of betalamic acid with an amino acid, are yellow to orange (Fig. 15.6). The early and late steps of betalain synthesis are catalysed by enzymes, but many of the intermediate reactions in the synthetic pathway appear to occur spontaneously (Strack *et al.* 2003). A tyrosinase, a form of phenoloxidase, is known to catalyse both the first step of the pathway and at least one later step (Grotewold 2006). Like flavonoids, betalains are stored in the vacuole of plant cells in a glycosylated form. In further support of the idea that the betalains and anthocyanins share some evolutionary heritage, phylogenetic analyses indicate that the genes encoding the betalain-specific glucosyltransferases are very closely related to those encoding flavonoid-specific glucosyltransferases (Strack *et al.* 2003). The evolutionary history of these pigments may involve only a few mutations to genes encoding key enzymes, following the loss of anthocyanins or betalains within a lineage.

CHAPTER 16

Enhancing Flower Colour

The production of coloured tissues, particularly insect-attracting petals, depends upon the synthesis of the pigments described in Chapter 15. However, very few flowers are coloured simply by the synthesis of one single pigment, and equally few petals are composed of a single block of unchanging colour. In the same way as an artist's palette can be used to produce an almost infinite variety of colours, plants are able to mix, modify, and enhance pigments to produce a vast array of final petal colours. These colours are usually distributed across the flower in patterns, which vary in their degree of regularity and complexity between different species. While colour contrast is much more important than pattern for attracting pollinators from a distance, pattern becomes important at close range and allows animals to distinguish between flowers of different species and to learn to 'handle' flowers (extract the reward) with the minimum possible expenditure of energy (Kevan *et al*. 1996). In this chapter we will consider the effects of mixing pigments together, the regulation of pigment distribution in the flower, and a number of other ways in which plants are able to modify the final colour of the flower. While much of the work described in this section has been conducted using human-visible pigments, it should be borne in mind throughout that the patterning we perceive on a flower may be very different from that perceived by an insect, and therefore that the UV-absorbing pigments visible to the insect eye are also likely to be subject to the same regulatory mechanisms.

16.1 Mixing pigments

The colour a flower appears is very rarely the colour of a single pure pigment, either carotenoid or flavonoid. In almost all species investigated, multiple pigments are present within petals and it is their combination which gives the final colour. This combination will vary according to whether the pigments are mixed within the vacuole (i.e., multiple flavonoids), or are produced in different plastids, or a combination of these factors. In particular, plastid distribution may vary across the flower, giving very different colour outcomes. Red rose petals contain no plastids, and so the red colour is purely a result of anthocyanin deposition in vacuoles. In contrast, yellow carotenoid-containing plastids are present throughout the petals of wall-flower, *Chiranthus cheiri*, blending visually with the magenta anthocyanin in the vacuole to produce the orangey-brown colour of the mature flowers. The importance of plastid distribution in determining pattern has been shown in cultivated primroses, where the central eye of the flower is composed of carotenoid-containing pigments, while the rest of the petal is pigmented by flavonoids.

Some combinations of pigments have more dramatic effects than others. In particular, a combination of a yellow flavonoid (aurone) with a yellow carotenoid may appear very little different from the carotenoid alone, because aurones are rarely produced in significant quantities. In general, carotenoids impart much stronger and heavier colours to petals than do apparently similarly coloured flavonoids. For example, the bright yellow of garden *Forsythia* flowers is due to the yellow carotenoids present, while loss of the yellow flavonoids also produced in the petals has little effect on final flower colour.

16.2 Co-pigmentation

A co-pigment is a compound which modifies the apparent colour of a pigment despite having little or no colour itself. Co-pigments are believed to exist with the ability to modify all varieties of anthocyanin, and this modification may involve a shift in absorption spectrum or simply an enhancement of the apparent brightness of the pigment. This enhancement may be a result of stabilization of the structural form of the anthocyanin molecule (Markovic *et al.* 2005).

Many flavonols and flavones act as co-pigments, and are thought to form stacked complexes with anthocyanin, causing a shift in the absorption spectrum of the anthocyanin molecule (Mol *et al.* 1998). Flavonols are usually colourless or pale yellow, and are derived from dihydroflavonols through enzymatic conversion by flavonol synthase. There are therefore four ways in which flavonols may affect final flower colour: by being yellow, and mixing with reddish anthocyanins to give a final orange colour; by absorbing UV and thus changing final flower colour to an insect's eye; by competing with the anthocyanin synthetic pathway for dihydroflavonol (see Fig. 15.5 in Chapter 15) and thus reducing the final concentration of anthocyanin produced; and by acting as a co-pigment, where they enhance anthocyanin colour. Analysis of Petunia mutants with lesions in flavonol synthase activity, along with investigation of the effects of ectopic expression of the gene encoding flavonol synthase, have shown that flavonols play roles in both substrate competition and co-pigmentation in this species (Holton *et al.* 1993a).

Cinnamic acids have also been shown to play an important role in co-pigmentation. They often constitute the main acyl groups in acylated anthocyanins and their stabilizing effects on the pigments enhance the final petal colour, which is otherwise subject to bleaching when exposed to acidic pH (Figueiredo *et al.* 1996). Aliphatic acyl groups such as malanyl are particularly effective at stabilizing anthocyanins, while aromatic acyl groups such as sinapayl shift the pigment towards blue. Recent evidence indicates that cinnamic acids also play a role in co-pigmentation of non-acylated anthocyanins, changing the absorption spectrum of common pigments such as malvin (Markovic *et al.* 2005).

16.3 Regulation of pigment distribution

Since the type of pigments and co-pigments produced has such a range of effects on flower colour it is hardly surprising that the pigmentation pathways should be under tight regulatory control. This regulation is both spatial and temporal, generating colour patterns on almost all flowers, and in some species changing the colours of petals during their development. For example, the flowers of *Viola cornuta* cultivar Yesterday, Today, and Tomorrow (a garden variety of pansy) change from white through pale lilac to purple over the space of 5–8 days (Fig. 16.1a). This change in colour has been shown to be the result of a steady increase in anthocyanin production over the time period, presumably as a result of transcriptional regulation of the genes encoding the enzymes of anthocyanin biosynthesis. While such a colour change may seem extraordinary, its dependence upon pollination (without which the petals remain white) suggests that plant growth regulator-mediated signals from germinating pollen tubes trigger the change in anthocyanin regulation, perhaps to change the attractiveness or visibility of the flower to potential pollinators (Farzad *et al.* 2002).

It has been shown that the activities of the genes encoding the enzymes of anthocyanin biosynthesis are predominantly regulated at the transcriptional level. It can be inferred from this that the majority of petal pigmentation patterns are specified by the expression patterns of regulatory genes which control the activity of the biosynthetic genes (Mol *et al.* 1996). The analysis of pigment regulation has focused on anthocyanins rather than carotenoids, since carotenoids are essential to many other aspects of plant growth and so cannot be manipulated as easily. By convention, anthocyanin regulatory loci are divided into those which specifically control anthocyanin deposition (by regulating 'late' biosynthetic genes, some way down the pathway), and those which control the synthesis of other flavonoids (by regulating 'early' biosynthetic genes). A number of genes have been shown to be important in both pathways, and three key groups of proteins have been shown to be involved. The most important of these are transcriptional activators of the MYB and basic helix-loop-helix (bHLH) families. In maize the

Figure 16.1 Pigment regulation. (a) *Viola cornuta* Yesterday, Today, and Tomorrow is fully purple 5–8 days after pollination (left), but opens as a white flower (middle) in which pigmentation steadily increases (right). Image kindly provided by Martha Weiss (Georgetown University, Washington, DC). (b) The *delila* mutant of Antirrhinum lacks pigmentation in the tube as a result of loss of activity of a BHLH transcription factor. (c) The *Venosa* locus produces pigmentation over the petal veins in a pale Antirrhinum flower. *VENOSA* encodes a MYB transcription factor. (d) The *an11* mutant of petunia lacks pigmentation as a result of loss of activity of a WD40 protein. The transposon in the *AN11* locus excises somatically, generating patches of wild type red tissue. Image kindly provided by Ronald Koes (Vrije Universiteit, Amsterdam). See also Plate 19.

MYB protein C1 and the bHLH protein R have been shown to activate expression of the structural genes encoding all enzymes of flavonoid synthesis, from chalcone synthase onwards, in kernal aleurone cells (reviewed in Martin *et al*. 1991). However, it appears that the evolution of floral pigmentation patterns in Eudicots has involved subfunctionalization of the roles of multiple members of these families, so that genes encoding enzymes required early in the anthocyanin synthetic pathway can be activated independently of genes encoding enzymes that act late in the pathway.

16.3.1 bHLH proteins

The best characterized example of a floral pigment regulatory protein is the bHLH transcription factor DELILA (DEL) of *Antirrhinum majus*. Mutations at the *DEL* locus result in wild type red lobes, but the corolla tube of the flower is white (Fig. 16.1b Goodrich *et al.* 1992). The DELILA protein is very similar in sequence to R of maize. The bHLH domain which defines members of this family is also common to the MYC family of transcription factors in animals, and is believed to function as a DNA-binding domain. Extensive analysis of DEL function has determined that the protein activates the late anthocyanin biosynthetic genes, specifically those encoding F3H, DFR, and UFGT (Martin *et al.* 1991). The regulation of DFR by DEL has been analysed in detail. In the wild type Antirrhinum flower, pigmentation is first deposited in a ring at the base of the tube, then in the lobes, and finally in the body of the tube. Mutation of the promoter region of *PALLIDA* (encoding DFR, see Chapter 15) releases it from the control of DEL. Analysis of a number of promoter mutations has established that DEL is responsible for PALLIDA activity in each of the regions of the flower at the correct developmental stage (Almeida *et al.* 1989). DEL has also been shown to be functional in heterologous hosts, activating anthocyanin synthesis in flowers of tobacco and leaves of tomato. Interestingly, although these experiments were conducted using a constitutive promoter, and *DEL* transcript was detected in all tissues, DFR was only activated in a subset of plant organs. This implies that DEL functions in conjunction with other, developmentally regulated, factors (Mooney *et al.* 1995).

The DEL orthologue in Petunia is known to be JAF13, another bHLH protein, very similar in sequence to DEL, which has also been shown to activate DFR expression (Quattrocchio *et al.* 1998). However, recent reports indicate that another bHLH protein, less similar to DEL and R, is also required for anthocyanin regulation in Petunia. AN1 (ANTHOCYANIN1) encodes a bHLH protein with only limited sequence similarity to JAF13 (and DEL and R), but a closer relationship to another maize protein, INTENSIFIER (IN). The IN locus is an inhibitor of anthocyanin synthesis in maize

kernals, but AN1 has been shown to activate DFR expression, in the same way that JAF13 does, and also to interact with a MYB protein, AN2 (Spelt *et al.* 2000, 2002). It also appears that AN1 may play a role in cell pH, as the *an1* mutant is allelic to the *ph6* mutant (see Section 16.5).

A similar situation seems to exist in Arabidopsis, where 2 of the 139 bHLH transcription factors show particularly sequence similarity to IN and AN1. Both GLABRA3 and MYC-146 (also known as AtMYC2) were shown to activate anthocyanin synthesis when transiently expressed in white varieties of *Matthiola incana* (Ramsay *et al.* 2003).

16.3.2 MYB proteins

The joint function of the bHLH factor R and the MYB protein C1 in anthocyanin activation in maize has led to a general expectation that MYB proteins would also be important in the regulation of anthocyanin synthesis in Eudicot flowers. Although the data are not as complete as for bHLH factors, there is convincing evidence that MYB proteins do play important roles. All MYB proteins share 1, 2, or 3 repeats of a helix-helix-turn-helix domain, which binds DNA. The protein family is ancient, with MYB transcription factors also known to play important roles in animal cell proliferation.

In Petunia, a MYB protein, AN2 (ANTHO-CYANIN2), is necessary for anthocyanin synthesis. *AN2* is not particularly similar in sequence to *C1*, except for the MYB domain, but is nonetheless able to complement a *c1* mutant (Quattrocchio *et al.* 1999). AN2 activates late biosynthetic genes, particularly *DFR*, and is hypothesized to function in concert with AN1. Subspecies of *Petunia axillaria* which do not produce anthocyanin have been shown to have frameshift mutations in the gene encoding AN2 (Quattrocchio *et al.* 1999).

Three lines of Antirrhinum, *venosa*, *rosea^colorata*, and *rosea^dorsea*, show variation in pigment patterning through the activities of genes encoding MYB transcription factors (Schwinn *et al.* 2006). The dominant *Venosa* allele is only visible in an Antirrhinum background which is not fully pigmented, but in a paler flower it results in the production of magenta anthocyanin in the epidermal tissue overlying the petal veins (Fig. 16.1c). The VENOSA protein is very similar to ROSEA1 and 2 (see below), suggesting duplication within the Antirrhinum lineage, but is also reasonably similar to AN2 of Petunia. The protein contains the necessary amino acid motifs for interaction with a bHLH protein, so may well function in a similar way to AN2. Analysis of transcription of biosynthetic genes in different genetic backgrounds revealed that VENOSA induces expression of *CHI, F3H, F3'H*, and *UFGT*. The *ROSEA* locus has been shown to comprise two genes, *ROSEA1* and *ROSEA2*, and a number of mutations in each coding sequence contribute to the different patterning seen in the *rosea*^{colorata} and *rosea*^{dorsea} mutant lines. ROSEA1 and ROSEA2 activate the expression of overlapping subsets of the anthocyanin synthetic genes, and also contain the appropriate amino acid motif for interaction with a bHLH protein. Double mutant analysis has confirmed that *ROSEA1, ROSEA2*, and *VENOSA* all interact with *DELILA*. Analysis of pigmentation patterns in wild species of Antirrhinum has revealed that most of the natural variation in pigmentation patterning is attributable to the activities of these three genes, confirming the importance of regulating anthocyanin synthesis for final flower colour (Schwinn *et al.* 2006).

The genes encoding two more MYB transcription factors, MYB305 and MYB340, are also known to be expressed in Antirrhinum flowers, although mutant phenotypes have not been described for either locus (Moyano *et al.* 1996). The primary function of these genes appears to be in the regulation of flavonols in the carpel and stamens, and they have only a limited role in petal pigmentation. The MYB305 and MYB340 proteins are virtually identical across their DNA-binding domains, differing in only four amino acids, and they both bind to the promoters of the genes encoding PAL, CHI, and F3H and activate transcription of the genes. As such they represent a link between the late biosynthetic genes (F3H) and the earlier parts of the pathway (PAL and CHI). MYB340 is the stronger activator of the two proteins, producing more transcript of all three of the structural genes than MYB305. However, MYB305 has a much stronger binding affinity for biosynthetic gene promoters than does MYB340. Since the weaker activator binds more strongly, and the stronger activator binds much more weakly,

coexpression of the two genes results in 50% less biosynthetic gene transcript than does expression of *MYB340* alone, because the proteins are in competition with each other (Moyano *et al.* 1996). These data show that flavonoids are subject to very precise regulation by gearing mechanisms, and that similar proteins may interact to regulate petal pigmentation. A gene with very high sequence similarity to *MYB305* and *MYB340* has been described from pea (*Pisum sativum*). *Myb26* is expressed in developing pea flowers, with transcript levels increasing as anthocyanin deposition occurs. However, the Myb26 protein is not able to complement the Petunia *an2* mutant, suggesting that Myb26, like MYB340 and MYB350, activates a set of early biosynthetic genes (Uimari and Strommer 1997).

16.3.3 WD40 proteins

The third group of proteins with an important role in pigment regulation are the WD40 repeat proteins. These proteins produce a seven-bladed propeller motif and act as surfaces for the interactions of other proteins, such as the transcription factors previously discussed (Ramsay and Glover 2005). The first WD40 protein shown to play a role in anthocyanin regulation was AN11 (ANTHOCYANIN11) of Petunia. The *an11* mutant has reduced activity of the late biosynthetic genes (from DFR onwards) and produces flowers with white petals as a result (Fig. 16.1d). WD40 proteins are so conserved across enormous evolutionary distances that the human sequence with most similarity to *AN11* was able to complement an *an11* mutant phenotype in a transient expression experiment (de Vetten *et al.* 1997). While no WD40 protein has yet been shown to be involved in pigment regulation in Antirrhinum, the characterization of the Arabidopsis protein TTG1 as a WD40 protein required for anthocyanin synthesis and trichome development suggests that the role of a WD40 protein in this regulatory pathway may be ubiquitous to the Eudicots (Walker *et al.* 1999).

16.4 The effects of metal ions

Interactions between floral pigments and metal ions can also alter the final colour of the petals. For example, the bright blue colour of cornflowers is not simply due to the presence of a blue pigment. Instead it stems from an interaction between the magenta anthocyanin, cyanidin, flavone co-pigments, and ions of the metals iron, magnesium, and calcium, absorbed by the plant's roots from the soil. The combination of the pigment and co-pigments with the metals results in a molecule that gives the flower a very bright blue colour , even though the anthocyanin alone is red/magenta (Shiono *et al.* 2005). Equally remarkably, the bright blue colour of Himalayan poppy, *Meconopsis grandis*, is due to the interaction between iron and the magenta anthocyanin cyanidin (Fig. 16.2a). Yoshida *et al.* (2006) were able to reproduce the blue colour of Himalayan blue poppy petals *in vitro* by mixing cyanidin with flavonol co-pigments, iron, and magnesium under low pH conditions. The iron was the most crucial factor besides the cyanidin. Another good example of this sort of interaction is the variable colour of the flowers of *Hydrangea*, which is commonly grown in gardens (Fig. 16.2b). *Hydrangea* flowers are blue if there is aluminium in the soil, as aluminium and the pigment delphinidin form a very stable, very blue complex. If there is less aluminium available in the soil, and more molybdenum, the same pigment interacts with the molybdenum ions, causing the flowers to appear light pink instead. The presence of this changeable anthocyanin pigment in *Hydrangea* is exploited by gardeners, who may water the soil around their plants with a solution containing the appropriate ion to generate the final flower colour of their choice. The aluminium–delphinidin complex is known to be acid-stable, giving the very blue colour even in an acidic pH which would normally make the delphinidin appear red. Magnesium, iron, and aluminium ions have all been shown to interact with anthocyanins to shift their absorption spectrum towards blue.

16.5 The importance of pH

The pH of petal cells can also affect the final colour of the flower, as pH determines anthocyanin structure and absorption spectrum. For example, the light blue petals of *Ipomoea tricolor*, morning glory, owe their colour to the effect of a high petal pH on

their anthocyanin (Fig. 16.2c). The closed buds of these flowers are purplish red and their cells have a pH of 6.6. However, when the flowers open the petal cell pH increases to 7.7, and the pigment changes colour to sky blue (Yoshida *et al.* 1995). To confirm that this change in colour was solely due to a change in pH, Yoshida *et al.* (1995) also showed that aluminium and iron had no role in morning glory colouration, and that the anthocyanin was stabilized by a caffeic acid co-pigment in both the bud and mature flower stages. Treatment of open flowers with an acidifying environment (carbon dioxide) resulted in a reduction in petal cell pH and a reversion of the colour to purple. To investigate the mechanism of this *in vivo* pH change, Yoshida *et al.* (2005) isolated proteins from vacuolar membranes of mature petals and immature buds. They observed that a Na^+/H^+ exchanger was present only in petal vacuolar membranes and was present at particularly high levels in mature flowers. This suggests that the increased pH is due to active transport of Na^+ and/or K^+ from the cytosol to the vacuoles. Further support for this mechanism comes from the characterization of a mutant of *Ipomoea nil*, which fails to undergo colour change on maturity and remains purple. Fukada-Tanaka *et al.* (2000) confirmed that the difference between purple mutant and blue wild type tissue was entirely due to a difference in pH, with the mutant tissue retaining a pH about 0.7 lower than wild type petals. Isolation of the gene perturbed in the purple mutant, *InNHX1*, confirmed that it encoded a protein which had strong sequence similarity to Arabidopsis and rice vacuolar Na^+/H^+ exchangers. The *InNHX1* transcript is most abundant in petals 12 hours prior to bud opening, an expression pattern strongly correlated with the timing of change in colour (Yamaguchi *et al.* 2001). A similar gene, *InNHX2*, was isolated from floral cDNA and also found to increase in expression levels in the 12 hours before bud opening. Using mutant strains of yeast sensitive to cation presence, Ohnishi *et al.* (2005) demonstrated that both InNHX1 and InNHX2 catalyse the transport of both Na^+ and K^+ into vacuoles, confirming their importance in regulating vacuolar pH and thus flower colour.

The ability of vacuolar pH, controlled by membrane transporters, to alter flower colour without

Figure 16.2 Metals and pH both affect flower colour. (a) The Himalayan blue poppy owes its blue colour to an interaction between anthocyanin and iron. Photograph kindly supplied by Cambridge University Botanic Garden. (b) Hydrangea flowers can be blue or pink, depending on the metal ions present in the soil. (c) Morning glory flowers have high vacuolar pH. Image kindly provided by Felix Jaffe (CSIRO, Australia). (d) An unstable *pH4* mutant of petunia, with revertant wild type red (acidic) sectors on a mutant bluey-pink (more alkaline) background. Image kindly provided by Ronald Koes (Vrije Universiteit, Amsterdam). See also Plate 20.

any change in the types of pigment produced, gives plants the flexibility to alter their final petal colour after pigments have been made. Stewart *et al.* (1975) observed that the colours of several wild flower species became bluer as they aged, and that this blueing was correlated with an increase in petal pH.

Petal pH has been shown to be under the control of genetic factors in a number of species, including *Primula sinensis, Papaver rhoeas, Tropaeolum majus, Lathyrus odoratus,* and the clover *Trifolium pratense,* and is likely to be in many other species besides. Petunia has proven the best model system in which to investigate these effects, and seven loci which determine petal cell pH have been identified (de Vlaming *et al.* 1983; Mol *et al.* 1998). In Petunia these loci work to maintain the acid status of the vacuole and thus the red colour of the anthocyanin. Wild type Petunia flowers have a vacuolar pH of 5.5, while it rises to as much as 6 in some of the *ph1–7* mutants (de Vlaming *et al.* 1983). *PH6* has been isolated, and shown to be allelic to the bHLH-encoding locus *AN1* (Spelt *et al.* 2002). More recently, *PH4* has also been isolated, and shown to encode a MYB-related transcription factor (Quattrocchio *et al.* 2006). The *pH4* mutant shows an increased vacuolar pH, from the wild type 5.5 to around 6.0, and as a result the flower appears a slightly bluer shade (Fig. 16.2d; see also Plate). The *PH4* gene is expressed in the petal epidermis, and the MYB transcription factor it encodes can interact with the bHLH proteins AN1 and JAF13. These data suggest that AN1 interacts with different MYB proteins to activate different pathways—with AN2 to activate anthocyanin synthesis but with PH4 to regulate cell pH. Mutation of *PH4* had little effect on the expression of anthocyanin synthetic genes, but did result in strong downregulation of a protease which might be involved in regulating vacuolar pH (Quattrocchio *et al.* 2006*)*. Yamaguchi *et al.* (2001) isolated the Petunia orthologue of the *InNHX1* gene, responsible for increased pH of the *Ipomoea nil* flower, but found no particularly strong expression of the gene in petals. This is unsurprising, considering that Petunia employs an opposite mechanism to that of *Ipomoea,* namely an active reduction in petal pH.

16.6 The role of petal cell shape

A very subtle way in which the colour of a petal can be enhanced is by distribution of specialized cell types in the epidermal layers. In a study that examined the surface structures of 201 species of flower from 60 families, 79% were found to have some form of cone-shaped cells on the epidermis oriented towards potential pollinators, described as conical-papillate cells (Kay *et al.* 1981). In many species these cells are further ornamented by ridges of thickened cuticle. The frequency of this specialized cell morphology within the flowering plants, and its almost universal restriction to the petal, argue for an adaptive explanation involving the function of the petal in pollinator attraction. It was proposed that conical-papillate cells increased the amount of light absorbed by the pigments in flowers, enhancing the perceived colour of the petal (Kay *et al.* 1981). Conical-papillate cells might also scatter light reflected back from the inner layers of the petal more evenly than would flat cells, resulting in a sparkling effect, or velvety texture to the petal.

In Antirrhinum, the conical-papillate cells (Fig. 16.3a) are found only on the adaxial epidermis of the petal lobes (where they will be seen by potential pollinators as they approach the flower). The *mixta* mutant of Antirrhinum fails to develop conical-papillate petal cells and instead has flat petal cells, more similar to those of the leaf epidermis (Fig. 16.3b). The significance of the conical-papillate cells in enhancing colour is shown by the fact that the *mixta* mutant was originally identified in a screen of mutagenized plants because it was paler in colour than wild type flowers (Fig. 16.3c). The mutant petal looks as though it has a matt texture, unlike the velvety sparkle of the wild type petal (Noda *et al.* 1994).

By comparing the ability of epidermal cells to focus light in the wild type and *mixta* mutant lines, conical-papillate cells have been shown to enhance visible pigmentation. Conical-papillate cells focus light approximately twice as well as the mutant flat cells, and they focus it into the region of the epidermis where the pigment is contained (Gorton and Vogelmann 1996). It has also been shown that the wild type Antirrhinum petals reflected

(a)

(b)

(c)

Figure 16.3 Petal cell shape affects flower colour. (a) Wild type Antirrhinum petal epidermis, composed of conical cells. (b) *mixta* mutant petal epidermis, composed of flat cells. (c) Wild type (left) and *mixta* (mutant) right flowers, showing the difference in colour attributable to the cell shape. See also Plate 21.

significantly less light away from the flower than *mixta* mutant petals did, and absorbed significantly more light. These differences can be attributed to the focusing of the light onto the light-absorbing pigments in the epidermal cells, and to the reduction in reflection of light at low angles of incidence, resulting in the greater depth of colour of wild type conical-celled flowers (Gorton and Vogelmann 1996). It is likely that this trick is used by many plant species to enhance petal colour, considering the wide distribution of conical cells among the angiosperms.

(a)

(c)

(b)

(d)

(e)

Plate 10.

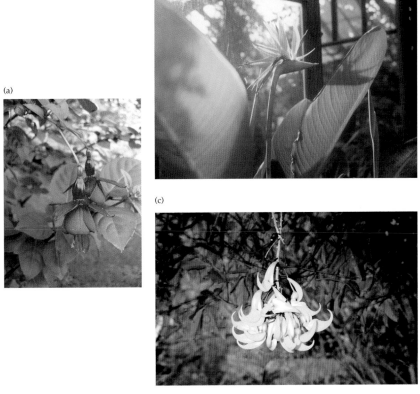

(b)

(a)

(c)

Plate 11.

(a) (b)

(c)

Plate 12.

a

b

Plate 13.

Plate 14.

Plate 15.

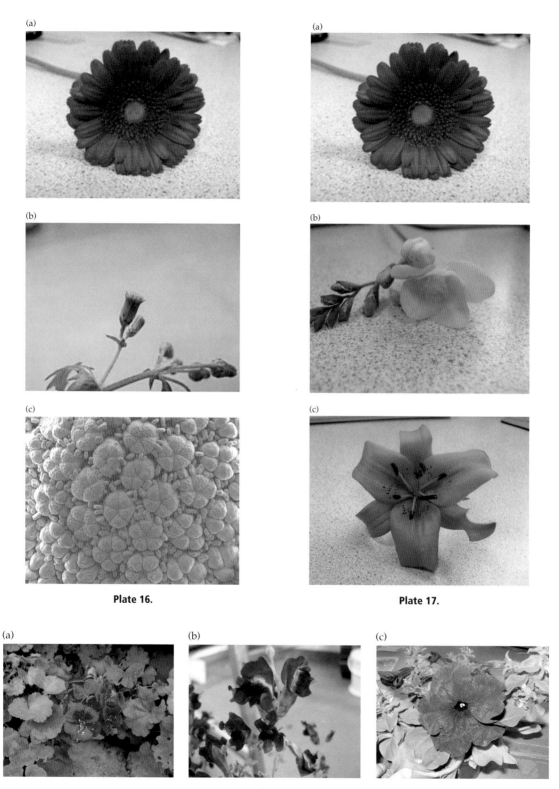

(a)

(b)

(c)

Plate 16.

(a)

(b)

(c)

Plate 17.

(a)

(b)

(c)

Plate 18.

(a)

(b)

(c)

(d)

Plate 19.

(a)

(b)

(c)

(d)

Plate 20.

Plate 21.

Plate 22.

(a)

(b)

Plate 23.

Plate 24.

Plate 25.

Plate 26.

Isolation of the *MIXTA* gene revealed that it encodes a MYB transcription factor, presumably responsible for the activation of structural genes involved in cytoskeletal arrangement and cell wall deposition (Noda *et al.* 1994). Transgenic experiments have demonstrated that the MIXTA protein is sufficient to activate the conical-papillate cell form, even when expressed in other tissues, such as leaf epidermis (Glover *et al.* 1998).

The *mixta* mutant has been used to test the idea that conical-papillate petal cells and the brighter colour they generate enhance the pollination success of the plant. Plants of the wild type and *mixta* mutant lines were grown in a field plot and the stamens were removed from the flowers to prevent self-pollination. Fruit set was then scored one month later and the presence of a fruit used as an indicator of a pollinator visit. The shape of the petal cells had a highly significant effect on the likelihood of fruit set (and thus a pollinator visit). Flat-celled flowers set significantly fewer fruits than conical-papillate celled flowers (Glover and Martin 1998). Interestingly, a similar result was observed when conical-papillate celled white Antirrhinum flowers were compared to flat-celled/white flowers (both generated by crossing wild type and *mixta* mutant lines to the *nivea* mutant, lacking chalcone synthase and thus all of flavonoid synthesis). Although conical-papillate and flat-celled surfaces were identical to the human eye in the absence of pigmentation, the flat-celled flowers experienced reduced fruit set and a reduction in pollinator visits (Glover and Martin 1998). These data indicate that enhanced colour is unlikely to be the main explanation for the evolutionary success of conical-papillate cells, and more recent experiments have confirmed that bumblebees have no intrinsic preference for the wild type Antirrhinum colour compared to the *mixta* mutant (Dyer *et al.* 2007).

An alternative, although not mutually exclusive, explanation for the specialized shape of petal epidermal cells is that the texture of a flower may provide tactile cues to pollinator position on the flower, sensed through the antennae of insects. Moreover, the shape of petal epidermal cells may be used as a tactile cue by insects in their discrimination between particular flowers. In support of this idea, bees have been trained to recognize the petals of different species by the shape of the cells. Bees were provided with a food reward when they touched epidermal layers composed of certain cell types, but no reward when they touched other shaped cells. They learned very quickly to associate the reward with the texture of the petal and would search for food only when presented with the epidermal tissue which usually accompanied the reward (Kevan and Lane 1985). To investigate the influence of the texture of epidermal tissues on pollinator behaviour the response of bumblebees to the wild type Antirrhinum and *mixta* mutant lines was analysed in a field experiment. Analysis of bee behaviour indicated that bees distinguished between the two genotypes both before and after landing (Comba *et al.* 2000). It is therefore unlikely that epidermal cell shape is used here solely as a tactile cue since cell shape evidently provides information that is available before landing. However, this does not exclude the possibility that cell shape is also used by pollinators as a tactile cue post-landing, and recent experiments with resin casts of wild type and *mixta* mutant Antirrhinum petals have confirmed that bees can discriminate between them using touch alone (H. Whitney and B. J. Glover, unpublished).

A third way in which conical cells might be perceived as adaptations to enhance the attractiveness of the petal is through a role in regulating intrafloral microclimate. The temperature within a flower has been shown to influence nectar secretion rate, nectar evaporation rate, and nectar concentration. It may also influence scent diffusion. There is also some evidence to suggest that a high intrafloral temperature is itself attractive to insects, particularly in cooler climates and at dawn (Dyer *et al.* 2006). Analysis of intrafloral temperature demonstrated a small but significant difference between the wild type and *mixta* mutant lines of Antirrhinum. The intrafloral temperature of wild type, conical-celled, flowers increased directly with increase in the warming effect of direct solar radiation. However, the intrafloral temperature of mutant flat-celled flowers remained constant, however strong the warming effect of solar radiation (Comba *et al.* 2000).

Finally, conical petal epidermal cells can also influence the degree of petal reflexing and thus the apparent surface area of the petal. This change in apparent surface area may play a role in enhancing flower visibility at a distance.

It is likely that conical petal cells influence pollinator behaviour through each of these different methods in different species. However, in the context of flower colour, it is certainly clear that conical petal epidermal cells do enhance the appearance of anthocyanin by focusing light into the cell vacuoles. The complexity of the many ways of altering final flower colour has added support to the idea that colour is intricately linked to the attraction of specific pollinators. In the next section of this book we consider whether pollinators and plants are really linked in precise relationships.

The Influence of Pollinators on Floral Form

CHAPTER 17

Are Flowers under Selective Pressure to Increase Pollinator Attention?

In the previous section of this book we discussed a variety of ways in which flower visibility and appearance can be enhanced, through changes to size, shape, structure, and colour. The underlying assumption of much of the work on flower development and morphology is that these features serve to increase the attractiveness of the flower to pollinating animals, thus maximizing pollinator attention, and consequently seed set and fitness. As we discussed in Chapter 13, it has long been believed that these elaborations are the consequence of adapting to attract particular pollinating animals, resulting in pollination syndromes. In Chapter 19 we will analyse in detail the evidence for the existence of these syndromes. However, before we consider whether pollination syndromes do exist, and how frequently, we need to consider whether there is evidence that two underlying assumptions are met. In Chapter 18 we will look at the evidence that pollinating animals actually discriminate between the traits described in the previous section—if they cannot, then those traits cannot form part of pollination syndromes unless they enhance pollen transfer indirectly. In this chapter we will address an even more fundamental issue—do plants actually benefit from increased pollinator attention and should floral attractiveness therefore be expected to increase across generations?

17.1 Competition for pollinator attention

The pollination syndrome concept rests on the idea that individual plants are competing for pollinator attention. If there is no pollination competition in a given habitat, then the limits to the reproductive success of an individual plant are not imposed by the pollination system. If this is the case, increasing the showiness of the flower, or specializing in some way to attract a subset of pollinators, will have no positive effect on the plant's fitness. Indeed, since pigments and elaborate petal structures are costly in terms of energy and key nutrients such as nitrogen, in the absence of a beneficial increase in seed set such structures will actually have a negative effect on plant fitness. Elaboration of floral attractants only enhances the reproductive success of a plant if by doing so more of that plant's pollen reaches an appropriate stigma, or more of its egg cells are fertilized by sperm from incoming pollen grains. If the plant is already dispersing pollen and being pollinated quite successfully, then there is no pressure to change.

Pollination competition falls into two different categories, which have different evolutionary consequences. The variety that we are most concerned with in this chapter is exploitation competition. Exploitation competition occurs when a plant receives insufficient visits by pollinators because another plant or plants nearby is more attractive. This situation results in the first plant setting less seed, or contributing fewer male gametes to the gene pool. The fitness of the plant is therefore compromised. Exploitation competition can be intraspecific or interspecific. When it is intraspecific it may be predicted to drive elaboration of floral structures and increase showiness of flowers. When it is interspecific it is more likely to result in pressure to alter habitat or flowering time, or else to specialize in attracting a smaller subset of pollinators. Interspecific exploitation competition is anecdotally of commercial importance, with many fruit farmers reporting that flowering weed crops decrease fruit set of their orchard trees. Free (1968) observed honeybee behaviour in fruit orchards containing dandelions, and confirmed that most bees foraged preferentially on dandelions, both for nectar and for pollen, at the expense of the fruit trees.

The second form of pollination competition, interference competition, is only ever interspecific. It occurs when pollen of the wrong species lands on a stigma, where it can begin to germinate and clog up the surface, reducing the chances of the right pollen grains germinating successfully and thereby decreasing female fitness. At the same time placement of pollen on the stigmas of plants of another species effectively wastes that pollen, resulting in reduced male fitness. Interference competition can have more significant effects than exploitation competition, particularly in situations where density of a particular plant species is low relative to the densities of other species (Kunin 1993). Interference competition may be expected to drive changes which minimize the number of pollinating species visiting a plant or the number of plant species visited by each pollinator, preventing interspecific pollen transfer. Such changes might include alterations in habitat or flowering time, but might also include increased specialization towards the foraging preferences of a limited subset of pollinators. This latter set of changes is only possible if a sufficient number of specialist pollinators are available, and is heavily dependent on the density of both plant and pollinator.

17.2 Facilitation of pollination

Attempts to identify and quantify the importance of pollination competition are made more difficult by the complicating factor of facilitation of pollination. In some cases, the presence of more flowers nearby, whether of the same or different species, may actually increase the pollination success of a particular plant. Such an increase in pollination success is likely to be frequency dependent, and when a certain number of nearby plants is reached the balance will tip towards a competitive situation (illustrated in Fig. 17.1). For example, a single individual plant flowering in an otherwise barren environment may simply not be spotted by passing pollinators or may be judged insufficiently rewarding to be worth stopping for. The addition of other plants in bloom to the same locality will increase the chances of a pollinator visiting the area and increase the chances of the first plant receiving pollination services. However, when the number of nearby plants increases sufficiently, the original plant will find itself competing with its neighbours, and its fitness may decline accordingly. Some of the methods used to identify pollination competition take into account the importance of facilitation, but other methods may not fully register its contribution.

17.3 Techniques for investigating the role of pollinator attention in limiting fitness

There are three main methods used to investigate the role of pollination competition in limiting plant fitness. The aim of each method is primarily to

Figure 17.1 Facilitation and competition in mixed species plots. Addition of flowers of species B (white) to a plot of flowers of species A (grey) initially facilitates pollination of species A. However, when relatively high proportions of species B are present, competition for pollination occurs and fitness of species A is compromised.

establish whether such competition occurs or has occurred in a particular situation, and then, if possible, to quantify the extent or importance of that competition. These methods only provide data on particular plants growing in particular habitats, and it has been difficult to extrapolate from these data, as they are themselves highly variable. Each method has advantages and disadvantages, but each provides useful information and is therefore used in current studies.

17.3.1 Fruit and seed set following hand pollination

Perhaps the simplest method to identify pollination competition is to hand-pollinate flowers and observe the effects on fruit and seed set, relative to control flowers not hand-pollinated but open to natural pollinators. In theory, such an experiment will reveal what proportion of ovules in the control plants is unfertilized because of an insufficiency of suitable pollen. This may be due to a shortage of pollinator attention, or else to an excess of heterospecific pollen clogging the stigma. In practice, this method is open to several criticisms. Firstly, it takes no account of the male component of fitness. It is not sufficient to assume that male fitness is directly related to female fitness, even though in some cases it might be. If the reduced seed set of the control flowers is attributable to insufficient pollinator attention, then male fitness might well be similarly affected. However, if the limitation is due to heterospecific pollen clogging the stigma, then variations in pollinator foraging habits may, for example, mean that the pollinator has moved on only to plants of the same species and has actually caused the experimental plant's male fitness to be extremely high. Since this experimental method makes no observation of pollinator behaviour, and male fitness is notoriously variable, it is likely that erroneous conclusions about the degree of pollination competition, if not about its absolute occurrence, are often drawn from its use. Secondly, this method must be used with great caution to ensure that female fitness is accurately recorded. For example, simply scoring fruit set in hand-pollinated and control plants is insufficient, as many species can produce fruit even in the absence

of fertilization, a situation which obviously does not contribute to female fitness and will actually decrease it. Similarly, seed number within fruit is also variable within species, and so it is necessary to quantify seed set, not fruit set, even if the plant is known not to produce fruit entirely free from seed. Ideally, the set of viable seed should be scored, to ensure that non-viable hybrid seed is not being counted. A greater difficulty in correctly scoring female fitness is that increased seed set in one part of the plant, or one flowering season, may be offset by reduced seed set in other parts or future seasons. This reflects the costly nature of seed and fruit development, and the fact that seed set in many plants is limited not by pollination competition but by resource availability. It is therefore not sufficient to hand-pollinate all flowers on a single branch of a tree, and compare them to control branches. It is similarly inappropriate to hand-pollinate all the flowers on a tree in a single growing season, without reference to the subsequent seasons. Little is known about the effects of nutrient acquisition and use in one year on the reproductive success of a plant in subsequent years (Horvitz and Schemske 1988). However, it has been clearly shown in several tropical orchid species that increased fruit set in one season (as a result of hand pollination) can result in significant negative effects on vegetative growth, inflorescence production, and flower production in subsequent years (Montalvo and Ackerman 1987; Ackerman and Montalvo 1990). Ideally, then, these experiments should compare seed set of a plant on which every single flower has been hand-pollinated for its entire life to seed set of a control plant. Such experiments are most easily performed with short-lived annuals, and so many of the data generated using this method must be treated with some caution. Finally, several authors have pointed out that assuming that seed set is limited by either pollen limitation or female resource availability is an oversimplification, with many plants likely to be operating at an equilibrium between these two limiting factors (for a detailed discussion of the causes and consequences of pollen limitation see Ashman *et al.* 2004).

A variation on the method described above is to compare seed set of open-pollinated flowers under

PRESTON RIDGE LIBRARY
COLLIN COLLEGE
FRISCO, TX 75035

different natural pollinator regimes—for instance, at different times of the season or before and after events (such as disease or natural disasters) which dramatically alter pollinator availability. The same difficulties apply to this form of analysis.

17.3.2 Mixed species plots

A direct way to measure the effect of pollination competition on fitness is to compare artificial plots containing a single plant species with those containing a mix of species. If the plots are in the same environment then the only variables should be the number and species composition of plants in them. It is then possible to measure a number of factors relating to fitness, including pollinator visitation

rates, fruit set, and seed set. As before, seed set is a much more useful measure of female fitness than fruit set. Recording pollinator visitation will give a measure of the extent to which any negative consequences of mixed plots are due to competition for pollinator attention as opposed to the inhibitory effects of heterospecific pollination, and from that it may be possible to estimate the male component of fitness. It is also possible using such plots to assess where facilitation of pollination turns into competition. If a range of plots containing a decreasing proportion of the plant of interest is used, then it becomes possible to chart the increase in seed set associated with facilitation and the point at which seed set begins to fall, marking the onset of competition.

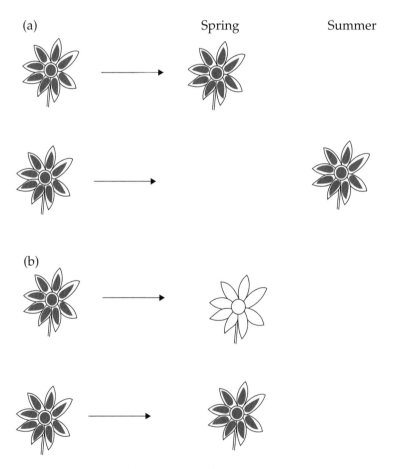

Figure 17.2 Character displacement to minimize pollination competition. (a) Two plants which would otherwise compete for pollinator attention can stagger their flowering time to prevent overlap. (b) Alternatively, other floral characters, such as colour, can be displaced to attract different pollinators.

The major drawback of this method of identifying pollination competition is the extent to which it is labour intensive. The difficulties associated with producing a range of plots and recording pollinator visitation as well as scoring seed set are such that only small plots are used, containing only two or three species, and data are only available from a very small number of habitats. Furthermore, it is usually only possible to use small fast-growing plants, or those that can be easily transplanted, in these artificial plots. It is accordingly difficult to extrapolate from these data to the likely importance of pollination competition for larger, long-lived plants such as trees, and for plants growing in extremely species-diverse habitats, such as tropical rainforests.

17.3.3 Character displacement

The third method used to identify pollination competition is both the least direct and probably the most used. In this method, plant species in a particular environment are analysed to see whether characters appear to have been displaced by the need to avoid pollination competition (illustrated in Fig. 17.2). For example, a study might consider all the species of a particular genus in one locality, predicting that, if pollination competition had been a problem in the past, present-day floral features would vary between species to attract different pollinators. Alternatively all the species in a habitat which attract a single type of pollinator might be studied, to assess whether the timing of their flowering is staggered to minimize competition. The power of these studies is that they can look at a large number of species in a wide range of habitats. Plant life habit does not matter in these studies, so trees can readily be compared to shrubs and annuals. Geographically distant localities can be compared, often using previously published data. An obvious null hypothesis is that flower form or flowering time (or whatever feature is studied) is simply a matter of phylogeny, and thus that closely related species should have similar characters. Deviation from this null hypothesis is easy to spot and can be readily analysed using statistical methods. However, the major drawback of this approach is that it can only ever be correlative. No direct proof

of current pollination competition is produced, and its previous presence can only ever be indirectly inferred with caution. Without knowing whether the species under analysis evolved sympatrically or allopatrically it can be very difficult to draw conclusions from simple observations of current characters. Indeed, some authors have argued that plants seldom coexist under stable conditions for long enough for pollination competition to drive character displacement, suggesting that even where such displacement is observed it may be too simplistic to attribute it to previous competition for pollinator attention (Feinsinger 1987).

17.4 Evidence based on fruit and seed set following hand pollination

One of the best studies of pollination limitation using hand pollination was published in 1988 by Horvitz and Schemske, who concluded that their system under investigation was not significantly limited for pollinator attention. Plants of *Calathea ovandensis* growing in a Mexican forest were examined. This species is herbaceous, flowering in multiple years, but only usually producing a single inflorescence per plant per year. Pairs of plants at similar developmental stages and within a few metres of one another were chosen for analysis, and all flowers in the inflorescence of one plant were hand-pollinated while no flowers on the control plant were pollinated. Since all flowers within an inflorescence were treated, and most plants produce only a single inflorescence per year, these authors effectively hand-pollinated entire plants for comparison to controls. They found that hand-pollinated inflorescences initiated 24% more fruits than controls, but since not all initiated fruit developed fully, they only set 6.8% more mature fruit. Total seed set did increase by 12.6% in the hand-pollinated inflorescences, but this difference was not statistically significant. The authors could not therefore conclude that there was direct evidence that this species was suffering pollination limitation in this locality. Instead, they concluded that seed set was likely to be resource limited, a conclusion strongly supported by the abortion of fruit early in development in the hand-pollinated plants. It should be noted here that pollination limitation and resource limitation are not mutually exclusive conditions for a plant, and that it is possible that the interaction between these limiting factors actually determines reproductive success. Campbell and Halama (1993) showed that pollen availability limited seed number per flower of *Ipomopsis aggregata* but that nutrient availability limited the number of flowers produced.

In a more recent example of this approach, Ghazoul (2006) hand-pollinated four flowers on each of 50 *Raphanus raphanistrum* (wild radish) plants in experimental plots, and merely tagged four flowers at the same developmental stage on 50 control plants. Pollen limitation was analysed by comparing fruit and seed set of the hand-pollinated plants versus the controls. While this experiment lacks the thoroughness of the previously described study, it was conducted with plants in defined plots and could control for environmental differences. The author found that hand-pollinated flowers set 50% more seed than control flowers, and that this difference was statistically significant.

A recent review of pollen limitation analysed all the published literature describing hand pollination as an attempt to supplement pollen supply. Knight *et al.* (2005) found that most studies recorded fruit set, rather than seed set. Of the 482 published reports on the effect of hand pollination on fruit set, 304 (63%) showed significant pollen limitation in the control plants.

Studies observing fruit and seed set in wild situations with variable pollinator profiles have also found evidence for pollination limitation. Bumblebee-pollinated *Antirrhinum majus* plants in an experimental field plot in Norfolk flowered all summer but the bees were most active in June and early July. Early in the summer as many as 88% of the flowers were pollinated (measured by emasculating flowers to prevent self-pollination, tagging, and then scoring fruit set). Later in the summer when the bees were less numerous only 41% of flowers were pollinated (B. J. Glover, unpublished).

A more striking example of this approach was provided by Rathcke (2000), who analysed fruit set of a bird-pollinated shrub in the Bahamas in response to hurricanes. *Pavonia bahamensis* (Bahama Swamp-bush, a member of the Malvaceae and endemic to the Bahamas) is pollinated by two bird

species, the Bahama Honeycreeper (or Bananaquit) and the Bahama Woodstar. During the winter flowering season of 1994–5, both bird species were frequently observed on Bahama Swamp-bush flowers, 98% of stigmas monitored received pollen, and fruit set was close to 100%. Following a severe hurricane in October 1996 (Hurricane Lili, a category 2 storm), both bird species were virtually absent in the flowering season of 1996–7. At the same time only 49% of stigmas received pollen and fruit set dropped to only 11%. Since almost all fruit in this species contain the same number of seeds, fruit set is a reliable indicator of seed set and thus total female reproductive fitness. It is clear from these data that this particular species was limited for pollinator attention when its pollinator population declined following severe weather conditions. Other circumstances that might be expected to impact on pollinator populations include changes in predator populations, disease epidemics, and alteration to habitats, including habitat loss. It is likely that many different sets of circumstances frequently result in pollination limitation for a range of different plant species growing in variable habitats all over the world.

17.5 Evidence from mixed species plots

Analysing pollination competition in mixed species plots has an advantage over other methods in that it allows facilitation of pollination to be monitored at the same time. A recent report by Ghazoul (2006) provides a textbook example of how relative population densities determine whether mixing flowering species together results in competition or facilitation of pollination. This result is consistent with the proposition that facilitation and competition are simply opposite ends of a continuum of interactions. For our purposes, the existence of such a continuum means that, even where facilitation seems to be occurring in a particular habitat, it is possible that at other times in the flowering season or in other localities with variable population densities, competition for pollination is occurring between the same groups of plant species. Such competition may be expected to result in adaptations to increase floral attractiveness or to make flowers more distinctive than those of other members of the community. At the same time, facilitation may actually increase intraspecific competition for pollinator attention, by ensuring that the maximum seed is set by those flowers of each species that are most attractive.

Ghazoul (2006) established seven field plots containing 50 *R. raphanistrum* (wild radish) plants and 50 *Cirsium arvense* (creeping thistle) plants. The number of inflorescences on the *C. arvense* plants varied naturally, and was recorded twice daily over eight days. At the same time the number of pollinating insects visiting the *R. raphanistrum* flowers within a two-minute time window was also recorded. The total number of pollinating insects within the plot was then recorded in a subsequent two-minute window, and the proportion of insects visiting the *R. raphanistrum* flowers calculated from these two observations. These data thus allowed a comparison of proportion of pollinators visiting species 1 with number of inflorescences of species 2 present. At relatively low numbers of *C. arvense* inflorescences (fewer than 32), presence of extra inflorescences increased pollinator visitation to *R. raphanistrum*. Therefore small numbers of flowers of a second species facilitated pollination of one plant species. However, as the number of *C. arvense* inflorescences increased above 32, a decreasing proportion of pollinators visited the *R. raphanistrum* flowers. When over 300 *C. arvense* inflorescences were present in the plot, pollinator visits to *R. raphanistrum* ceased almost entirely. Competition for pollinator attention thus became extreme when *R. raphanistrum* was at a lower relative density within the mixed plot. This study suffers from a number of problems, notably that presence of insect visitors does not necessarily equate to subsequent seed set. However, previous experiments reported within the same paper had shown a correlation between flower visitation and seed set, so it is likely that the same correlation held true in this particular study. It is certainly the case that an absence of insect visitors, as seen when large numbers of competing inflorescences were present, will result in a significant reduction in seed set.

In a study using similar methods, Brown *et al.* (2002) investigated the consequences of pollination competition from an invasive Eurasian species, *Lythrium salicaria* (purple loosestrife), for seed set in the American *Lythrium alata* (winged loosestrife). Over two years they measured pollinator visits in 15-minute intervals and seed set per fruit for 15 fruits per plant, comparing plants in single species plots and mixed species plots. They found that increasing the number of *L. alata* plants in a single species plot had no significant effect on seed set per fruit. It did cause a reduction in visitor numbers per flower per 15-minute interval, significantly so when the number of plants within a plot was tripled. However, since there was no effect on seed set it seems likely that this increased competition for pollination attention was of no fitness consequence to the plants. In contrast, when *L. salicaria* was added to the plots, *L. alata* plants experienced reductions in both seed set and number of visitors per flower. The reduction in pollinator visits was more significant than that observed in the single species plots, suggesting that the reduction in seed set might reflect insufficient pollinator visits, as well as the transfer of heterospecific pollen. This is another example of a specific situation in which competition for pollinator attention may occur. However, the mobility of most plant species, often aided by the activity of mankind, makes it very likely that large numbers of plants regularly come into contact with invasive species and suffer pollination competition as a result.

Using genetic markers to assess outcrossing in the progeny of plants in mixed species plots led Bell *et al.* (2005) to conclude that pollination competition affected seed quality as well as seed quantity. In plots containing *Lobelia siphilitica* as well as *Mimulus ringens* (plants with similar blue flowers that coexist in American meadows) the *Mimulus* plants set significantly fewer seeds per fruit than in single species plots. This reduction in seed set was also associated with a reduction in outcrossing rate, from 0.63 to 0.43, indicating that pollination limitation had resulted in an increased rate of self-pollination.

17.6 Analysis of character traits potentially displaced by pollination competition

Studies which search for evidence for pollination competition through displacement of character traits have primarily focused on flowering time (commonly referred to as flowering phenology in the ecological literature, although not in the literature concerned with the molecular regulation of flowering-time control, perhaps because the latter focuses on shorter term effects than does the former). The suggestion that competition for pollinator attention might lead to displacement of flowering time in plants growing together is based on two processes. Firstly, it is possible that evolutionary displacement of flowering times may occur, with those individuals flowering at times slightly offset from their competitors being most successful and setting most seed. Alternatively, an ecological sorting process may serve to eliminate species from a locality if their flowering competes too directly with other species, resulting in assemblages of plants that do not flower together. However, several factors will also limit the extent to which flowering time can be displaced. To start with, flowering time is under strict genetic control, as discussed in Section IIA of this book. The many different environmental and endogenous factors that interact to regulate flowering time are not easily perturbed, as there are so many checkpoints and fail safes within the complex network of interactions. For displacement of flowering time to occur, then, it is necessary for mutations to occur in one or more key genes which are sufficient to influence the end point of the entire network. Such mutations do occur in crop plants, where they have allowed breeders to manipulate flowering time to commercial advantage, and so it is likely that they will also occur from time to time in wild species. It is also necessary for competition for pollination to be sufficiently strong that these mutants are at a selective advantage strong enough to enable them to persist and eventually dominate a population. At the same time, the environment in which a plant finds itself will have a significant effect on flowering time. No amount of pollination competition should be sufficient to drive a species to flower under extremely cold conditions, or with insufficient time to set seed before frost hinders metabolic and developmental activity. In tropical environments variations in rainfall pattern will also be important, and flowering may be constrained to those seasons with adequate water availability, unless the plant has adaptations

that allow water storage. However, even with these limitations, a number of studies have found apparent displacement of flowering time in a manner which strongly suggests the existence of competition for pollinator attention.

Aizen and Vázquez (2006) considered the flowering time of a group of unrelated bird-pollinated species in South America. They used data for 13 different species, each from a different genus, but found growing together at three different sites. Each plant produced ornthophilous-appearing flowers, usually red and tubular with large quantities of nectar. For each species the hummingbird *Sephanoides sephanoides* had been recorded as the main flower visitor, and had also been shown to transport pollen of each species. The hummingbird-pollinated species were shown to have a much broader range of flowering times than was usual for plants flowering in these localities. Other, non-bird-pollinated, plants tended to flower in late spring, while the hummingbird-pollinated species flowered throughout spring and summer. Flowering times were compared to a number of models, and the authors concluded that flowering times of these 13 species were overdispersed, flowering over a broader range of times than would be predicted based on the behaviour of other plants in the same environment. They therefore concluded that this system provided evidence that competition for the attention of the hummingbird pollinator had occurred, resulting in displacement of flowering times by either evolutionary or ecological processes.

In a similar study using a group of more closely related plant species and their bat pollinators, Lobo *et al.* (2003) also showed that flowering time appeared to be displaced to minimize pollination competition. Flowering patterns of trees from the family Bombacaceae were compared in three different geographic locations which differed in rainfall patterns. All the trees studied have been reported to be pollinated by necteverous bats, and also by moths. Within each locality, the flowering time of the trees was staggered so that maximum flower production of each species occurred sequentially, rather than together. This did not seem to be related to rainfall, but did correlate with phylogeny (the same species tended to flower at the same time in different habitats) and with the presence of other bat-pollinated species and the need to avoid concurrent flowering. Interestingly, analysis of pollen content of bat faeces suggested that the pollinating bats specialized on whichever tree was at the peak of its flowering in a single location, essentially feeding on only one pollen type at a time, for periods of up to a month. This specialization by a major group of pollinators on the most abundantly flowering species will act to reinforce selective pressures inhibiting concurrent flowering, by strongly punishing trees which flower earlier or later than the mean for their species. In this way clear displacement of flowering time can be maintained and further pollination competition prevented.

A recent study confirms that concurrent flowering of species that usually flower sequentially can have detrimental effects on fitness. The organ pipe cactus, *Stenocereus thurberi*, has a peak flowering time approximately one month later than two other species of cactus that grow alongside it in Mexico. All three species are bat-pollinated. Fleming (2006) showed that the fruit set by those individual organ pipe cacti that flower early, and therefore concurrently with the competing species, contained sterile seeds. He further showed that pollination with heterospecific pollen caused the development of fruit, but that the seed within lacked embryos. Selective pressure will therefore act to maintain displacement of flowering time in these species, as concurrent flowering reduces viable seed set. A similar study by Waser (1978) drew similar conclusions for the interaction between *Delphinium nelsonii* and *Ipomopsis aggregata* in Colorado. These species flower sequentially, and are both pollinated by broad-tailed hummingbirds. Waser reported that plants that flowered at the margins of the flowering period for their species, and that therefore overlapped with the other species, set 30–45% fewer seeds per flower than plants flowering at displaced times. Similarly, plants grown in mixed species plots set 25–50% fewer seeds than those in single species plots, and plants hand-pollinated with pollen from the other species set significantly fewer seeds than control plants. These studies provide evidence for the importance of displaced flowering time in maintaining reproductive success in these species.

An extremely detailed analysis by Stone *et al.* (1998) demonstrated that assemblages of closely related species can show displacement of a number of characters to minimize competition for pollination. Working with ten species of *Acacia* growing together in mixed habitats in Tanzania, they firstly showed that, while most species flowered during the rainy season, two species flowered in the dry season. This restriction of flowering to a less favourable time of year can be interpreted as a mechanism to avoid competition. They also showed that there was some partitioning of pollinator species between tree species, but that this partitioning was only ever partial and substantial overlap in pollinating species existed. However, they were able to show that a novel character trait had been displaced to minimize competition for pollination and heterospecific pollen transfer. In habitats where multiple *Acacia* trees flowered together after the autumn rains, the timing of pollen release within a single day had been partitioned. Thus, all trees of each *Acacia* species released their pollen together, but the timing of pollen release was different for each species. Some species had peak pollen availability as early as 6am, while others released maximum pollen as late as 4pm. The window of peak pollen availability was usually little more than an hour, allowing ample time for all co-flowering species to independently release pollen each day. The local pollinator populations had learned to follow this daily synchronous release of pollen, foraging from each species at the most rewarding time. Megachilid bees, honeybees, and flies showed temporal patterns of flower visitation that ensured they visited each tree species at the time of maximum pollen availability. This partitioning of pollen release ensured that the pollinators served only one tree species at a time, minimizing both competition for pollinator attention and the interfering effects of heterospecific pollen transfer.

A number of studies have considered displacement of characters other than timing of flowering and pollen release. For example, Armbruster *et al.* (1994) considered length of nectar tube and position of pollen placement on the pollinators of a group of species of *Stylidium* flowering together in Western Australia. They reported that overlap in pollination system (determined by nectar tube length and the position and length of the pollen-depositing column) was virtually absent from groups of co-flowering species, suggesting that character displacement had occurred.

Not all studies of character displacement have found evidence to support the idea that pollinator attention is limiting. Ollerton *et al.* (2007a) reported that the parasitic plant, *Orobanche elatior*, and its host, *Centaurea scabiosa*, flower concurrently in England and share a common pollinator, the bumblebee *Bombus pascuorum*. The host species is also pollinated by a number of other insects, but the authors postulated that the use of a common pollinator species might drive character displacement. However, they were unable to find any differences in flowering time, placement of pollen on the bee's body, or daily timing of pollinator visits, suggesting that competition between these two plant species was insufficient to drive such character displacement.

In this chapter we have considered evidence, collected through a number of different methods, that shows the presence of competition for pollinator attention, either in the present or in the evolutionary past of a group of plants. At least one of these methods, the analysis of displaced characters such as flowering time, in fact shows an absence of competition for pollination in the current environment. The data considered show that the interaction between flowering plants of the same or different species can range from positive facilitation of pollination through very minimal effects to strong competition for pollination, severely limiting seed set in affected individuals. The nature of this interaction varies with the frequency of different plants within a habitat and can operate either through the direct attraction of pollinators or through the consequences of pollinator fidelity for heterospecific pollen transfer, or through both mechanisms

together. An understanding of how these interactions between concurrently flowering plants operate can inform agricultural practice, as well as strategies to conserve and restore species, particularly in response to climate change. However, the purpose of this chapter was simply to ask whether competition for pollinator attention sometimes occurs. If plants compete for pollination services, and those individuals within a species with showier flowers attract more or better pollinators (or make better use of each visit), we can expect showier flowers to become the norm. Indeed, the strength of such natural selection on floral traits may increase in the near future. Recent models of climate change forecast a smaller number of pollinators globally. These models suggest that selection on floral traits will be strongest in species-rich habitats, where competition for pollination services is likely to be strongest if animal species diversity declines. It is therefore predicted that plants in biodiversity hotspots (such as tropical rainforests or the South African Cape flora) will be at increased risk of extinction if they are not able to specialize on particularly abundant pollinators or substantially enhance their floral attractiveness (Vamosi *et al.* 2006). In the uncertain and constantly changing circumstances experienced by most flowering plants, it is a daunting task to quantify the frequency and importance of the selection imposed by pollination limitation relative to the many other selection pressures plants face. However, the fact that pollination competition occurs is sufficient to allow flowers to evolve into more elaborate structures and for pollination syndromes to exist, at least in theory. Chapter 18 will consider another key concept underlying the pollination syndrome theory, before Chapter 19 assesses whether specialized pollination systems are the rule or the exception.

Do Pollinators Discriminate between Different Floral Forms?

In Chapter 17 we concluded that some plants, at some times, experience competition for pollinator attention. In such circumstances natural selection will favour traits which enhance the attractiveness of flowers to their pollinators. In this chapter we will investigate whether the different shapes, structures, and colours that flowers produce have the potential to enhance pollinator visitation. To do this, they must fulfil two criteria. First, they must be visible to the appropriate pollinator, or detectable using some other sense. Secondly, the pollinator must discriminate between different floral forms. Simply because a change in floral form is detectable to an animal, it does not necessarily follow that the animal will discriminate between the original and the novel form. Such discrimination will only occur if one form provides an advantage to the animal, in terms of being easier to detect or handle in some way, or in terms of providing a superior reward of nectar or pollen. In this chapter, then, we will begin by discussing the current evidence on what different pollinating animals can see and detect in other ways. Then we will consider the experimental evidence that pollinators do discriminate between different floral forms, focusing on colour, shape, and scent. An understanding of which floral features are discriminated by pollinators will allow us to assess the usefulness of the pollination syndrome concept in Chapter 19.

18.1 What pollinators see

Animal vision depends on two components—the presence of structural features which allow objects to be detected, and the presence of the neural processing ability to translate such signals. Our understanding of animal vision varies greatly between different types of pollinators. In this section we will consider two comparatively well-studied examples, bees and birds.

18.1.1 Bees

Bees, like all other insects, possess compound eyes. This means that the eye is composed of several thousand independent units, known as ommatidia. Each ommatidium contains its own photoreceptor cells and its own lens, and so functions as a mini eye in its own right.

Bumblebee and honeybee colour vision, like human colour vision, is trichromatic, based on three colours. The ommatidia contain photoreceptors, containing opsin pigments, which perceive green (540 nm), blue (430 nm), and ultraviolet (UV, 340 nm) light. They do not contain photoreceptors for red, leading to a common misconception that bees cannot see red. However, the spectral sensitivity curve of the green receptor has an extended tail, and reaches to around 650 nm, well beyond the orange/red boundary (Chittka and Waser 1997). This allows bees to perceive red light, although it may be difficult to distinguish from green light of a lower intensity. Bee colour vision is therefore very different from human colour vision, being weaker with regard to one of our primary colours but containing an extra colour, UV, which we cannot see. As a result, it is essential to accurately measure colour using spectrophotometry at a range of wavelengths when considering how a flower looks to a bee, rather than relying on the human eye (Fig. 18.1). Indeed, bee colour vision has recently been shown to be more complex

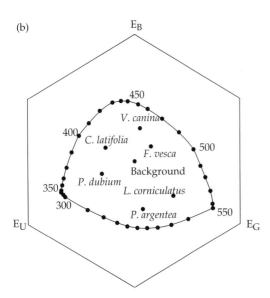

Figure 18.1 Bee colour vision. (a) A set of typical floral spectral reflectance functions. The flowers measured are yellow *Potentilla argentea* (with UV reflectance); red *Papaver dubium* (with UV reflectance); blue *Viola canina* (without UV reflectance); violet *Campanula latifolia* (with UV reflectance), and white *Fragaria vesca* (without UV reflectance). (b) The colour hexagon shows how colours look to a bee. The continuous curve denotes the spectrum locus, in 10-nm steps from 300 to 550 nm. The bottom segment of the spectrum locus connects the loci of 300 and 550 nm in nine mixtures of the two lights in ratios of 0.9:01; 0.8:0.2, and so forth. Colour loci for flowers whose reflectance are given in (a) are marked here. Both parts of this figure kindly provided by Lars Chittka (QMUL).

than previously thought. All the ommatidia of *Bombus impatiens* contain green receptor cells, but they do not all contain blue and UV. There are three types of ommatidia in the eye—those containing green and blue photoreceptors, those containing green and UV photoreceptors, and those containing all three types of photoreceptor, shown by *in situ* hybridization of probes for the UV absorbing pigment to sections of the eye (Spaethe and Briscoe 2005). This means that different ommatidia may perceive a flower as different colours, adding to the complex task of the neural processing systems in decoding the signals they receive. It is also likely that colour vision varies between individuals within a population, as well as between species of bee.

Compound eyes do not provide the same degree of resolution as the simple eyes of vertebrates, and it has been estimated that the honeybee eye has a resolving ability around 100 times worse than the human eye (Chittka and Raine 2006). Combined with this relatively poor resolution, insects also appear to have relatively poor visual processing and it is necessary for at least 15 ommatidia to perceive a flower for the bee to recognize it by its colour alone (Chittka and Raine 2006). It is clearly the case that larger flowers should be recognizable at greater distances than smaller flowers on the basis of their colour. Spaethe *et al.* (2001) confirmed that it took longer for bees to find flowers as their size decreased, even though their colour remained constant.

However, bees do not only use colour to identify flowers, but rely at greater distances on the contrast between flowers and green vegetation (Chittka and Raine 2006). This so-called 'green contrast' provides a reliable means of identifying a target (the flower) against the almost universal background colour of plants. It greatly increases the distance at which flowers are visible to bees, although larger flowers will still be visible as coloured objects at greater distances than will smaller flowers. A consequence of the use of green contrast by bees is that the difference between a flower's colour and green can be of more importance in bee discrimination than the intensity or brightness of the colour itself, particularly when the flower is small (Spaethe *et al.* 2001).

18.1.2 Birds

Birds have camera eyes like those of other verte-brates, including humans. This means that their resolving power at various distances is more similar to our own than is that of insects. However, bird colour vision is quite complex, and may vary significantly between different species of bird. Essentially, bird eyes contain cone cells with pig-ments, opsins, which absorb light of different colours. Bird vision is tetrachromatic, based on the presence of four or five different types of cone cell (Bennett and Thery 2007). However, the colours that these absorb are modified by the presence of a drop of oil through which light must pass to reach the pigments. The oil contains one of a number of dif-ferent carotenoids, and the combination of different carotenoids in the oil with different opsin pigments results in a wide range of colour sensitivities (Varela *et al.* 1993). All birds have red photoreceptors, and many birds are most sensitive to colours at the red end of the spectrum. However, the colour vision of most birds extends into the UV range, to light with a wavelength of around 300–370 nm (Chen *et al.* 1984), and so covers a wider spectrum of colours than that of either humans or insects. Hummingbirds are not known to have particularly strong sensitivity to red, and do not show a strong preference for red when offered a choice of coloured feeders, although many hummingbird-pollinated flowers produce red pigment (Goldsmith and Goldsmith 1979). One possible explanation for this is that red flowers are harder for bees to handle (because they show little contrast to green in insect vision) and are therefore preferentially visited by birds as their nectar is less likely to be depleted than that of more strongly bee-visible flowers (Raven 1972; Rodriguez-Girones and Santamaria 2004).

18.2 What pollinators sense in other ways

Pollinating animals may also be attracted to flowers over a range of distances by their scent. Scent func-tions as a long distance cue for moths, beetles, and some bees, and at more moderate distance also for butterflies. It is also known to act as a short-range landing cue or nectar guide for a large array of insects. Once on a flower, tactile cues may be important in allowing rapid handling of the flower, and may therefore make flowers with certain sur-face properties more attractive than those without them. Taste may also play a part in flower discrim-ination, if particular nectar qualities can be associ-ated with features which can be discriminated at longer ranges.

The scents produced by flowers can travel very great distances. For example, the amines volatal-ized by the Titan Arum (*Amorphophallus titanum*) inflorescence attract pollinating bees and flies over distances of hundreds of metres, as individual plants are very widely dispersed within the envir-onment. Scent make-up is very complex, and dif-ferent animals are able to detect different scent components to different degrees. They may also draw different information from different odours. Some scents carry very specific information (such as organic decay or sex pheromones) while others are more ubiquitous and can be learned by insects (particularly bees, moths, and butterflies) as associ-ated with floral rewards.

Insect odour perception has been best studied in the fruit fly, *Drosophila melanogaster*, which produces 62 odour receptor proteins, a family of G protein-coupled receptors with seven transmem-brane domains (Ache and Young 2005; Hallem *et al.* 2006). Although the general mechanisms of odour perception, and the types of proteins involved, are comparable in other insects studied, it is likely that *Drosophila* has a simpler olfactory system with regard to floral scents than do pollinating insects, which distinguish between many compounds in their foraging bouts. While bees have not been studied in as much detail, bee antennae are thought to contain around 130 types of odour receptor (Chittka and Raine 2006), making the understand-ing of bee response to different floral scents a much more complex problem than that of visual response to colour. Guerrieri *et al.* (2005) tested the ability of bees to distinguish between a range of arbitrarily chosen odours, and observed that alcohols, alde-hydes, and ketones were the most important chem-icals in bee scent perception, with the majority of odour receptors focused on discriminating between molecules of these types. These compounds are not all commonly released by flowers, and so bee discrimination of floral scents may differ somewhat from this general model. The situation is likely to be much more complex in many animals that do not use visual signals to the same degree, such as some beetles, moths, and bats.

The sense of touch is a great deal more similar between different animals than either vision or olfaction. However, differences in body size will significantly influence the range and scale of structures which an animal can perceive tactilely. The great variety of cell types present on petals suggests that insect feet and antennae can distinguish shapes and structures as small as a single cell in size, and behavioural experiments have confirmed this ability (Kevan and Lane 1985).

The sense of taste depends on receptors triggered by contact with particular molecules. Most animals can sense sweet and salt, with other taste receptors varying more between species. Electrophysiological techniques have revealed that insects typically have four different types of taste-sensitive neuron, responding to sugar, water, low salt concentrations, and high salt concentrations, and they may also have receptors for the presence of amino acids (Hallem *et al.* 2006). These are triggered by the activities of taste receptors, of which *D. melanogaster* has 68 (Hallem *et al.* 2006). The sweet receptors in bees are sensitive enough to allow discrimination between sucrose at concentrations varying by only 0.5%, although their preferred range is 20–30%. Analysis of hummingbird responses to sucrose feeders at various concentrations has shown that 40% sucrose is preferred over lower concentrations (Stromberg and Johnsen 1990). This discrimination between rewards at different concentrations could be used to select between flowers if it can be linked to a colour or shape difference which is remotely detectable. However, temperature may also influence the perceived taste of a substance if taste receptors are not buffered for temperature. For example, there is evidence that some insect sweet receptors perceive warmer sucrose as sweeter than cooler sucrose of the same concentration (Uehara and Morita 1972). Another effect of temperature on reward relates to the need for warm-blooded animals to heat ingested food to body temperature. For hummingbirds this metabolic cost may have a significant effect on their perception of the quality of reward offered by a flower (Lotz *et al.* 2003). These

factors may result in selective pressure to moderate flower temperature as well as nectar quality.

18.3 Discrimination between petals of different colours

Investigation of pollinator discrimination between different floral traits is experimentally difficult. It has traditionally been approached in two different ways. Bee behavioural scientists have tended to use artificial flowers, often composed of pieces of coloured card, to investigate behavioural responses to single factors such as colour, size, or shape. These experiments can be difficult to interpret from a plant evolution standpoint, because they do not necessarily mimic structures and colours that can easily be achieved in nature with the resources available to the plant. Plant evolutionary biologists have tended to compare pollinator responses or seed set in flowers with different forms. However, unless the differences between the two floral forms are very carefully described and controlled, this can result in equally problematic data, as the flowers may vary in more than one feature. Such experiments are also constrained by the availability of individual species with variation in an experimentally interesting trait, such as flower colour.

The literature contains many examples of attempts to assess pollinator discrimination between petals of different colours. In some cases discrimination is clear, in others the animals showed no discrimination, and in many cases some animals discriminated while others did not. For example, experiments with white and violet alfalfa flowers in a growth room indicated that honey bees did not discriminate between flower colours. When a similar trial was conducted in the field, honey bees again showed no discrimination, but pollen-gathering leaf-cutter bees much preferred the purple flowers (Pederson 1967). Similarly, experiments with equal numbers of purple- and white-flowered plants of *Vigna sinensis* (cow pea) showed that bumblebees visited purple flowers three times as often as white flowers, but honey bees visited white flowers twice as often as purple flowers (Leleji 1973). These experiments indicate that the attraction of multiple pollinators can result in mixed selective pressures, even where one animal shows very clear discrimination. For example, *Raphanus raphanistrum*, wild radish, has yellow or white flowers, controlled by a single locus. The frequency of the yellow morph varied from 7 to 60% in the populations studied by Kay (1976), but in all these populations the butterfly *Pieris rapae* much preferred the yellow form to the white. On the site with 60% yellow flowers 307 visits to wild radish flowers by the butterflies were observed—and 306 of the 307 were to yellow flowers. However, honey bees show no preference for yellow flowers over white, maintaining the polymorphism within the population. In 1986 Stanton *et al.* followed this work up by investigating the consequences of pollinator preferences for maternal and paternal reproductive success in *R. raphanistrum*. Since the white allele is dominant over the yellow they were able to assess male reproductive success by growing progeny of yellow parents and noting whether they had white or yellow flowers. They discovered that, despite pollinator preference for the yellow morph, female reproductive success of white and yellow flowered plants was equal, suggesting that resource limitation is more important than pollinator limitation in female fitness of this system. However, nearly three-quarters of the progeny of yellow-flowered plants were themselves yellow-flowered, indicating that their male parent was also yellow. In this example pollinator discrimination in favour of the yellow morph has clear consequences for the male fitness of yellow-flowered plants. In contrast, Schemske and Bierzychudek (2001) showed that the pollinating beetle, *Trichochorous* sp., showed no preference for blue- or white-flowered forms of the desert annual *Linanthus parryae*, suggesting that other factors maintain this polymorphism within the population.

One factor which has been shown to complicate matters when analysing pollinator discrimination between different floral colours is the tendency of many animals to behave in a frequency-dependent manner. In theory, pollinators should preferentially visit more common floral forms, because they are easier to learn to handle and easier to remember. In wild populations, then, particular colour morphs can be selected against simply because they are rare. This has been shown in particular detail with the tall morning glory, *Ipomoea purpurea*, and its native American pollinators, mainly bumblebees. The wild type flower is blue-purple, but a white-flowered

form arises commonly in natural populations, and is present at a range of frequencies. When white flowers are present at frequencies below around 25%, bumblebees discriminate against white flowers more than would be expected based solely on their frequency. However, when the frequency of white flowers increases to 50%, bumblebees no longer show any colour discrimination (Epperson and Clegg 1987; Clegg and Durbin 2003). This positive-frequency-dependent discrimination may serve to limit the extent to which novel flower colour mutations can spread through populations.

Despite the different preferences shown by different animals the studies described above, and many similar studies, provide evidence that pollinating animals do sometimes discriminate between flowers of different colours, as we would expect if they can link colour to quality or availability of reward. This should be sufficient for flower colour to be under selection in some circumstances, and thus to form part of pollination syndromes. However, these studies all suffer from the classic difficulty of the unknown nature of the polymorphism. The observed difference in flower colour may not be the only difference between two naturally occurring morphs of a plant species, and pollinators may actually be discriminating on the basis of other differences besides colour. This is particularly likely where flower colour is genetically linked to a trait such as nectar quality or flower size. Recent attempts to address this difficulty have focused on using near isogenic lines (NILs) produced by extensive backcrossing, or using molecularly characterized mutant lines. These allow an assessment of the strength of direct selection on isolated phenotypic traits, although such data are only applicable to wild situations where the trait in question can segregate independently of other aspects of pollinator attraction.

Bradshaw and Schemske (2003) provided clear evidence that both bumblebees and hummingbirds distinguish between different coloured forms of *Mimulus*, using NILs. This followed on from their previous work (Schemske and Bradshaw 1999) using a segregating F2 population resulting from a cross between *Mimulus lewisii* and *Mimulus cardinalis*. *M. lewisii* is normally pink, as a result of anthocyanin deposition, and is primarily pollinated by bumblebees (Fig. 18.2a). *M. cardinalis* is normally orange/red, as a result of both anthocyanin and carotenoid deposition, and is primarily pollinated by hummingbirds (Fig. 18.2b). These two types of animal rarely visit the 'incorrect' flower type—some observations of pollinator behaviour in mixed plots have shown that hummingbirds only visit *M. cardinalis* and that 98.9% of bee visits are to *M. lewisii* (Ramsey *et al.* 2003), although other studies suggest that the incidence of crossover visits is somewhat higher than this. In the segregating F2 population, bumblebee visits were shown to increase when flower size increased, and to be negatively correlated with carotenoid and anthocyanin levels, while hummingbird visits increased with increased anthocyanin levels and increased nectar volume (Schemske and Bradshaw

1999). To analyse these data further, Bradshaw and Schemske (2003) introgressed the *YUP* locus, responsible for carotenoid deposition, from each species into the other background, through four generations, ensuring 97% genetic identity between the new lines and their most similar parent. This resulted in orange coloured *M. lewisii* flowers and deep pink *M. cardinalis* flowers. Nectar production (a highly significant factor in pollinator attraction) was compared between parent and new lines, and found not to be significantly different. Mixed plots containing 50 wild type and 50 novel plants of a species were grown, with the two species being investigated separately. Pollinator visits were recorded, and revealed that orange-flowered *M. lewisii* received 68-fold more

(a)　(b)

(c)　(d)

Figure 18.2 Flowers for which pollinators have been shown to discriminate between colour morphs. (a) *Mimulus lewisii*. (b) *Mimulus cardinalis*. Images (a) and (b) kindly provided by Toby Bradshaw (Washington State University). (c) *Antirrhinum majus* wild type and *sulfurea* lines. (d) *A. majus* wild type and *incolorata* lines. Images (c) and (d) kindly provided by Enrico Coen (John Innes Centre). See also Plate 22.

visits from hummingbirds than the wild type pink, but a significant reduction in bumblebee visits. Similarly, the pink-flowered *M. cardinalis* received 74-fold more visits from bumblebees than the wild type orange (although little reduction in humming-bird visits). These experiments show that both bumblebees and hummingbirds exhibit strong discrimination on the basis of petal colour. The near isogenic nature of the lines used in this study makes it likely, although not certain, that colour is the only significant factor in the choices made by pollinators. However, since the pollinators were not flower naïve, but accustomed to foraging among wild varieties of both species in which colour is linked with a number of other traits, including nectar availability, it is possible that the animals had learned that colour was a cue that they could usually expect to be associated with a desirable reward.

The use of mutant lines which have been sub-jected to extensive molecular characterization allows the assessment of pollinator discrimination between different colour morphs with absolute confidence. *Antirrhinum majus* has proved to be an extremely useful model for this kind of work, being both bumblebee-pollinated and a classic molecular genetic model. Jones and Reithel (2001) investi-gated the responses of wild American bumblebees to experimental plots containing wild type red, white *incolorata* mutant (Fig. 18.2d; *INC* encodes flavanone 3-hydroxylase, see Chapter 15), and yellow *sulfurea* mutant lines of Antirrhinum (Fig. 18.2c). Different bees exhibited different pref-erences, with individual animals expressing strong preferences for either red or yellow flowers within a foraging bout. Similarly, Glover and Martin (1998) analysed pollinator responses to the *nivea* mutant of Antirrhinum, which contains a deletion of the gene encoding chalcone synthase (the first step of the anthocyanin synthetic pathway, see Chapter 15) and is therefore white. The petals contain no ultraviolet absorbing pigment, besides no visible pigment, and are therefore white to the bee eye as well as to the human eye. Analysis of seed set of emasculated flowers in mixed plots indicated that the *nivea* mutant flowers set significantly less seed than wild type red flowers, even though they were equally fertile when hand-pollinated. The extent of

pollinator discrimination varied with the number of pollinators present in the plot, but in all circum-stances *nivea* mutants were discriminated against. Dyer *et al.* (2007) conducted a detailed analysis of bumblebee behaviour in response to the *nivea* mutant, and concluded that it was significantly less visible than red flowers at a range of distances, as a result of the similarity of the white flower colour to green vegetation when viewed through a bee's visual system. This similarity reduced the 'green contrast' of the *nivea* mutant relative to wild type flowers.

18.3.1 Discrimination in favour of nectar guides

Many flowers produce markings on their petals believed to act as nectar guides, directing pollin-ators towards the reward. These are predicted to reduce the time taken to handle a flower, and there-fore increase the attractiveness of the plant. A very few studies have investigated pollinator discrim-ination between flowers with differences in this specialized subset of petal colours.

Scora (1964) showed that bees and wasps use the nectar guides on *Monarda punctata* flowers, consist-ing of 3 to 5 spots or lines of pigment in the back of the throat, to position themselves correctly to access the nectar. Scora used colchicine treatment to pro-duce mutant plants without the nectar guides, but with otherwise normal flowers. These initially appeared to attract bees and wasps to the same degree as wild type flowers, but the pollinators rejected the flowers after landing, without trying to probe for nectar.

A particularly well-studied system where nectar guides have been shown to be important is that of *Delphinium nelsonii*, which has two sets of pollin-ators, hummingbirds and bumblebees (Waser and Price 1981). The flowers are usually deep blue but occasional albino flowers arise in natural popula-tions. The seed set of white flowers and blue flowers is very similar if they are hand-pollinated, suggesting that they are equally fertile. However, the seed set of open-pollinated albino flowers is as much as 45% lower than that of blue flowers and white flowers in artificial plots receive 24% fewer visits from pollinators than do blue flowers. Both bees and hummingbirds released into cages containing both blue and white flowers preferred the blue flowers. This appeared to be due to a difference in handling time between the two flower morphs, resulting from the 'disappearance' of the nectar guides in the white flowers. The blue flowers have a nectar guide, consisting of two small white petals which frame the entrance to the concealed nectaries (Fig. 18.3a). These are not visible against the white background in the albino flowers. It takes up to 1.8 times as long for an animal to extract nectar from the white flowers as from the blue flowers, and up to 2.2 times as long to fly between white flowers as blue flowers. This results in a significant difference in the total time taken to process a white flower and a blue flower, even though they provide identical nectar rewards. In this case pollinator discrimination between different colours of flower is directly attributable to the visibility of the nectar guides against the different backgrounds and the consequent effects on handling time (Waser and Price 1981).

Frequency-dependent behaviour may also influence pollinator discrimination between flowers with and without nectar guides. *Clarkia xantiana* ssp. *xantiana* produces pale pink flowers and is polymorphic for the presence of nectar guides. These consist of dramatic spots of deep red pigment, surrounded by a white area, at the base of each petal (Fig. 18.3b). Eckhart *et al.* (2006) showed that the pollinators of this plant (several species of bee) showed no discrimination between flowers with spots and flowers without, unless they were presented at different frequencies. When flower frequencies were varied, one bee species preferentially visited the more common form, whether that was spotted or spotless, while two other bee species preferentially visited the rarer form. As with petal colour in general, it appears that frequency-dependent foraging strategies can serve to complicate our understanding of pollinator discrimination between different forms of nectar guide.

18.3.2 Discrimination between colours as a cue to other differences

Pollinating animals may discriminate between different petal colours not only because some colours

Figure 18.3 Flowers with significant nectar guides. (a) *Delphinium nelsonii*. Image kindly supplied by Nick Waser (University of California—Riverside). (b) *Clarkia xantiana* ssp. *xantiana*. Image kindly provided by Vince Eckhart (Grinnell College, USA). See also Plate 23.

are more visible, or provide more handling information than others, but also through learned associations. Dyer *et al.* (2006) showed that bumblebees could learn to associate particular flower colours with particular rewards, in this case warmer sucrose. Bees were presented with two colours of artificial flower, pink and purple. The colour difference was very slight, and bees had previously been shown not to discriminate between the two colours when both were equally rewarded. The purple flowers were then heated to several degrees Centigrade above ambient, while the pink were not, and both were provided with sucrose solution. The bees quickly learned only to land on the purple flowers, which contained the warmer reward. This ability to learn an association between flower colour and a particular reward may explain pollinator discrimination between apparently equally visible flower colours, if one colour is associated with other beneficial properties such as enhanced nectar supply or reduced handling time.

18.4 Discrimination between corollas of different sizes

There have been fewer experimental demonstrations of pollinator discrimination between flowers of different sizes, despite the prediction, based on our knowledge of bee vision, that insects should strongly favour larger flowers because of their enhanced visibility. One reason for the shortage of convincing experimental data may be that flower size is not as variable as might be expected, as perturbations in cell division or cell expansion are compensated for by reciprocal changes to the other process (see Chapter 14). However, where experimental systems have been identified in which flower size varies significantly and reproducibly within a population, then it has been possible to show that both bees and hummingbirds discriminate on the basis of various aspects of corolla size. There is some evidence in each case to suggest that reward production increases with flower size, so such discrimination may be due to a learned association between flower size and reward quality or quantity.

Some of the most thorough and convincing work has been conducted over a number of years on *Polemonium viscosum*, the alpine skypilot, by Candace Galen. The alpine skypilot produces flowers with different degrees of corolla width or flare at different elevations in Colorado. The flowers of upland plants, growing in tundra habitats, have corollas on average 12% wider than those of lowland plants growing at the timberline. The upland plants are bumblebee-pollinated, while the lowland plants are pollinated by a whole range of different insects, but mostly by flies. Both sets of plants are self-incompatible, so effective pollination is essential for seed set. Bumblebees entering a mixed plot visit the wide flowered plants first, and wide-flowered plants set more seed than narrow-flowered plants. Pollen removal by bumblebees has also been shown to be correlated with corolla flare (Galen and Stanton 1989), indicating that the male component of fitness is also related to corolla size through bumblebee discrimination. Estimates of the selective advantage of wider corollas in attracting bumblebees predicted that the 12% difference in corolla flare could be achieved in only 1–3 generations (Galen 1996a, b). Pollinators are not the only animals to distinguish between alpine skypilots with different degrees of corolla flare. Nectar-gathering ants, which often sever the stigma and cause female infertility, preferentially visit short flowers with wide corollas. To test whether reduced corolla flare would protect against this damage, Galen and Cuba (2001) constricted floral growth using glue, so that flowers attained their normal size but had corollas 36–46% narrower than usual. When ants were presented with mixed arrays of untreated control flowers and the manipulated narrower flowers they chose the wider controls 68% of the time. However, such protection is not selectively advantageous in the wild, because when bumblebees were allowed to pollinate a mixed array of the two flower types the narrower manipulated flowers set 62% fewer seeds than the controls. These experiments confirm that pollinating bumblebees, as well as predatory ants, can and do discriminate between flowers of the same species with different widths.

Work with *Ipomopsis aggregata*, a hummingbird-pollinated plant growing in the Rocky Mountains of America, has shown that birds also discriminate between flowers with different widths and lengths. Campbell *et al.* (1991) showed that wider flowers in their study site produced more nectar than narrower flowers, and that there was a relationship between nectar production, corolla width, and hummingbird visitation. Flowers with wide corollas exported significantly more pollen as a consequence. *I. aggregata* is shown in Fig. 18.4 (see also Plate 24), being visited by its other main pollinator, a large bumblebee.

The selection pressure imposed by pollinators is not only by direct floral discrimination and choice. Fulton and Hodges (1999) showed that the hawkmoths which are the major pollinator of *Aquilegia pubescens* exert selective pressure on floral position through their choices—when flowers were positioned pendantly, as opposed to their normal upright position, they were visited ten times less frequently than flowers in the normal position. However, artificial shortening of the nectar spurs of *A. pubescens* had no effect on hawkmoth visitation rates. It did have a strong effect on pollination efficiency, with hawkmoths removing significantly fewer pollen grains from flowers with short nectar spurs than from flowers with long nectar spurs.

Figure 18.4 Flowers for which pollinators have been shown to discriminate on the basis of size. *Ipomopsis aggregata*. Image kindly provided by Nick Waser (University of California—Riverside).

18.5 Discrimination between zygomorphic and actinomorphic flowers

Although zygomorphy (bilateral symmetry) has evolved multiple times in the angiosperms, and is often described as a key specialization to attract a particular subset of pollinators (Cubas 2003), there have been few experimental attempts to investigate pollinator discrimination between floral morphs with different symmetries. This is mainly due to the difficulty of identifying suitable model systems. Very few species exhibit naturally occurring variation in floral symmetry to any great degree. However, the use of actinomorphic mutants, such as those of *Antirrhinum majus*, is also problematic as their phenotypes may be so extreme as to prohibit pollination at all. For example, the radially symmetrical *cycloidea/dichotoma* double mutant (see Section 14.2) produces a tightly closed corolla so narrow that bumblebees, the normal pollinator, cannot enter the flower to probe for nectar (B. J. Glover, unpublished). Many studies have shown that bees can distinguish between artificial shapes with different symmetries (Giurfa *et al.* 1999), but it is only very recently that evidence for pollinator discrimination between differently symmetrical floral forms of the same species has arisen. A recent study using *Erysimum mediohispanicum*, a member of the Brassicaceae, has shown pollinator discrimination in favour of zygomorphy. This plant, growing in the Sierra Nevada, shows continuous variation in flower shape from purely actinomorphic to quite strongly zygomorphic with the two abaxial petals considerably larger than the two adaxial ones. One of its major pollinators in the study site is the beetle *Meligethes maurus*. Gomez *et al.* (2006) studied flower morphology and a number of components of fitness in a wild population of *E. mediohispanicum* and reported that zygomorphic flowers received more pollinator visits than actinomorphic flowers and also produced more surviving juvenile offspring.

18.6 Discrimination between flowers with different petal cell shapes

In Section 16.6 we discussed in some detail the variety of roles that petal epidermal cell shape can play in modifying flower colour, texture, and temperature. That bumblebees can discriminate between wild type Antirrhinum plants with conical petal epidermal cells and isogenic *mixta* mutants with flat petal cells was shown by Glover and Martin (1998). Fruit set was recorded in mixed field plots of emasculated flowers, and the conical-celled form was shown to set significantly more fruit, reflecting more pollinator visits. These data were confirmed by Comba *et al.* (2000), who observed that bumblebees failed to land on flat-celled *mixta* flowers as often as on conical-celled wild type flowers, and that they also rejected flat-celled flowers without probing them more frequently than they did conical-celled flowers. Recent data suggest that naïve bumblebees have no intrinsic preference for the slightly deeper colour of wild type flowers compared to *mixta* mutants, although they can be trained to discriminate them if one genotype is rewarded with nectar and the other punished with quinine (Dyer *et al.* 2007). Work is ongoing to investigate whether the discrimination observed in the wild is related to tactile effects on handling time or on the consequences of cell shape for floral temperature. However, it is clear that some pollinating animals can discriminate between flowers with different shaped petal epidermal cells.

18.7 Discrimination between flowers on the basis of scent

Analysis of pollinator response to floral scent has been considerably more difficult than analysis of visual responses, because variation in scent production is less obvious to the investigator than is variation in appearance. However, a number of studies have established that pollinating animals do distinguish between flowers on the basis of their scent. For example, Dobson *et al.* (1999) showed that the scent released by the pollen of *Rosa rugosa* flowers had a significant effect on their attractiveness to bumblebees. These flowers produce no nectar and attract pollen-gathering bees. Emasculated flowers received fewer bee visits than controls, but attractiveness was significantly enhanced by applying eugenol, an odour normally released by rose pollen, to the emasculated flowers. Similarly, Ashman *et al.* (2005) showed that odour extracts of male flowers, applied to the female flowers of the gynodiecious wild strawberry *Fragaria virginiana*, increased the frequency of approaches by several species of small bees to those flowers. While these two studies both focus on pollen-released scents, it is likely that petal scents also play a significant role in allowing pollinator discrimination between floral forms.

In this chapter we have considered how pollinators sense various features of flowers, and whether there is convincing evidence that they do sometimes discriminate between such features. Although it is by no means clear from the literature that all diversity in floral form is of significance to pollinating animals, there are certainly data to suggest that polymorphisms in flower size, flower shape, flower colour, and flower scent are all sometimes noted and responded to by pollinators. These data provide the evidence of natural selection operating on floral form that is necessary before any attempt can be made to classify floral diversity with regard to the preferences of particular pollinators. In the next chapter we will look in more detail at pollination syndromes, considering the other assumptions underlying the concept and the evidence as to whether the concept is useful.

CHAPTER 19

Pollination Syndromes—The Evidence

In Chapter 13 we discussed the concept of a pollination syndrome, a suite of floral characters such as those described in Chapters 14, 15, and 16, which have evolved together because they enhance pollination by a particular taxonomic or functional group of animals. This concept has underlain much of floral biology for many years, and our purpose in this chapter is to assess the usefulness of the concept in understanding flowers and flowering. We will begin by considering why and how the pollination syndrome concept has become so entrenched in the literature on flowering, and then assess whether the key assumptions that underlie it are met. Finally we will assess the experimental evidence that pollination syndromes do exist, and the experimental evidence which shows them to be false—those cases where the major pollinator in the native habitat is not that which the flower's morphology would lead us to predict. This chapter will also provide a brief overview of the relative importance of generalization and specialization in pollination ecology.

19.1 Historical context

The recognition that pollinating animals are essential for reproductive success in many flowering plants came in 1761, in the work of Koelreuter (discussed and partially translated by Waser 2006). By 1793, Sprengel was already assigning likely pollinators to plants on the basis of their floral form and attempting to classify flowers into groups according to their rewards and the numbers of their floral parts (Waser 2006). A century or so later came the first attempt to classify flowers according to the

morphological and structural features which suited them for pollination by different animals. Delpino's scheme was the forerunner of the pollination syndromes we know today (Waser 2006). However, much of the scientific enthusiasm for pollination syndromes can be traced to Charles Darwin and the controversy surrounding his theories. Darwin's own work on orchid pollination was one of the earlier essays on the relationships between suites of floral characters and their animal pollinators. In the rush to find evidence to support Darwin's new evolutionary theory, the visually obvious similarities between flower size and pollinator size, flower colour and pollinator preferences, and flower structure and pollinator feeding mechanism were seized upon by many authors as clear examples of the consequences of natural selection. The thorough description of the different pollination syndromes by Faegri and van der Pijl (1966), along with several other authors, with reference to corolla size and shape, colour, scent, and reward production, formally established the pollination syndrome as the paradigm within which pollination ecology was conducted. It was only in the 1980s that experimental approaches were first taken to assess whether particular floral traits were under selection by pollinators. Although these experiments (some of which are described in Chapter 18) did reveal the existence of pollinator selection on floral form, they did not address the frequency with which such selection occurs. In 1996 two critical essays were published, both questioning the utility of the pollination syndrome concept, and setting out the difficulties with its entrenchment in the botanical literature (Herrera 1996; Waser et al. 1996). Both papers suggested that generalization, with flowers receiving pollinator service from more than one type of animal, was more frequent in nature than the pollination syndrome-dominated literature might lead us to expect. One concern was that positive cases of pollinator selection were over-reported in relation to negative cases, generating a situation where we can say qualitatively that such selection occurs but nothing quantitative about its relative frequency or importance. Most studies also fail to take account of constraints that might prevent plants from adapting, even when selection by pollinators is strong. This requires an understanding of the genetic control of flower development as well as an overview of the other ecological constraints on a flower, such as the need to avoid predators (such as the nectar-robbing ants studied in *Polemonium viscosum* by Galen and Cuba 2001). It also requires an appreciation of the relationship between the traits of the flower and the lifetime fitness of the plant. It is only in recent years that a more integrative approach to analysing pollination syndromes has been reported, and the data from such studies now allow us to make initial comments on the utility of the pollination syndrome concept.

19.2 Putting the assumptions together

The pollination syndrome concept rests on a number of assumptions, although they are rarely stated explicitly. They are:

1. That plant reproductive success is limited by a lack of pollinator attention;
2. That pollinators discriminate between different floral morphs within a species, exerting selection on floral traits;
3. That plants can respond to that selective pressure, having the necessary genetic variability with respect to key traits;
4. That plants are not penalized in other ways for responding to that selective pressure;
5. That this sequence of events has occurred in a majority, or even a substantial majority, of the plant–pollinator interactions we see today; and
6. That this has resulted in a degree of specialization in most plant–pollinator interactions.

1. That plant reproductive success is limited by a lack of pollinator attention.

We discussed assumption 1 in Chapter 17, and concluded that plant reproductive success is sometimes limited by lack of pollinator attention (or too general pollinator attention, resulting in interference competition). What is not as easy to assess is how frequently such limitations occur. The best study giving any quantitative data is the review by Knight *et al.* (2005) of published reports of pollen limitation. As much as 63% of published studies did record pollen limitation of control plants, although the variety of methods used in the different studies makes even this conclusion tentative. There are also vastly insufficient data for us to conclude whether pollinator attention limits the fitness of plants in all or a few particular habitats, and whether it occurs occasionally in a plant lifespan, for all of the lifespan, or something in between. In short, assumption 1 is qualitatively true, but we do not know very much about it at a quantitative level.

2. That pollinators discriminate between different floral morphs within a species, exerting selection on floral traits.

We discussed assumption 2 in Chapter 18, and concluded that some pollinators certainly discriminate between some differences in floral form, including those of colour and size. While assessment of pollinator discrimination does require appropriate consideration of pollinator sensory capabilities, there is sufficient evidence that it does occur to allow us to predict its occurrence between many other floral morphs.

It is a great deal harder to demonstrate selective pressure than it is to demonstrate pollinator discrimination, so many authors assume one follows from the other. It is likely that this is often, although not always, the case. Studies such as those of Candace Galen with the alpine skypilot have demonstrated changes in floral morphology over generations to be consistent with selective pressure imposed by pollinators, making it likely that such selective pressure does occur in other systems. What is not yet known is how significant different floral traits are for different pollinators. For example, it is possible that traits that directly affect reward availability are under stronger selective pressure than those that affect the colour of the flower.

3. That plants can respond to that selective pressure, having the necessary genetic variability with respect to key traits.

In several earlier chapters we discussed the genetic control of floral size, shape, symmetry, colour, and pattern. It is clear that many of these traits are under the control of multiple loci, often acting in a coordinated fashion as a result of the activity of transcription factors. Mutations to the coding or regulatory regions of these transcription factors, or to the regulatory regions of the genes they interact with, may result in morphological changes on which selective pressure can act. There is certainly no shortage of genetic material with the potential to be variable, and most of the traits under discussion are quantitative (such as spur length, pigment content, or petal size) and should therefore be developmentally labile. Indeed, much evidence exists from a number of species that variation within floral traits that influence pollinators is heritable, although it may also have an environmental component (Mitchell and Shaw 1993; Campbell 1996; Galen 1996a, b).

4. That plants are not penalized in other ways for responding to that selective pressure.

In Chapter 18 we discussed the possibility that traits that enhance attractiveness to pollinators might also have detrimental consequences. For example, Galen and Cuba (2001) showed that increased corolla flare in the alpine skypilot made the flowers more attractive to bumblebee pollinators but also more vulnerable to nectar-robbing ants which damaged carpels and reduced female fertility. Similarly, changes to pigmentation might be expected to affect floral temperature, heat load, and ultraviolet damage to tissues. The most obvious cost of an increase in floral size, pigmentation, or reward production is that of increased use of resources. Therefore, we should not expect that all floral traits will respond to pollinator-imposed

selection, as some will be constrained by other parameters. A good deal of further research is needed to establish how interactions with the abiotic environment, and with predators, constrain floral responses to pollinator-imposed selection.

5. That this sequence of events has occurred in a majority, or even a substantial majority, of the plant–pollinator interactions we see today.

Assumption 5 cannot be tested experimentally, as it considers events which have occurred during evolutionary time. At best we can hope to extrapolate from experimental evidence obtained on extant plants. Since we know that this sequence of assumptions can be true for some plants today, we can certainly assume that it was true for some plants in the past. However, we are again faced with the difficulty of making any quantitative statement.

6. That this has resulted in a degree of specialization in most plant–pollinator interactions.

This final assumption must be true for the organizing principle of the pollination syndrome to be useful. If it is not true, then pollination syndromes represent tidy ideas that are not reflected by what we see in nature, and should be treated with a great deal of caution. While the statement that a flower exhibits a particular pollination syndrome does not imply that it is only visited by a single species, it does imply that a substantial proportion of its pollination is achieved by a single functional (usually taxonomic) group of animals. Hence, a pollination syndrome implies some degree of specialization and is inconsistent with the idea of flowers as general advertisers of rewards to all passing animals. The assumption that plant–pollinator interactions are specialized has pervaded the literature for many years, but recently more empirical approaches have been taken to test this assumption. These approaches have been based on the assessment of number of pollinating species per plant species, and number of plant species visited per pollinator, and they have found that generalization is much more common than expected. These approaches are reviewed extensively elsewhere (see Waser *et al.*

1996, Waser and Ollerton 2006). Waser *et al.* (1996) concluded that moderate generalization was the rule in most published plant–pollinator systems, with the variation in pollinator availability (and, less commonly, efficiency) limiting the extent to which plants could specialize on single pollinators. Similarly, they pointed out that the limited flowering seasons of many plant species, and the variability in numbers of short-lived plants of any one species in each habitat each year, would limit the extent to which pollinators could specialize on single plant species. In all the studies Waser *et al.* cited, the great majority of plant species were visited by (and, where this was tested, actually pollinated by) more than one animal species. These species were often of multiple genera or even orders.

In contrast, Fenster *et al.* (2004) argued that specialization and pollination syndromes should not be thought of as acting at the level of species, but at the level of 'functional groups'. They argued that the precise species which pollinated a flower was of no relevance, but that what mattered in terms of floral evolution was the selective pressure that this animal would impose, which would depend on its general morphological and behavioural characters. They reassessed some of the literature that Waser *et al.* (1996) had reviewed, but instead of counting number of animal species visiting a plant, they counted the number of functional groups (such as long-tongued bees, or pollen-gathering bees) recorded. They concluded that around 75% of flowering plant species for which data were available were specialized on a single functional group, even if that group contained multiple species. For example, *Silene vulgaris* is pollinated by 26 species, but all are nocturnal moths, acting as a single functional group (Fenster *et al.* 2004). We do not currently know to what extent behavioural patterns and discriminatory abilities vary between species within a single pollinator functional group, making it difficult to assess how valid such groups are as units of specialization.

Specialization between plants and pollinators is frequently asymmetric, adding further complexity to the identification of pollination syndromes. Specialist flowers pollinated by only a small number of animals are often using animals which are themselves great generalists, foraging from very many different plant species. Similarly, specialist pollinators which forage on a limited number of plant species are often foraging from plants that experience pollination service from a great many other animal species. This asymmetry in pollination biology can provide increased stability, increasing the survival chances of rare or highly specialist species (Bascompte *et al.* 2003).

This final assumption, then, is either true or false depending on how specialization is defined. If it is true that most flowers are specialized at the level of functional groups, then the pollination syndrome concept has the potential to be a useful organizing principle for pollination ecology. We will next consider some of the published evidence that plants and pollinators are linked in pollination syndromes, and some of the published examples where they are not.

19.3 Evidence for pollination syndromes

Evidence that the pollination syndromes are useful tools comes from studies in which floral morphology is described, pollinator predicted from floral morphology, and then the plant observed in its native habitat to see which animals do provide pollination service. Very few such studies have been reported, and almost all date to the past 20 years. It can be difficult to perform these studies well, because they should ideally measure pollination, not visitation. Many animals may visit a flower without providing much, if any, pollination service. However, measuring pollination efficiency is difficult in a wild situation. It is also important to note the full range of visitors to the flower, not just the most common or 'expected' ones. A number of such studies have provided support for the pollination syndrome concept.

Wilson *et al.* (2004) described the morphology of and visitors to 49 different species of *Penstemon*, *Keckiella*, and *Nothochelone*, which are collectively thought of as garden penstemons (Figs. 19.1a and b). Animals were recorded as pollinating if they visited a flower and left with pollen on their body. They found that flower colour segregated very neatly with pollinator—bird-pollinated flowers were shades of red or pink while bee-pollinated flowers were purple, blue, yellow, or white. They

(a) (b)

Figure 19.1 Plants for which pollination syndromes hold true. (a) *Penstemon centranthifolius* (hummingbird-pollinated). (b) *Penstemon heterophyllus* (bee-pollinated). Images (a) and (b) kindly supplied by Scott Armbruster (University of Portsmouth). See Plate 25.

also found that bee-pollinated flowers differed in floral morphology according to the size of their pollinator. Large open corollas were pollinated by large bees, while flowers with long thin corollas were pollinated by smaller bees. This distinction was less clear between bird- and bee-pollinated flowers, with some intermediate forms present. Overall, they concluded that it was possible to correctly predict the main pollinator of a penstemon on the basis of its floral morphology, and that the pollination syndromes are useful tools in this system.

Machado and Lopes (2004) conducted a similar study of unrelated plants in a particular tropical dry forest habitat in Brazil. They recorded floral morphology and floral visitors for 99 species, and were able to identify trends along the lines of the pollination syndromes. For example, hummingbird-pollinated flowers were usually red, most bat-pollinated flowers were white, and actinomorphy was common in flowers pollinated by small insects. Although these data are less convincing than the multivariate analysis of Wilson *et al.* (2004), it is certainly reassuring to be able to trace patterns which reflect the expected syndromes.

The true test of pollinator efficiency is reproductive success, and Hargreaves *et al.* (2004) were able to show that seed set was very strongly dependent on visitation by the predicted pollinator, a sunbird, in the African sugarbush, *Protea roupelliae*. The flowers have dark pink bracts, no scent, but large quantities of nectar. They are visited by a range of nectarivorous birds, but also by a large number of different insect species. Their floral morphology would suggest that their main pollinator is a bird, but it is possible that any of these visitors is the actual pollinator. Hargreaves *et al.* (2004) excluded birds but not insects from inflorescences, using large mesh cages. They found that 75% of caged inflorescences set less than 5% of the maximum possible seed, while less than a third of uncaged inflorescences performed as badly, confirming that birds were the major pollinators of these plants.

A different approach to investigate pollination syndromes is to ask which animals pollinate a number of plant species with very similar flowers. The pollination syndrome concept predicts the pollinator would be the same in such cases. Pauw (2006) observed pollinator visits to six species of South African orchids with very similar flowers. All the flowers studied were greenish-yellow, with shallow corollas, had a pungent scent, and secreted oil. The only animal found to visit these flowers was the female of the oil-collecting bee *Redivivia peringueyi*, and analysis of captured animals confirmed that they carried pollinia from the orchids on their bodies. This is a relatively specialized pollination syndrome, and it is in these specialized cases that the concept is likely to be of most value in predicting pollinator species.

19.4 Evidence against pollination syndromes

Similar sorts of studies have also found against the pollination syndrome concept. By investigating the reproductive behaviour of species which supposedly show a particular pollination syndrome, some authors have found that the real pollinator is not as expected. Similarly, some studies of groups of plant species have failed to find correlations between floral morphology and pollinator. These studies are usually very similar in design to those discussed earlier, suggesting that their failure to match flowers to pollination syndromes reflects the inconsistency with which the concept is useful, rather than flaws in any particular study.

Hingston and McQuillan (2000) identified the animals visiting 114 native Tasmanian plant species, and found that pollination syndromes did not predict floral visitors. Although they did not record pollination, only visitation, their results were still rather discouraging for the pollination syndrome concept. Some flowers appeared to have many characters associated with a particular syndrome, but were not visited by the appropriate animals. Almost all flowers had visitors from more than one order of animals, and flowers of the same species had different visitors in different habitats. The authors therefore concluded that pollination syndromes were unreliable indicators of pollinating species, at least in the Tasmanian flora.

One of the most damaging studies on the pollination syndrome concept was published by Carlos Herrera (1993), who studied the floral morphology and female reproductive success of a Spanish endemic violet, *Viola cazorlensis*, over 5 years

(a)

(b)

Figure 19.2 Plants for which pollination syndromes do not hold true. (a) *Viola cazorlensis*. Image kindly supplied by Carlos Herrera (Seville). (b) *Microloma sagittatum* with its sunbird pollinator. Image kindly supplied by Anton Pauw (Stellenbosch). See Plate 26.

(Fig. 19.2a). This violet is pollinated by a single species, the diurnal hawkmoth, *Macroglossum stellatarum*. The flowers are characterized by a particularly long thin nectar spur (averaging around

25 mm in length), which had been supposed to be part of the plant's pollination syndrome, fitting the long proboscis of the moth. However, Herrera showed that spur length had no effect at all on female reproductive success. While most other variable flower morphological traits studied did show some correlation with seed set, spur length did not show any correlation at all, indicating that it is not currently under selective pressure by the hawk-moth and is in no sense part of the current pollination syndrome of this plant. Even more strikingly, Herrera showed that fruit production, flower production, plant size, and herbivory all affected seed set to a greater degree than all accumulated differences in floral morphology. He therefore concluded that selection on floral morphology may be virtually irrelevant in comparison to other ecological factors which determine plant fitness.

Analysis of plant reproductive behaviour has undermined even some of the most specialized of pollination syndromes. Zhang *et al.* (2005) showed that the apparently sapromyiophilous inflorescences of *Tacca chantrieri* were predominantly self-pollinated in China, rather than pollinated by flies. The inflorescences contain dark purple bracts and flowers which can be nearly black. However, the anthers were found to dehisce before the flowers opened, depositing pollen very effectively on the adjacent stigmas. Selfing rates were estimated using molecular markers (isozymes) and found to vary from 76 to 94%. Therefore, although this species has a floral morphology strongly suggestive of one particular pollination syndrome, in its native habitat it is primarily self-pollinated. A similarly surprising result was reported by Pauw (1998), who discovered that the supposedly insect-pollinated milkweed, *Microloma sagittatum*, was in fact pollinated by sunbirds (Fig. 19.2b). The plant produces pollen in pollinaria, which had previously only been thought to attach to insect bodies. However, observation in the native habitat revealed that sunbirds collect nectar from the flower, and in the process the pollinaria attach to their tongues, from where they are transferred to the next flower. The pollinaria attach to the tongue using a microscopic groove which clips to the edge of the tongue. Pauw (1998) also showed that caged plants, visited by a single sunbird but no insects, were successfully

pollinated. This is another striking example in which the floral morphology would have predicted an entirely different pollinator from the one observed (Ollerton 1998).

A recent and extremely wide-ranging study has concluded that pollination syndromes do not allow us to correctly predict pollinators for the majority of plants. Ollerton *et al.* (2007b) conducted a multivariate analysis of the floral traits associated with traditional pollination syndromes, generating three-dimensional spaces associated with each syndrome. They then recorded the same floral traits as present or absent from flowers from six different communities on different continents. Their results showed that the majority of plant species did not produce flowers which fit within one of the spaces associated with the traditional pollination syndromes, but that flowers were generally more variable. Finally, they recorded the most frequent effective pollinator of each of the plants under study, using pollen removal and contact between animal body and floral reproductive organs to ensure only effective pollinators were recorded. On average, in fewer than 25% of species was this pollinator the one that the flower's appearance would have led them to predict, basing their predictions on the pollinator of the traditional syndrome nearest in space to the flower's actual morphology. The authors concluded that the traditional pollination syndromes largely failed to describe the real variation in floral morphology and also failed to identify the true associations between plants and their pollinators.

19.5 The most effective pollinator?

The literature tells us that the assumptions which underlie the pollination syndrome concept are sometimes true, and that authors in some cases have found evidence for the existence of pollination syndromes—that is, that plants are sometimes pollinated by the animals that their floral morphology would lead us to predict. However, there are also many recorded cases where pollinator identity does not match floral morphology or where floral morphology cannot be assigned to a specific traditional syndrome, and the recent data of Ollerton *et al.* (2007b) indicate that this may be the more frequent situation globally. The arguments about the relative importance of generalization and specialization in pollination ecology are likely to continue for some time, but their resolution may depend on a change in terminology so that flowers are described with reference to the full range of their pollinators, rather than to one 'ideal' pollinator. Indeed, many authors have pointed out the importance of considering the effects of multiple pollinators on a species. Flowers have often been described as having one 'major' pollinator and one or more 'minor' pollinators. However, this terminology is insufficiently precise, and other authors have proposed the term 'most effective pollinator'. Stebbins (1970) proposed that pollination ecologists should focus on those pollinators that visit a flower 'most frequently and effectively', as they would be the likeliest to impose selective pressure. Mayfield *et al.* (2001) reported that these two elements were not necessarily linked. They showed that the most frequent visitors to flowers of *Ipomopsis aggregata* were hummingbirds, but that bumblebees were more effective at both pollen removal and ensuring seed production. A single bumblebee visit resulted in the removal of an average of 112.7 pollen grains, and the production of 4.36 seeds, while a single hummingbird visited resulted in the removal of 38.1 pollen grains and the production of 1.08 seeds. However, bumblebees visited wild flowers only one-third to one-quarter as frequently as hummingbirds did. The flowers have the morphology supposedly typical of bird-pollinated flowers, being red with long corollas. However, it is known that bumblebees can see these red flowers, and visit them particularly when hummingbirds are rare and the nectar available is abundant. It may be that in years where few hummingbirds visit a population, bumblebees can learn to associate the colour and shape of *I. aggregata* flowers with a rich reward. This would imply that the plant is under selective pressure in two different directions. Its most frequent pollinator, the hummingbird, may be expected to impose one set of pressures while its most efficient pollinator, the bumblebee, will impose another. The floral form seen in the wild might be thought of as a compromise, and one which must ensure it does not exclude either pollinator. It would be interesting to observe floral morphology in a population

pollinated by only one of these two animals over several generations. More importantly, it would be extremely useful to have similar data for other species. If it is often the case that one pollinator is more efficient but another is more abundant, it is hardly surprising that many flowers fail to match the expectations of traditional pollination syndromes. Indeed, it may be that in some of those studies that recorded the most frequent pollinator as not that expected by the flower's morphology, a less frequent pollinator was more efficient, and the flower was actually responding to selective pressure imposed by the more effective animal.

Aigner (2004, 2006) pointed out that it is not only pollination frequency and pollinator effectiveness which determine what selective pressure is imposed on a flower, but also whether trade-offs occur between fitness for pollination by different animals. A situation where two different pollinators are applying opposing selective pressures and increased fitness for one reduces the effectiveness of the other may, in fact, be quite rare. Perhaps more common would be a situation in which one pollinator applies pressure in one direction, but this has no consequences for other pollinators. In Aigner's own work with species of *Dudleya*, hummingbirds were shown to remove dye (as a marker for pollen) more effectively and deposit pollen more successfully in flowers with corollas which had been artificially narrowed. However, although bumblebees, the other major pollinator of these plants, were more effective than hummingbirds at pollinating wider flowers, they were not themselves more effective with flowers of any particular width. In this case hummingbirds might be expected to apply selective pressure in favour of reduced corolla flare, but such a reduction in corolla flare has no consequence for the other major pollinators, bumblebees. If the same is true for the *Ipomopsis* example described above, it may be that the flowers have the appearance of hummingbird-pollinated flowers simply because the traits favoured by the hummingbirds are selectively neutral with regard to the bumblebees. Indeed, it is not difficult to imagine situations in which a pollinator which is very clearly neither the most common nor the most effective applies selective pressure which does result in a specialized floral morphology, simply because the trait under selection does not affect pollination by the major pollinators. Such examples would certainly cause deviations from the classical pollination syndromes.

This chapter has considered the evidence that pollination syndromes are real—that is, that flowers are pollinated by the animal that their morphology would lead us to expect. It is difficult to draw firm quantitative conclusions, because the literature is patchy and the quality of the observations varies greatly. However, it is clear that in some cases the pollination syndrome is a useful concept, particularly if applied with caution. It is also clear that much more work remains to be done on the frequency of pollination by different animals, the effectiveness of that pollination, and the selective pressures imposed by each animal. Only then can we hope to fully describe floral morphology with respect to pollinating animals, and only then can we really link the interactions of plants with their pollinators to the genes controlling flower development and floral form.

Epilogue

This book was conceived with a single clear aim: to link what we know of the molecular and genetic control of how flowers come to look as they do with what we know from evolutionary and ecological perspectives about why they look as they do. My impetus was a firm belief that the traditional analysis of plant–pollinator interactions would be enhanced by a better understanding of the molecular processes underpinning floral morphology, and that the traditional molecular and developmental analysis of floral morphology would also benefit from an appreciation of the consequences of that morphology for a real plant in its natural habitat. Nineteen chapters later I am no less convinced of the need for an integrative approach, but I have discovered that to take such an approach is not simple when so much ambiguity remains in the various specialist fields of floral biology. In particular, the current debates about the utility of the pollination syndrome concept, and the relative frequencies and symmetry of specialization and generalization in interactions between plants and pollinators, make it far from easy for a molecular developmentalist to contextualize their own research. It becomes tempting in these circumstances to oversimplify to a degree that will only antagonize the true pollination biologist. I hope that, if nothing else, this book will serve to show each sort of specialist the complexity of the rest of the field, and perhaps encourage collaboration and discussion as the way forward.

Despite the difficulties in presenting an integrated picture of floral biology, some major themes have come to the fore. In particular, flexibility seems to be as significant in plant development as it is in plant breeding system. Despite the apparent complexity of the developmental pathways leading to floral induction, floral organ development, and final flower morphogenesis, plant development nevertheless has the flexibility to respond to environmental inputs. The characteristic flexibility of plant development is usually interpreted as an adaptation to survival as an immobile organism in a world populated by mobile animals (although it could equally be argued that the mobility of animals is an adaptation to survival in the absence of flexible development!). In the same way, most plants exhibit great flexibility of breeding system, tolerating self-fertilization under marginal conditions, varying their reproductive output as males or females in response to environmental pressures, and recruiting different pollen vectors in different habitats. This flexibility can also be interpreted as an adaptation to a sessile life cycle in a changeable world populated by mobile animals.

In the same way that developmental flexibility is mirrored by breeding system flexibility, the presence of redundant mechanisms operating as fail-safes in development is mirrored by the likely general nature of plant–pollinator interactions. Almost uniquely among the eukaryotes, plant development is characterized by an enormous degree of redundancy. The proportion of genes in the Arabidopsis genome that encode transcriptional regulators is several times the proportion of similar genes found in the Drosophila genome. These transcriptional regulators act in overlapping cascades to regulate the expression of redundant structural genes found in large multigene families. The resulting redundancy acts as a safety net, ensuring successful plant development even in the absence of one or several component factors. Redundancy may be a necessary adaptation of an organism continually exposed to damaging ultraviolet light, ensuring survival and reproduction even when cells and the genes they contain are damaged by daily exposure to sunlight. In the same way, the generalist nature of plant–pollinator inter-actions also provides safety nets, ensuring plant reproduction in the absence of the usual or most effective pollinator. This generalism may be at the level of functional groups, with multiple species of animal functioning in the same way with regard to pollen deposition and removal, or it may be at a higher level, with several clearly distinct types of animal all attracted to subsets of the same set of flo-ral characters. The ability to self-fertilize, or repro-duce asexually through vegetative organs, provides the ultimate safety net in plant reproduction.

The importance of both flexibility and fail-safe mechanisms in floral development and pollination underlines the vital nature of both processes to plant reproductive success. The angiosperms are the most speciose plant group by some distance,

and they also represent the bulk of terrestrial bio-mass (both plant and animal). They have colonized more habitats and occupied more niches than any other plant group. And this success can be traced to the reproductive innovations made by the first angiosperms—the development of flowers. Flowers, and the flexibility and redundancy inbuilt in their development and function, are responsible for the astonishing radiation of the angiosperms and their continued success today. To fully under-stand that success we will need to continue our exploration of both flower development and polli-nation biology in an open and integrative spirit, encouraging the disparate fields of floral biology to cross-fertilize each other with ideas just as the angiosperms encourage cross-fertilization through flowers themselves.

References

Aarts, M., R. Hodge, K. Kalantidis, D. Florack, Z. Wilson, B. Mulligan, W. Stiekema, R. Scott, and A. Pereira. (1997). The Arabidopsis *MALE STERILITY 2* protein shares similarity with reductases in elongation/condensation complexes. *Plant Journal* **12**:615–623.

Abbott, R. J., P. A. Ashton, and D. G. Forbes. (1992). Introgressive origin of the radiate groundsel, *Senecio vulgaris* L. var. *hibernicus* Syme: *Aat3* evidence. *Heredity* **60**:295–299.

Abe, M., Y. Kobayashi, S. Yamamoto, Y. Daimon, A. Yamaguchi, Y. Ikeda, H. Ichinoki, A. Notaguchi, K. Goto, and T. Araki. (2005). FD, a bZIP protein mediating signals from the floral pathway integrator FT at the shoot apex. *Science* **309**:1052–1056.

Achard, P., H. Cheng, L. de Grauwe, J. Decat, H. Schoutteten, T. Moritz, D. V. D. Straeten, J. Peng, and N. P. Harberd. (2006). Integration of plant responses to environmentally activated phytohormonal signals. *Science* **311**:91–94.

Ache, B., and J. Young. (2005). Olfaction: diverse species, conserved principles. *Neuron* **48**:417–430.

Ackerman, J., and A. Montalvo. (1990). Short- and long-term limitations to fruit production in a tropical orchid. *Ecology* **71**:263–272.

Ahmad, M., and A. R. Cashmore. (1993). *HY4* gene of *A. thaliana* encodes a protein with characteristics of a blue-light photoreceptor. *Nature* **366**:162–166.

Ahn, J. H., D. Miller, V. J. Winter, M. J. Banfield, J. H. Lee, S. Y. Yoo, S. R. Henz, R. L. Brady, and D. Weigel. (2006). A divergent external loop confers antagonistic activity on floral regulators FT and TFL1. *EMBO Journal* **25**:605–614.

Aigner, P. (2004). Floral specialisation without trade-offs: optimal corolla flare in contrasting pollination environments. *Ecology* **85**:2560–2569.

Aigner, P. (2006). The evolution of specialised floral phenotypes in a fine-grained pollination environment. In N. Waser and J. Ollerton (eds), *Plant–Pollinator Interactions*, pp. 23–46. University of Chicago Press, London.

Ainsworth, C. (2000). Boys and girls come out to play: the molecular biology of dioecious plants. *Annals of Botany* **86**:211–221.

Ainsworth, C., S. Crossley, V. Buchanan-Wollaston, M. Thangavelu, and J. Parker. (1995). Male and female flowers of the dioecious plant sorrel show different patterns of MADS box gene expression. *Plant Cell* **7**:1583–1598.

Aizen, M., and D. Vázquez. (2006). Flowering phonologies of hummingbird plants from the temperate forest of southern South America: is there evidence of competitive displacement? *Ecogeography* **29**:357–366.

Alabadi, D., T. Oyama, M. Yanovsky, F. Harmon, P. Mas, and S. Kay. (2001). Reciprocal regulation between *TOC1* and *LHY/CCA1* within the *Arabidopsis* circadian clock. *Science* **293**:880–883.

Al-Babili, S., J. Lintig, H. Haubruck, and P. Beyer. (1996). A novel, soluble form of phytoene desaturase from *Narcissus pseudonarcissus* chromoplasts is Hsp70-complexed and competent for flavinylation, membrane association and enzymatic activation. *Plant Journal* **9**:601–612.

Alfenito, M., E. Souer, C. Goodman, R. Buell, J. Mol, R. Koes, and V. Walbot. (1998). Functional complementation of anthocyanin sequestration in the vacuole by widely divergent glutathione S-transferases. *Plant Cell* **10**:1135–1149.

Almeida, J., R. Carpenter, T. P. Robbins, C. Martin, and E. S. Coen. (1989). Genetic interactions underlying flower colour patterns in *Antirrhinum majus*. *Genes and Development* **3**:1758–1767.

Andersen, C. H., C. S. Jensen, and K. Petersen. (2004). Similar genetic switch systems might integrate the floral inductive pathways in dicots and monocots. *Trends in Plant Science* **9**:105–107.

Anderson, M., E. Cornish, S. Mau, E. Williams, R. Hoggart, A. Atkinson, I. Boenig, B. Grego, R. Simpson, P. Roche, J. Haley, J. Penschow, H. Niall, G. Treager, J. Coughlan, R. Crawford, and A. Clarke. (1986). Cloning of cDNA for a stylar glycoprotein associated with expression of self-incompatibility in *Nicotiana alata*. *Nature* **321**:38–44.

Arber, A. (1937). The interpretation of the flower: a study of some aspects of morphological thought. *Biological Reviews* **12**:157–184.

Arber, A. (1946). Goethe's botany. *Chronica Botanica* **10**:63–126.

Armbruster, W. S., M. Edwards, and E. Debevec. (1994). Floral character displacement generates an assemblage structure of Western Australian triggerplants (*Stylidium*). *Ecology* **75**:315–329.

Ashman, T., T. Knight, J. Steets, P. Amarasekare, M. Burd, D. Campbell, M. Dudash, M. Johnston, S. Mazer, R. Mitchell, M. Morgan, and W. Wilson. (2004). Pollen limitation of plant reproduction: ecological and evolutionary causes and consequences. *Ecology* **85**:2408–2421.

Ashman, T., M. Bradburn, D. Cole, B. Blaney, and R. Raguso. (2005). The scent of a male: the role of floral volatiles in pollination of a gender dimorphic plant. *Ecology* **86**:2099–2105.

Aubert, D., L. Chen, Y. Moon, D. Martin, L. Castle, C. Yang, and Z. R. Sung. (2001). EMF1, a novel protein involved in the control of shoot architecture and flowering in Arabidopsis. *Plant Cell* **13**:1865–1875.

Auckerman, M. J., I. Lee, D. Weigel, and R. M. Amasino. (1999). The Arabidopsis flowering-time gene *LUMINIDEPENDENS* is expressed primarily in regions of cell proliferation and encodes a nuclear protein that regulates *LEAFY* expression. *Plant Journal* **18**:195–203.

Ausin, I., C. Alonso-Blanco, J. Jarillo, L. Ruiz-Garcia, and J. Martinez-Zapater. (2004). Regulation of flowering time by FVE, a retinoblastoma-associated protein. *Nature Genetics* **36**:162–166.

Ayasse, M., F. P. Schiestl, H. F. Paulus, F. Ibarra, and W. Francke. (2003). Pollinator attraction in a sexually deceptive orchid by means of unconventional chemicals. *Proceedings of the Royal Society B: Biological Sciences* **270**:517–522.

Barak, S., E. Tobin, C. Andronis, S. Sugano, and R. Green. (2000). All in good time: the *Arabidopsis* circadian clock. *Trends in Plant Science* **5**:517–522.

Barrett, P. M., and K. J. Willis. (2001). Did dinosaurs invent flowers? Dinosaur-angiosperm coevolution revisited. *Biological Reviews of the Cambridge Philosophical Society* **76**:411–447.

Barrett, S. C. H. (2002). The evolution of plant sexual diversity. *Nature Reviews Genetics* **3**:274–284.

Barrett, S. C. H., L. K. Jesson, and A. M. Baker. (2000). The evolution and function of stylar polymorphisms in flowering plants. *Annals of Botany* **85**:A253–A265.

Bascompte, J., P. Jordano, C. Melian, and J. Olesen. (2003). The nested assembly of plant-animal mutualistic networks. *Proceedings of the National Academy of Sciences, USA* **100**:9383–9387.

Bastow, R., J. S. Mylne, C. Lister, Z. Lippman, R. A. Martienssen, and C. Dean. (2004). Vernalization requires epigenetic silencing of *FLC* by histone methylation. *Nature* **427**:164–167.

Battey, N. H., and F. Tooke. (2002). Molecular control and variation in the floral transition. *Current Opinion in Plant Biology* **5**:62–68.

Baum, D. A. (1998). The evolution of plant development. *Current Opinion in Plant Biology* **1**:79–86.

Becker, A., and G. Theissen. (2003). The major clades of MADS-box genes and their role in the development and evolution of flowering plants. *Molecular Phylogenetics and Evolution* **29**:464–489.

Beld, M., C. Martin, H. Huits, A. Stuitje, and A. G. M. Gerats. (1989). Flavonoid synthesis in *Petunia*; partial characterisation of dihydroflavonol 4-reductase genes. *Plant Molecular Biology* **13**:491–502.

Bell, J., J. Karron, and R. Mitchell. (2005). Interspecific competitionfor pollination lowers seed production and outcrossing in *Mimulus ringens*. *Ecology* **86**:762–771.

Bennett, A., and M. Thery. (2007). Avian colour vision and colouration: multidisciplinary evolutionary biology. *American Naturalist* **169**:S1–S6.

Bernier, G., and C. Perilleux. (2003). A physiological overview of the genetics of flowering time control. *Plant Biotechnology Journal* **3**:3–16.

Bertin, R. (1993). Incidence of monoecy and dichogamy in relation to self-fertilization in angiosperms. *American Journal of Botany* **80**:557–560.

Beveridge, C., and I. Murfet. (1996). The *gigas* mutant in pea is deficient in the floral stimulus. *Physiologia Plantarum* **96**:637–645.

Bey, M., K. Stuber, K. Fellenberg, Z. Schwarz-Sommer, H. Sommer, H. Saedler, and S. Zachgo. (2004). Characterisation of Antirrhinum petal development and identification of target genes of the class B MADS box gene *DEFICIENS*. *Plant Cell* **16**:3197–3215.

Blazquez, M., and D. Weigel. (2000). Integration of floral inductive signals in *Arabidopsis*. *Nature* **404**:889–892.

Blazquez, M., R. Green, O. Nilsson, M. Sussman, and D. Weigel. (1998). Gibberellins promote flowering of Arabidopsis by activating the *LEAFY* promoter. *Plant Cell* **10**:791–800.

Blazquez, M., C. Ferrandiz, F. Madueno, and F. Parcy. (2006). How floral meristems are built. *Plant Molecular Biology* **60**:855–870.

Borner, R., G. Kampnann, J. Chandler, R. Gleissner, E. Wisman, K. Apel, and S. Melzer. (2000). A MADS domain gene involved in the transition to flowering in *Arabidopsis*. *Plant Journal* **24**:591–599.

Bowman, J., J. Alvarez, D. Weigel, E. Meyerowitz, and D. Smyth. (1993). Control of flower development in *Arabidopis thaliana* by *APETALA1* and interacting genes. *Development* **119**:721–743.

Bradley, D., O. Ratcliffe, C. Vincent, R. Carpenter, and E. Coen. (1997). Inflorescence commitment and architecture in *Arabidopsis*. *Science* **275**:80–83.

Bradshaw, H., and D. Schemske. (2003). Allele substitution at a flower colour locus produces a pollinator shift in monkeyflowers. *Nature* **426**:176–178.

Brenner, G. J. (1996). Evidence for the earliest stages of angiosperm pollen evolution: a paleoequatorial section from Israel. In D. W. Taylor and L. J. Hickey (eds) *Flowering Plant Origin, Evolution and Phylogeny*, pp. 91–115. New York: Chapman and Hall.

Breuil-Broyer, S., P. Morel, J. Almeida-Engler, V. Coustham, I. Negrutiu, and C. Trehin. (2004). High-resolution boundary analysis during *Arabidopsis thaliana* flower development. *Plant Journal* **38**:182–192.

Brown, B. J., R. J. Mitchell, and S. A. Graham. (2002). Competition for pollination between an invasive species (purple loosestrife) and a native congener. *Ecology* **83**:2328–2336.

Brugliera, F., G. Barr-Rewell, T. Holton, and J. Mason. (1999). Isolation and characterisation of a flavonoid 3′-hydroxylase cDNA clone corresponding to the *Ht1* locus of *Petunia hybrida*. *Plant Journal* **19**:441–451.

Cabrillac, D., J. Cock, C. Dumas, and T. Gaude. (2001). The S-locus receptor kinase is inhibited by thioredoxins and activated by pollen coat proteins. *Nature* **410**: 220–223.

Campbell, D. (1996). Evolution of floral traits in a hermaphroditic plant: field measurements of heritabilities and genetic correlations. *Evolution* **50**:1442–1453.

Campbell, D., and K. Halama. (1993). Resource and pollen limitations to lifetime seed production in a natural plant population. *Ecology* **74**:1043–1051.

Campbell, D., N. Waser, M. Price, E. Lynch, and R. Mitchell. (1991). Components of phenotypic selection: pollen export and flower corolla width in *Ipomopsis aggregata*. *Evolution* **45**:1458–1467.

Carpenter, R., and E. Coen. (1990). Floral homeotic mutations produced by transposon-mutagenesis in *Antirrhinum majus*. *Genes and Development* **4**:1483–1493.

Cashmore, A., J. Jarillo, Y. Wu, and D. Liu. (1999). Cryptochromes: blue light receptors for plants and animals. *Science* **284**:760–765.

Causier, B., R. Catillo, J. Zhou, R. Ingram, Y. Xue, Z. Schwarz-Sommer, and B. Davies. (2005). Evolution in action: following function in duplicated floral homeotic genes. *Current Biology* **15**:1508–1512.

Chandler, J., J. M. Martinez-Zapater, and C. Dean. (2000). Mutations causing defects in the biosynthesis and response to gibberellins, abscisic acid and phytochrome B do not inhibit vernalization in *Arabidopsis fca-1*. *Planta* **210**:677–682.

Chanvivattana, Y., A. Bishopp, D. Schubert, C. Stock, Y. Moon, Z. R. Sung, and J. Goodrich. (2004). Interaction of Polycomb-group proteins controlling flowering in *Arabidopsis*. *Development* **131**:5263–5276.

Chappell, J. (1995). Biochemistry and molecular biology of the isoprenoid biosynthetic pathway in plants. *Anuual Review of Plant Physiology and Plant Molecular Biology* **46**:521–547.

Charlesworth, D. (2002). Self-incompatibility: how to stay incompatible. *Current Biology* **12**:R424–R426.

Charlesworth, D. (2006). Evolution of plant breeding systems. *Current Biology* **16**:R726–R735.

Chen, D., J. Collins, and T. Goldsmith. (1984). The ultraviolet receptor of bird retinas. *Science* **285**:337–340.

Chittka, L. (1996). Does bee colour vision predate the evolution of flower colour? *Naturwissenschaften* **83**:136–138.

Chittka, L., and N. Raine. (2006). Recognition of flowers by pollinators. *Current Opinion in Plant Biology* **9**:428–435.

Chittka, L., and N. M. Waser. (1997). Why red flowers are not invisible to bees. *Israel Journal of Plant Sciences* **45**:169–183.

Citerne, H., R. T. Pennington, and Q. C. B. Cronk. (2006). An apparent reversal in floral symmetry in the legume *Cadia* is a homeotic transformation. *Proceedings of the National Academy of Sciences, USA* **103**:12017–12020.

Clegg, M., and M. Durbin. (2003). Tracing floral adaptations from ecology to molecules. *Nature Reviews Genetics* **4**:206–215.

Clough, S. J., and A. F. Bent. (1998). Floral dip: a simplified method for Agrobacterium-mediated transformation of *Arabidopsis thaliana*. *Plant Journal* **16**:735–743.

Coe, E. H., S. McCormick, and S. A. Modena. (1981). White pollen in maize. *Journal of Heredity* **72**:318–320.

Coen, E., and E. Meyerowitz. (1991). The war of the whorls: genetic interactions controlling flower development. *Nature* **353**:31–37.

Coen, E. S., R. Carpenter, and C. Martin. (1986). Transposable elements generate novel patterns of gene expression in *Antirrhinum majus*. *Cell* **47**:285–296.

Coen, E., J. M. Romero, S. Doyle, R. Elliott, G. Murphy, and R. Carpenter. (1990). *floricaula*: a homeotic gene required for flower development in Antirrhinum majus. *Cell* **63**:1311–1322.

Colcombet, J., A. Boisson-Cernier, R. Ros-Palau, C. Vera, and J. Schroeder. (2005). Arabidopsis SOMATIC EMBRYOGENESIS RECEPTOR KINASES 1 and 2 are essential for tapetum development and microspore maturation. *Plant Cell* **17**:3350–3361.

Comba, L., S. A. Corbet, H. Hunt, S. Outram, J. S. Parker, and B. J. Glover. (2000). The role of genes influencing the corolla in pollination of *Antirrhinum majus*. *Plant, Cell and Environment* **23**:639–647.

Corbesier, L., and G. Coupland. (2005). Photoperiodic flowering of *Arabidopsis*: integrating genetic and physiological approaches to characterisation of the floral stimulus. *Plant, Cell and Environment* **28**:54–66.

Corbesier, L., Vincent, C., Jang, S., Fornara, F., Fan, Q., Searle, I., Giakountis, A., Farrona, S., Gissot, L., Turnbull, C. and Coupland, G. (2007). FT protein movement contributes to long-distance signaling in floral induction of *Arabidopsis*. *Science* **316**, 1030–1033.

Coupland, G. (1995). Genetic and environmental control of flowering time in *Arabidopsis*. *Trends in Genetics* **11**:393–397.

Corley, S. B., R. Carpenter, L. Copsey, and E. Coen. (2005). Floral asymmetry involves an interplay between TCP and MYB transcription factors in *Antirrhinum*. *Proceedings of the National Academy of Sciences, USA* **102**:5068–5073.

Cox, P. A. (1988). Hydrophilous pollination. *Annual Reviews in Ecology and Systematics* **19**:261–280.

Crane, P. R., E. M. Friis, and K. R. Pedersen. (1995) The origin and early diversification of angiosperms. *Nature* **374**:27–33.

Crawford, B. C. W., U. Nath, R. Carpenter, and E. S. Coen. (2004). *CINCINNATA* controls both cell differentiation and growth in petal lobes and leaves of Antirrhinum. *Plant Physiology* **135**:244–253.

Cronk, Q. C. B. (2001). Plant evolution and development in a post-genomic context. *Nature Reviews Genetics* **2**:607–619.

Cubas, P. (2003). Floral zygomorphy, the recurring evolution of a successful trait. *BioEssays* **26**:1175–1184.

Cubas, P., C. Vincent, and E. S. Coen. (1999) An epigenetic mutation responsible for natural variation in floral symmetry. *Nature* **401**:157–161.

Danyluk, J., N. Kane, G. Breton, A. Limin, D. B. Fowler, and F. Sarhan. (2003). TaVRT-1, a putative transcription factor associated with vegetative to reproductive transition in cereals. *Plant Physiology* **132**:1849–1860.

Darwin, C. (1859). *On the Origin of Species by Means of Natural Selection*. London: Murray.

Davis, S., J. Kurepa, and R. Vierstra. (1999). The *Arabidopsis thaliana HY1* locus, required for phytochrome-chromophore biosynthesis, encodes a protein related to heme oxygenases. *Proceedings of the National Academy of Sciences, USA* **96**:6541–6546.

Dellaporta, S., and A. Calderon-Urrea. (1994). The sex determination process in maize. *Science* **266**:1501–1505.

Deng, X., M. Matsui, N. Wei, D. Wagner, A. Chu, K. Feldmann, and P. Quail. (1992). *COP1*, and Arabidopsis regulatory gene, encodes a protein with both a zinc-binding motif and a G_β homologous domain. *Cell* **71**:791–801.

de Vetten, N., F. Quattrocchio, J. Mol, and R. Koes. (1997). The *an11* locus controlling flower pigmentation in petunia encodes a novel WD-repeat protein conserved in yeast, plants and animals. *Genes and Development* **11**:1422–1434.

de Vlaming, P., A. W. Scram, and H. Wiering. (1983). Genes affecting flower colour and pH of flower limb homogenates in *Petunia hybrida*. *Theoretical and Applied Genetics* **66**:271–278.

Dewitte, W., C. Riou-Khamlichi, S. Scofield, J. M. S. Healy, A. Jacqmard, N. J. Kilby, and J. A. H. Murray. (2003). Altered cell cycle distribution, hyperplasia, and inhibited differentiation in Arabidopsis caused by the D-type cyclin CYCD3. *Plant Cell* **15**:79–92.

Dilcher, D. (2000). Towards a new synthesis: major evolutionary trends in the angiosperm fossil record. *Proceedings of the National Academy of Sciences of the USA* **97**:7030–7036.

Disch, S., E. Anastasiou, V. Sharma, T. Laux, J. Fletcher, and M. Lenhard. (2006). The E3 ubiquitin ligase BIG BROTHER controls Arabidopsis organ size in a dosage-dependent manner. *Current Biology* **16**:272–279.

Di Stilio, V., E. Kramer, and D. Baum. (2005). Floral MADS box genes and homeotic gender dimorphism in *Thalictrum dioicum* (Ranunculaceae)—a new model for the study of dioecy. *Plant Journal* **41**:755–766.

Ditta, G., A. Pinyopich, P. Robles, S. Pelaz, and M. Yanofsky. (2004). The *SEP4* gene of *Arabidopsis thaliana* functions in floral organ and meristem identity. *Current Biology* **14**:1935–1940.

Dobson, H., E. Danielson, and I. van Wesep. (1999). Pollen odor chemicals as modulators of bumble bee foraging on *Rosa rugosa* Thunb. (Rosaceae). *Plant Species Biology* **14**:153–166.

Dooner, H. K. (1983). Coordinate genetic regulation of flavonoid biosynthetic enzymes in maize. *Molecular and General Genetics* **189**:136–141.

Dresselhaus, T. (2006). Cell-cell communication during double fertilization. *Current Opinion in Plant Biology* **9**:41–47.

Dyer, A. G., H. M. Whitney, S. E. J. Arnold, B. J. Glover, and L. Chittka. (2006). Bees associate warmth with floral colour. *Nature* **442**:525.

Dyer, A. G., H. M. Whitney, S. E. J. Arnold, B. J. Glover, and L. Chittka. (2007). Mutations perturbing petal cell shape and anthocyanin synthesis influence bumblebee perception of *Antirrhinum majus* flower colour. *Arthropod Plant Interactions* **1**:45–55.

Ebel, C., L. Mariconti, and W. Gruissem. (2004). Plant retinoblastoma homologues control nuclear proliferation in the female gametophyte. *Nature* **429**:776–780.

Eckardt, N. (2003). A component of the cryptochrome blue light signalling pathway. *Plant Cell* **15**:1051–1052.

Eckhart, V., N. Rushing, G. Hart, and J. Hansen. (2006). Frequency-dependent pollinator foraging in polymorphic *Clarkia xantiana* ssp. *xantiana* populations: implications for flower colour evolution and pollinator interactions. *Oikos* **112**:412–421.

Endress, P. K. (1987). The early evolution of the angiosperm flower. *Trends in Ecology and Evolution* **2**:300–304.

Endress, P. K. (1996). Structure and function of female and bisexual organ complexes in Gnetales. *International Journal of Plant Science* **157**:S113–S125.

Endress, P. K. (2001). Evolution of floral symmetry. *Current Opinion in Plant Biology* **4**:86–91.

Epperson, B., and M. Clegg. (1987). Frequency-dependent variation for outcrossing rate among flower colour morphs of *Ipomoea purpurea*. *Evolution* **41**:1302–1311.

Faegri, K., and L. van der Pijl. (1966). *The Principles of Pollination Ecology*. Pergamon Press, Oxford.

Fankhauser, C., K. Yeh, J. C. Lagarias, H. Zhang, T. Elich, and J. Chory. (1999). PKS1, a substrate phosphorylated by phytochrome that modulates light signalling in *Arabidopsis*. *Science* **284**:1539–1541.

Farzad, M., R. Griesbach, and M. R. Weiss. (2002). Floral colour change in *Viola cornuta* L. (Violaceae): a model system to study regulation of anthocyanin production. *Plant Science* **162**:225–231.

Favaro, R., A. Pinyopich, R. Battaglia, M. Kooiker, L. Borghi, G. Ditta, M. Yanofsky, M. Kater, and L. Colombo. (2003). MADS-box protein complexes control carpel and ovule development in Arabidopsis. *Plant Cell* **15**:2603–2611.

Feinsinger, P. (1987). Effects of plant species on each other's pollination: is community structure influenced? *Trends in Ecology and Evolution* **2**:123–126.

Feng, X., Z. Zhao, Z. Tian, *et al.* (2006). Control of petal shape and floral zygomorphy in *Lotus japonicus*. *Proceedings of the National Academy of Sciences, USA* **103**:4970–4975.

Fenster, C., W. S. Armbruster, P. Wilson, M. Dudash, and J. Thomson. (2004). Pollination syndromes and floral specialization. *Annual Reviews in Ecology, Evolution and Systematics* **35**:375–403.

Figueiredo, P., M. Elhabiri, K. Toki, N. Saito, O. Dangles, and R. Brouillard. (1996). New aspects of anthocyanin complexation. Intramolecular copigmentation as a means for colour loss? *Phytochemistry* **41**:301–308.

Finnegan, E. J., R. K. Genger, K. Kovac, W. J. Peacock, and E. S. Dennis. (1998). DNA methylation and the promotion of flowering by vernalization. *Proceedings of the National Academy of Sciences, USA* **95**:5824–5829.

Finnegan, E. J., K. A. Kovac, E. Jaligot, C. C. Sheldon, W. J. Peacock, and E. S. Dennis. (2005). The downregulation of *FLOWERING LOCUS C (FLC)* expression in plants with low levels of DNA methylation and by vernalization occurs by distinct mechanisms. *Plant Journal* **44**:420–432.

Fleming, T. (2006). Reproductive consequences of early flowering in organ pipe cactus, *Stenocereus thurberi*. *International Journal of Plant Sciences* **167**:473–481.

Foote, H., J. Ride, V. Franklin-Tong, E. Walker, M. Lawrence, and C. Franklin. (1994). Cloning and expression of a distinctive class of self-incompatibility (S) gene from *Papaver rhoeas* L. *Proceedings of the National Academy of Sciences, USA* **91**:2265–2269.

Fowler, S., K. Lee, H. Onouchi, A. Samach, K. Richardson, B. Morris, G. Coupland, and J. Putterill. (1999). *GIGANTEA*: a circadian clock-controlled gene that regulates photoperiodic flowering in *Arabidopsis* and encodes a protein with several possible membrane-spanning domains. *EMBO Journal* **18**:4679–4688.

Franklin-Tong, V., and C. Franklin. (2003). Gametophytic self-incompatibility inhibits pollen tube growth using different mechanisms. *Trends in Plant Science* **8**:598–605.

Franks, R. G., Z. Liu, and R. L. Fischer. (2006). *SEUSS* and *LEUNIG* regulate cell proliferation, vascular development and organ polarity in *Arabidopsis* petals. *Planta* **224**:801–811.

Free, J. (1968). Dandelion as a competitor to fruit trees for bee visits. *Journal of Applied Ecology* **5**:169–178.

Freeman, D. C., E. D. McArthur, K. Harper, and A. Blauer. (1981). Influence of environment on the floral sex ratio of monoecious plants. *Evolution* **35**:194–197.

Friedman, W. E. (2006). Embryological evidence for developmental lability during early angiosperm evolution. *Nature* **441**:337–340

Friis, E. M., K. R. Pedersen, and P. R. Crane. (1999). Early angiosperm diversification: the diversity of pollen associated with angiosperm reproductive structures in Early Cretaceous floras from Portugal. *Annals of the Missouri Botanical Garden* **86**:259–296.

Frohlich, M. W. (2002). The Mostly Male theory of flower origins: summary and update regarding the Jurassic pteridosperm *Pteroma*. In Q. C. B. Cronk, R. M. Bateman, and J. A. Hawkins (eds) *Developmental Genetics and Plant Evolution*, pp. 85–108. Taylor and Francis, London.

Frohlich, M. W., and D. S. Parker. (2000). The mostly male theory of flower evolutionary origins: from genes to fossils. *Systematic Botany* **25**:155–170.

Fukada-Tanaka, S., Y. Inagaki, T. Yamaguchi, N. Saito, and S. Iida. (2000). Colour-enhancing protein in blue petals. *Nature* **407**:581.

Fulton, M., and S. Hodges. (1999). Floral isolation between *Aquilegia formosa* and *Aquilegia pubescens*. *Proceedings of the Royal Society of London Series B* **266**:2247–2252.

Furner, I. J., J. Ainscough, J. Pumfrey, and L. Petty. (1996). Clonal analysis of the late flowering *fca* mutant of *Arabidopsis thaliana*: cell fate and cell autonomy. *Development* **122**:1041–1050.

Furness, C., P. Rudall, and F. B. Sampson. (2002). Evolution of microsporogenesis in angiosperms. *International Journal of Plant Science* **163**:235–260.

Galego, L., and J. Almeida. (2002). Role of DIVARICATA in the control of dorsoventral asymmetry in Antirrhinum flowers. *Genes and Development* **16**:880–891.

Galen, C. (1996a). Rates of floral evolution: adaptation to bumblebee pollination in an alpine wildflower, *Polemonium viscosum*. *Evolution* **50**:120–125.

Galen, C. (1996b). The evolution of floral form: insights from an alpine wildflower, *Polemonium viscosum* (Polemoniaceae). In D. Lloyd and S. Barrett (eds), *Floral Biology*, pp. 273–291. Chapman and Hall, New York.

Galen, C., and J. Cuba (2001). Down the tube: pollinators, predators and the evolution of flower shape in the alpine skypilot, *Polemonium viscosum*. *Evolution* **55**:1963–1971.

Galen, C., and M. Stanton. (1989). Bumble bee pollination and floral morphology: factors influencing pollen dispersal in the alpine sky pilot, *Polemonium viscosum* (Polemoniaceae). *American Journal of Botany* **76**:419–426.

Galizia, C. G., J. Kunze, A. Gumbert, A. K. Borg-Karlson, S. Sachse, C. Markl, and R. Menzel. (2005). Relationship of visual and olfactory signal parameters in a food-deceptive flower mimicry system. *Behavioural Ecology* **16**:159–168.

Garner, W., and H. Allard. (1920). Effect of the relative length of day and night and other factors of the environment on growth and reproduction in plants. *Journal of Agricultural Research* **18**:553–606.

Gendall, A. R., Y. Y. Levy, A. Wilson, and C. Dean. (2001). The *VERNALIZATION 2* gene mediates the epigenetic regulation of vernalization in *Arabidopsis*. *Cell* **107**:525–535.

Ghazoul, J. (2006). Floral diversity and the facilitation of pollination. *Journal of Ecology* **94**:295–304.

Gigord, L., M. Macnair, and A. Smithson. (2001). Negative frequency-dependent selection maintains a dramatic flower colour polymorphism in the rewardless orchid *Dactylorhiza sambucina* (L.) Soo. *Proceedings of the National Academy of Sciences, USA* **98**:6253–6255.

Gillies, A. C. M., P. Cubas, E. S. Coen, and R. J. Abbott. (2002). Making rays in the Asteraceae: genetics and evolution of radiate versus discoid flower heads. In Q. C. B. Cronk, R. M. Bateman, and J. A. Hawkins (eds), *Developmental Genetics and Plant Evolution*, pp. 233–246. Taylor and Francis, London.

Giurfa, M., A. Dafni, and P. Neal. (1999). Floral symmetry and its role in plant-pollinator systems. *International Journal of Plant Sciences* **160**:S41–S50.

Glover, B. J., and C. Martin. (1998). The role of petal cell shape and pigmentation in pollination success in *Antirrhinum majus*. *Heredity* **80**:778–784.

Glover, B. J., M. Perez-Rodriguez, and C. Martin. (1998). Development of several epidermal cell types can be specified by the same MYB-related plant transcription factor. *Development* **125**:3497–3508.

Gocal, G., C. Sheldon, F. Gubler, T. Moritz, D. Bagnall, C. MacMillan, S. Li, R. Parish, E. Dennis, D. Weigel, and R. King. (2001). *GAMYB-like* genes, flowering, and gibberellin signalling in Arabidopsis. *Plant Physiology* **127**:1682–1693.

Goldraij, A., K. Kondo, C. Lee, C. Hancock, M. Sivaguru, S. Vazquez-Santana, S. Kim, T. Phillips, F. Cruz-Garcia, and B. McClure. (2006). Compartmentalization of S-RNase and HT-B degradation in self-incompatible *Nicotiana*. *Nature* **439**:805–810.

Goldsmith, T., and K. Goldsmith. (1979). Discrimination of colours by the black-chinned hummingbird, *Archilochus alexandri*. *Journal of Comparative Physiology A* **130**:209–220.

Golz, J. F., E. J. Keck, and A. Hudson. (2002). Spontaneous mutations in *KNOX* genes give rise to a novel floral structure in Antirrhinum. *Current Biology* **12**:515–522.

Gomez, J., F. Perfectti, and J. Camacho. (2006). Natural selection on *Erysimum mediohispanicum* flower shape: insights into the evolution of zygomorphy. *American Naturalist* **168**:531–545.

Gomi, K., and M. Matsuoka. (2003). Gibberellin signalling pathway. *Current Opinion in Plant Biology* **6**:489–493.

Goodrich, J., R. Carpenter, and E. S. Coen. (1992). A common gene regulates pigmentation pattern in diverse plant species. *Cell* **68**:955–964.

Goodwin, T. W. (1980). *The Biochemistry of the Carotenoids Volume 1*. Chapman and Hall, New York.

Goremykin, V. V., K. I. Hirsch-Ernst, S. Wolfl, and F. H. Hellwig. (2003). Analysis of the *Amborella trichopoda* chloroplast genome sequence suggests that *Amborella* is not a basal angiosperm. *Molecular Biology and Evolution* **20**:1499–1505.

Gorton, H. L., and T. C. Vogelmann. (1996). Effects of epidermal cell shape and pigmentation on optical properties of *Antirrhinum* petals at visible and ultraviolet wavelengths. *Plant Physiology* **112**:879–888.

Goto, K., J. Kyozuka, and J. Bowman. (2001). Turning floral organs into leaves, leaves into floral organs. *Current Opinion in Genetics and Development* **11**:449–456.

Goto, N., T. Kumagai, and M. Koornneef. (1991). Flowering responses to light breaks in photomorphogenic mutants of *Arabidopsis thaliana*, a long-day plant. *Physiologia Plantarum* **83**:209–215.

Gray, J., B. McClure, I. Boenig, M. Anderson, and A. Clarke. (1991). Action of the style product of the self-incompatibility gene of *Nicotiana alata* (S-RNase) on in vitro-grown pollen tubes. *Plant Cell* **3**:271–283.

Green, R., and E. Tobin. (1999). Loss of the circadian clock-associated protein 1 in *Arabidopsis* results in altered clock-regulated gene expression. *Proceedings of the National Academy of Sciences, USA* **96**:4176–4179.

Gross, C. (2005). A comparison of the sexual systems in the trees from the Australian tropics with other tropical biomes—more monoecy but why? *American Journal of Botany* **92**:907–919.

Grotewold, E. (2006). The genetics and biochemistry of floral pigments. *Annual Review of Plant Biology* **57**:761–780.

Gu, T., M. Mazzurco, W. Sulaman, D. Matias, and D. Goring. (1998). Binding of an arm-repeat protein to the kinase domain of the S-locus receptor kinase. *Proceedings of the National Academy of Sciences, USA* **95**:382–387.

Guerrieri, F., M. Schubert, J. Sandoz, and M. Giurfa. (2005). Perceptual and neural olfactory similarity in honeybees. *PLoS Biology* **3**:e60.

Gyula, P., E. Schafer, and F. Nagy. (2003). Light perception and signalling in higher plants. *Current Opinion in Plant Biology* **6**:446–452.

Hallem, E., A. Dahanukar, and J. Carlson. (2006). Insect odor and taste receptors. *Annual Review of Entomology* **51**:113–135.

Hanzawa, Y., T. Money, and D. Bradley. (2005). A single amino acid converts a repressor to an activator of flowering. *Proceedings of the National Academy of Sciences, USA* **102**:7748–7753.

Hargreaves, A., S. Johnson, and E. Nol. (2004). Do floral syndromes predict specialisation in plant pollination systems? An experimental test in an 'ornithophilous' African *Protea*. *Oecologia* **140**:295–301.

Hayama, R., T. Izawa, and K. Shimamoto. (2002). Isolation of rice genes possibly involved in the photoperiodic control of flowering by a fluorescent differential display method. *Plant Cell Physiology* **43**:494–504.

Hayama, R., S. Yokoi, S. Tamaki, M. Yano, and K. Shimamoto. (2003). Adaptation of photoperiodic control pathways produces short-day flowering in rice. *Nature* **422**:719–722.

Hazebroek, J. P., J. D. Metzger, and E. R. Mansager. (1993). Thermoinductive regulation of gibberellin metabolism in *Thlaspi arvensei*. *Plant Physiology* **102**:547–552.

He, Y., and R. M. Amasino. (2005). Role of chromatin modification in flowering-time control. *Trends in Plant Science* **10**:30–35.

He, Y., S. D. Michaels, and R. M. Amasino. (2003). Regulation of flowering time by histone acetylation in *Arabidopsis*. *Science* **302**:1751–1754.

He, Y., M. Doyle, and R. M. Amasino. (2004). PAF1-complex-mediated histone methylation of *FLOWERING LOCUS C* chromatin is required for the vernalization-responsive, winter-annual habit in *Arabidopsis*. *Genes and Development* **18**:2774–2784.

Heidmann, I., N. Efremova, H. Saedler, and Z. Schwarz-Sommer. (1998). A protocol for transformation and regeneration of *Antirrhinum majus*. *Plant Journal* **13**:723–728.

Helliwell, C. A., C. C. Wood, M. Robertson, W. J. Peacock, and E. S. Dennis. (2006). The Arabidopsis FLC protein interacts directly *in vivo* with SOC1 and FT chromatin and is part of a high-molecular-weight protein complex. *Plant Journal* **46**:183–192.

Henderson, I. R., F. Liu, S. Drea, G. G. Simpson, and C. Dean. (2005). An allelic series reveals essential roles for FY in plant development in addition to flowering-time control. *Development* **132**:3597–3607.

Hepworth, S. R., F. Valverde, D. Ravenscroft, A. Mouradov, and G. Coupland. (2002). Antagonistic regulation of flowering-time gene SOC1 by CONSTANS and FLC via separate promoter motifs. *EMBO Journal* **21**:4327–4337.

Hepworth, S. R., J. E. Klenz, and G. W. Haughn. (2006). UFO in the *Arabidopsis* inflorescence apex is required for floral-meristem identity and bract suppression. *Planta* **223**:769–778.

Herrera, C. (1993). Selection on floral morphology and environmental determinants of fecundity in a hawk moth-pollinated violet. *Ecological monographs* **63**:251–275.

Herrera, C. (1996). Floral traits and plant adaptation to insect pollinators: a devil's advocate approach. In D. Lloyd and S. Barrett (eds), *Floral Biology*, pp. 65–87. Chapman and Hall, New York.

Hicks, K., A. Millar, I. Carre, D. Somers, M. Straume, D. Meeks-Wagner, and S. Kay. (1996). Conditional circadian dysfunction of the *Arabidopsis early-flowering 3* mutant. *Science* **274**:790–792.

Hicks, K., T. Albertson, and D. Wagner (2001). *EARLY FLOWERING 3* encodes a novel protein that regulates circadian clock function and flowering in Arabidopsis. *Plant Cell* **13**:1281–1292.

Hileman, L., E. Kramer, and D. Baum (2003). Differential regulation of symmetry genes and the evolution of flower morphologies. *Proceedings of the National Academy of Sciences, USA* **100**:12814–12819.

Hingston, A., and P. McQuillan. (2000). Are pollination syndromes useful predictors of floral visitors in Tasmania? *Austral Ecology* **25**:600–609.

Hirschberg, J. (1999). Production of high-value compounds: carotenoids and vitamin E. *Current Opinion in Biotechnology* **10**:186–191.

Hiscock, S., and S. McInnis. (2003). Pollen recognition and rejection during the sporophytic self-incompatibility response: *Brassica* and beyond. *Trends in Plant Science* **8**:606–613.

Hiscock, S., and D. Tabah. (2003). The different mechanisms of sporophytic self-incompatibility. *Philosophical*

Transactions of the Royal Society of London, Series B **358**:1037–1045.

Hiscock, S., S. McInnis, D. Tabah, C. Henderson, and A. Brennan. (2003). Sporophytic self-incompatibility in *Senecio squalidus* L. (Asteraceae)—the search for S. *Journal of Experimental Botany* **54**:169–174.

Holton, T. A., F. Brugliera, and Y. Tanaka. (1993a). Cloning and expression of flavonol synthase from *Petunia hybrida*. *Plant Journal* **4**:1003–1010.

Holton, T. A., F. Brugliera, D. Lester, Y. Tanaka, C. Hyland, J. Menting, C. Lu, E. Farcy, T. Stevenson, and E. Cornish. (1993b). Cloning and expression of cytochrome P450 genes controlling flower colour. *Nature* **366**:276–279.

Honys, D., and D. Twell. (2003). Comparative analysis of the Arabidopsis pollen transcriptome. *Plant Physiology* **132**:640–652.

Hu, Y., Q. Xie, and N. Chua. (2003). The Arabidopsis auxin-inducible gene *ARGOS* controls lateral organ size. *Plant Cell* **15**:1951–1961.

Hu, Y., H. Poh, and N. Chua. (2006). The Arabidopsis *ARGOS-LIKE* gene regulates cell expansion during organ growth. *Plant Journal* **47**:1–9.

Huang, T., H. Boehlenius, S. Eriksson, F. Parcy, and O. Nilsson. (2005). The mRNA of the *Arabidopsis* gene *FT* moves from leaf to shoot apex and induces flowering. *Science* **309**:1694–1696.

Hughes, N. F. (1994). *The Enigma of Angiosperm Origins*. Cambridge University Press, Cambridge.

Huijser, P., J. Klein, W. Lonnig, H. Meijer, H. Saedler, and H. Sommer. (1992). Bracteomania, an inflorescence anomaly, is caused by the loss of function of the MADS-box gene *squamosa* in *Antirrhinum majus*. *EMBO Journal* **11**:1239–1249.

Hunaca-Mamani, W., M. Garcia-Aguilar, G. Leon-Martinez, U. Grossniklaus, and J. Vielle-Calzada. (2005). CHR11, a chromatin-remodeling factor essential for nuclear proliferation during female gametogenesis in *Arabidopsis thaliana*. *Proceedings of the National Academy of Sciences, USA* **102**:17231–17236.

Huq, E., J. Tepperman, and P. Quail. (2000). GIGANTEA is a nuclear protein involved in phytochrome signalling in Arabidopsis. *Proceedings of the National Academy of Sciences, USA* **97**:9789–9794.

Ingram, G., J. Goodrich, M. Wilkinson, R. Simon, G. Haughn, and E. Coen. (1995). Parallels between *UNUSUAL FLORAL ORGANS* and *FIMBRIATA*, genes controlling flower development in Arabidopsis and Antirrhinum. *Plant Cell* **7**:1501–1510.

Ingram, G., S. Doyle, R. Carpenter, E. Schultz, Simon, and E. Coen. (1997). Dual role for *fimbriata* in regulating floral homeotic genes and cell division in *Antirrhinum*. *EMBO Journal* **16**:6521–6534.

Irish, V., and Y. Yamamoto. (1995). Conservation of floral homeotic gene function between Arabidopsis and Antirrhinum. *Plant Cell* **7**:1635–1644.

Ishikawa, R., S. Tamaki, S. Yokoi, N. Inagaki, T. Shinomura, M. Takano, and K. Shimamoto. (2005). Suppression of the floral activator *Hd3a* is the principle cause of the night break effect in rice. *Plant Cell* **17**:3326–3336.

Ito, T., H. Sakai, and E. Meyerowitz. (2003). Whorl-specific expression of the *SUPERMAN* gene of *Arabidopsis* is mediated by *cis* elements in the transcribed region. *Current Biology* **13**:1524–1530.

Izawa, T., T. Oikawa, S. Tokutomi, K. Okuno, and K. Shimamoto. (2000). Phytochromes confer the photoperiodic control of flowering in rice a short-day plant). *Plant Journal* **22**:391–399.

Izawa, T., T. Oikawa, N. Sugiyama, T. Tanisaka, M. Yano, and Shimamoto. (2002). Phytochrome mediates the external light signal to repress *FT* orthologues in photoperiodic flowering of rice. *Genes and Development* **16**:2006–2020.

Izawa, T., Takahashi, Y. and Yano, M. (2003) Comparative biology comes into bloom: genomic and genetic comparison of flowering pathways in rice and *Arabidopsis*. *Current Opinion in Plant Biology* **6**, 113–120.

Jack, T. (2001). Relearning our ABCs: new twists on an old model. *Trends in Plant Science* **6**:310–316.

Jack, T. (2004). Molecular and genetic mechanisms of floral control. *Plant Cell* **16**:S1–S17.

Jacobsen, S. E., and N. E. Olszewski. (1993). Mutations at the *SPINDLY* locus of Arabidopsis alter gibberellin signal transduction. *Plant Cell* **5**:887–896.

Jacobsen, S., K. Binkowski, and N. Olszewski. (1996). SPINDLY, a tetratricopeptide repeat protein involved in gibberellin signal transduction in *Arabidopsis*. *Proceedings of the National Academy of Sciences, USA* **93**:9292–9296.

Jaeger, K. and Wigge, P. (2007). FT protein acts as a long-range signal in *Arabidopsis*. *Current Biology* **17**, 1050–1054.

Jeon, J., S. Jang, S. Lee, J. Nam, C. Kim, S. Lee, Y. Chung, S. Kim, Y. Lee, Y. Cho, and G. An. (2000). *leafy hull sterile 1* is a homeotic mutation in a rice MADS box gene affecting flower development. *Plant Cell* **12**:871–884.

Jesson, L. K., and S. C. H. Barrett. (2002). Solving the puzzle of mirror-image flowers. *Nature* **417**:707.

Johanson, U., J. West, C. Lister, S. Michaels, R. Amasino, and C. Dean,. (2000). Molecular analysis of *FRIGIDA*, a major determinant of natural variation in *Arabidopsis* flowering time. *Science* **290**:344–347.

Johnson, S. D., and J. J. Midgley. (1997). Fly pollination in *Gorteria diffusa* (Asteraceae) and a possible mimetic function for dark spots on the capitulum. *American Journal of Botany* **84**:429–436.

Jones, A. M., K. H. Im, M. A. Savka, M. J. Wu, N. G. DeWitt, R. Shillito, and A. N. Binns. (1998). Auxin-dependent cell expansion mediated by overexpressed auxin binding protein 1. *Science* **282**:1114–1117.

Jones, K. N., and J. S. Reithel. (2001). Pollinator-mediated selection on a flower colour polymorphism in experimental populations of Antirrhinum (Scrophulariaceae). *American Journal of Botany* **88**:447–454.

Juergens, A., S. Doetterl, and U. Meve. (2006). The chemical nature of fetid floral odors in stapeliads (Apocynaceae-Asclepiadoideae-Ceropegieae). *New Phytologist* **172**:452–468.

Kajiwara, S., P. D. Fraser, K. Kondo, and N. Misawa. (1997). Expression of an exogenous isopentenyl diphosphate isomerase gene enhances isoprenoid biosynthesis in *Escherichia coli*. *Biochemical Journal* **324**:421–426.

Kandasamy, M., D. Paolillo, C. Faraday, J. Nasrallah, and M. Nasrallah. (1989). The S-locus specific glycoproteins of *Brassica* accumulate in the cell wall of developing stigma papillae. *Developmental Biology* **134**:462–472.

Kanno, A., H. Saeki, T. Kameya, H. Saedler, and G. Theissen. (2003). Heterotropic expression of class B floral homeotic genes supports a modified ABC model for tulip (*Tulipa gesneriana*). *Plant Molecular Biology* **52**:831–841.

Kardailsky, I., V. Shukla, J. Ahn, N. Dagenais, S. Christensen, J. Nguyen, J. Chory, M. Harrison, and D. Weigel. (1999). Activation tagging of the floral inducer *FT*. *Science* **286**:1962–1965.

Kay, Q. O. N. (1976). Preferential pollination of yellow-flowered morphs of *Raphanus raphanistrum* by *Pieris* and *Ersistralis* spp. *Nature* **261**:230–232.

Kay, Q. O. N., H. S. Daoud, and C. H. Stirton. (1981). Pigment distribution, light reflection and cell structure in petals. *Botanical Journal of the Linnean Society* **83**:57–84.

Keck, E., P. McSteen, R. Carpenter, and E. Coen. (2003). Separation of genetic functions controlling organ identity in flowers. *EMBO Journal* **22**:1058–1066.

Kempin, S., B. Savidge, and M. Yanovsky. (1995). Molecular basis of the *cauliflower* phenotype in *Arabidopsis*. *Science* **267**:522–525.

Kenrick, P. (1999). The family tree flowers. *Nature* **402**:358–359.

Kevan, P. G., and M. A. Lane. (1985). Flower petal microtexture is a tactile cue for bees. *Proceedings of the National Academy of Sciences of the USA* **82**:4750–4752.

Kevan, P., M. Giurfa, and L. Chittka. (1996). Why are there so many and so few white flowers? *Trends in Plant Science* **1**:280–284.

Kim, S., J. Koh, H. Yoo, H. Kong, Y. Hu, H. Ma, P. Soltis, and D. Soltis. (2005). Expression of floral MADS-box genes in basal angiosperms: implications for the evolution of floral regulators. *Plant Journal* **43**:724–744.

King, J. *Reaching for the Sun.* Cambridge University Press, Cambridge.

Knight, T., J. Steets, J. Vamosi, S. Mazer, M. Burd, D. Campbell, M. Dudash, M. Johnston, R. Mitchell, and T. Ashman. (2005). Pollen limitation of plant reproduction: pattern and process. *Annual Reviews of Ecology, Evolution and Systematics* **36**:467–497.

Kobayashi, Y., H. Kaya, K. Goto, M. Iwabuchi, and T. Araki. (1999) A pair of related genes with antagonistic roles in mediating flowering signals. *Science* **286**:1960–1962.

Koelewijn, H., and J. van Damme. (1996). Gender variation, partial male sterility and labile sex expression in gynodioecious *Plantago coronopus*. *New Phytologist* **132**:67–76.

Koes, R. E., C. E. Spelt, H. J. Reif, P. van den Elzen, E. Veltkamp, and J. N. M. Mol. (1986). Floral tissue of *Petunia hybrida* (V30) expresses only one member of the chalcone synthase multigene family. *Nucleic Acids Research* **14**:5229–5239.

Kohchi, T., K. Mukougawa, N. Frankenberg, M. Masuda, A. Yokota, and J. Lagarias. (2001). The Arabidopsis *HY2* gene encodes phytochromobilin synthase, a ferredoxin-dependent biliverdin reductase. *Plant Cell* **13**:425–436.

Kojima, S., Y. Takahashi, Y. Kobayashi, L. Monna, T. Sasaki, T. Araki, and M. Yano. (2002). *Hd3a*, a rice ortholog of the *Arabidopsis FT* gene, promotes transition to flowering downstream of *Hd1* under short-day conditions. *Plant Cell Physiology* **43**:1096–1105.

Koornneef, M., C. J. Hanhart, and J. H. van der Veen. (1991). A genetic and physiological analysis of late flowering mutants in *Arabidopsis thaliana*. *Molecular and General Genetics* **229**:57–66.

Koornneef, M., C. Alonso-Blanco, H. Blankestijn-DeVries, C. J. Hanhart, and A. J. Peeters. (1998). Genetic interactions among late-flowering mutants of *Arabidopsis*. *Genetics* **148**:885–892.

Kowyama, Y., K. Kakeda, K. Kondo, T. Imada, and T. Hattori. (1996). A putative receptor protein kinase gene in *Ipomoea trifida*. *Plant Cell Physiology* **37**:681–685.

Kramer, E. M., and V. F. Irish. (1999). Evolution of genetic mechanisms controlling petal development. *Nature* **399**:144–148.

Kramer, E. M., M. A. Jaramillo, and V. S. Di Stilio. (2004). Patterns of gene duplication and functional evolution during the diversification of the *AGAMOUS* subfamily of MADS box genes in angiosperms. *Genetics* **166**:1011–1023.

Kramer, E. M., H. Su, C. Wu, and J. Hu. (2006). A simplified explanation for the frameshift mutation that created a novel C-terminal motif in the *APETALA3* gene lineage. *BMC Evolutionary Biology* **6**:30.

Krassilov, V. A., and I. A. A. Dobruskina. (1995). Angiosperm fruit from the Lower Cretaceous of Israel and origins in rift valleys. *Paleontological Journal* 29:110–115.

Krizek, B. A., M. W. Lewis, and J. C. Fletcher. (2006). *RABBIT EARS* is a second-whorl repressor of *AGAMOUS* that maintains spatial boundaries in Arabidopsis flowers. *Plant Journal* 45:369–383.

Kunin, W. (1993). Sex and the single mustard: population density and pollinator behaviour effects on seed set. *Ecology* 74:2145–2160.

Kusaba, M., C. Tung, M. Nasrallah, and J. Nasrallah. (2002). Monoallelic expression and dominance interactions in anthers of self-incompatible *Arabidopss lyrata*. *Plant Physiology* 128:17–20.

Labandeira, C. C. (1997). Insect mouthparts: ascertaining the paleobiology of insect feeding strategies. *Annual Reviews in Ecology and Systematics* 28:153–193.

Lai, Z., W. Ma, B. Han, L. Liang, Y. Zhang, G. Hng, and Y. Xue. (2002). An F-box gene linked to the self-incompatibility (S) locus of *Antirrhinum* is expressed specifically in pollen and tapetum. *Plant Molecular Biology* 50:29–42.

Lariguet, P., I. Schepens, D. Hodgson, U. V. Pedmale, M. Trevisan, C. Kami, M. Carbonnel, J. M. Alonso, J. R. Ecker, E. Liscum, and C. Fankhauser. (2006). PHYTOCHROME KINASE SUBSTRATE 1 is a phototropin 1 binding protein required for phototropism. *Proceedings of the National Academy of Sciences, USA* 103:10134–10139.

Larson, R. L., and E. H. Coe. (1977). Gene-dependent flavonoid glucosyltransferase in maize. *Biochemical Genetics* 15:153–156.

Laux, T., K. Mayer, J. Berger, and G. Jurgens. (1996). The *WUSCHEL* gene is required for shoot and floral meristem integrity in *Arabidopsis*. *Development* 122:87–96.

Le Corre, V., F. Roux, and X. Reboud. (2002). DNA polymorphism at the *FRIGIDA* gene in *Arabidopsis thaliana*: extensive nonsynonymous variation is consistent with local selection for flowering time. *Molecular Biology and Evolution* 19, 1261–1271.

Lee, H., S. Suh, E. Park, E. Cho, J. Ahn, S. Kim, J. Lee, Y. Kwon, and I. Lee. (2000). The AGAMOUS-LIKE 20 MADS domain protein integrates floral inductive pathways in *Arabidopsis*. *Genes and Development* 14:2366–2376.

Lee, H., L. Xiong, Z. Gong, M. Ishitani, B. Stevenson, and J. Zhu. (2001). The *Arabidopsis HOS1* gene negatively regulates cold signal transduction and encodes a RING finger protein that displays cold-regulated nucleocytoplasmic partitioning. *Genes and Development* 15:912–924.

Lee, I., and R. Amasino. (1995). Effect of vernalization, photoperiod, and light quality on the flowering phenotype of Arabidopsis plants containing the *FRIGIDA* gene. *Plant Physiology* 108:157–162.

Lee, I., M. Auckerman, S. Gore, K. Lohman, S. Michaels, L. Weaver, M. John, K. Feldmann, and R. Amasino. (1994). Isolation of *LUMINIDEPENDENS*: a gene involved in the control of flowering time in Arabidopsis. *Plant Cell* 6:75–83.

Lee, S., J. Kim, J. Han, M. Han, and G. An. (2004). Functional analyses of the flowering time gene *OsMADS50*, the putative *SUPRESSOR OF OVEREXPRESSION OF CO 1/AGAMOUS LIKE 20 (SOC1/AGL20)* ortholog in rice. *Plant Journal* 38:754–764.

Leleji, O. (1973). Apparent preference by bees for different flower colours in cowpeas (*Vigna sinensis* (L) Savi ex Hassk. *Euphytica* 22:150–153.

Levin, J., and E. Meyerowitz. (1995). *UFO*: an Arabidopsis gene involved in both floral meristem and floral organ development. *Plant Cell* 7:529–548.

Levy, Y. Y., S. Mesnage, J. S. Mylne, A. R. Gendall, and C. Dean. (2002). Multiple roles of *Arabidopsis VRN1* in vernalization and flowering time control. *Science* 297:243–246.

Lim, M., J. Kim, Y. Kim, K. Chung, Y. Seo, I. Lee, J. Kim, C. Hong, H. Kim, and C. Park. (2004). A new *Arabidopsis* gene, *FLK*, encodes an RNA binding protein with K homology motifs and regulates flowering time via *FLOWERING LOCUS C*. *Plant Cell* 16:731–740.

Lin, C. (2000). Plant blue-light receptors. *Trends in Plant Science* 5:337–342.

Lin, C., H. Yang, H. Guo, T. Mockler, J. Chen, and A. Cashmore. (1998). Enhancement of blue-light sensitivity of Arabidopsis seedlings by a blue light receptor cryptochrome 2. *Proceedings of the National Academy of Sciences, USA* 95:2686–2690.

Liu, J., S. Gilmour, M. Thomashow, and S. van Nocker. (2002). Cold signalling associated with vernalization in *Arabidopsis thaliana* does not involve CBF1 or abscisic acid. *Physiologia Plantarum* 114:125–134.

Lobo, J., M. Quesada, K. Stoner, E. Fuchs, Y. Herrerias-Diego, J. Rojas, and G. Saborio. (2003). Factors effecting phonological patterns of Bombacaceous trees in seasonal forests in Costa Rica and Mexico. *American Journal of Botany* 90:1054–1063.

Lotz, C., C. Martinez del Rio, and S. Nicolson. (2003). Hummingbirds pay a high cost for a warm drink. *Journal of Comparative Physiology B* 173:455–462.

Luo, D., R. Carpenter, C. Vincent, L. Copsey, and E. Coen. (1996). Origin of floral asymmetry in Antirrhinum. *Nature* 383:794–799.

Luu, D., X. Qin, D. Morse, and M. Cappadocia. (2000). S-RNase uptake by compatible pollen tubes in gametophytic self-incompatibility. *Nature* 407:649–651.

Ma, H. (2005). Molecular genetic analyses of microsporogenesis and microgametogenesis in flowering plants. *Annual Reviews of Plant Biology* **56**:393–434.

McClure, B., and V. Franklin-Tong. (2006). Gametophytic self-incompatibility: understanding the cellular mechanisms involved in 'self' pollen tube inhibition. *Planta* **224**:233–245.

McCormick, S. (2004). Control of male gametophyte development. *Plant Cell* **16**:S142–S153.

Machado, I., and A. Lopes. (2004). Floral traits and pollination systems in the Caatinga, a Brazilian tropical dry forest. *Annals of Botany* **94**:365–376.

Macknight, R., I. Bancroft, T. Page, C. Lister, R. Schmidt, K. Love, L. Westphal, G. Murphy, S. Sherson, C. Cobbett, and C. Dean. (1997). *FCA*, a gene controlling flowering time in Arabidopsis, encodes a protein containing RNA-binding domains. *Cell* **89**:737–745.

Macknight, R., M. Duroux, R. Laurie, P. Dijkwel, G. Simpson, and C. Dean. (2002). Functional significance of the alternative transcript processing of the Arabidopsis floral promoter *FCA*. *Plant Cell* **14**:877–888.

McSteen, P., C. Vincent, S. Doyle, R. Carpenter, and E. Coen. (1998). Control of floral homeotic gene expression and organ morphogenesis in *Antirrhinum*. *Development* **125**:2359–2369.

Maizel, A., M. A. Busch, T. Tanahashi, J. Perkovic, M. Kato, M. Hasebe, and D. Weigel. (2005). The floral regulator LEAFY evolves by substitutions in the DNA binding domain. *Science* **308**:260–263.

Makino, S., T. Kiba, A. Imamura, N. Hanaki, A. Nakamura, T. Suzuki, M. Taniguchi, C. Ueguchi, T. Sugiyama, and T. Mizuno. (2000). Genes encoding pseudo-response regulators: insight into his-to-asp phosphorelay and circadian rhythm in *Arabidopsis thaliana*. *Plant Cell Physiology* **41**:791–803.

Malcomber, S., and E. Kellog. (2005). *SEPALLATA* gene diversification: brave new whorls. *Trends in Plant Science* **10**:427–435.

Mandel, M. A., and M. Yanofsky. (1995). A gene triggering flower formation in *Arabidopsis*. *Nature* **377**:522–524.

Mandel, M. A., C. Gustafson-Brown, B. Savidge, and M. Yanofsky. (1992). Molecular characterisation of the *Arabidopsis* floral homeotic gene *APETALA1*. *Nature* **360**:273–277.

Markham, K. R., K. S. Gould, C. S. Winefield, K. A. Mitchell, S. J. Bloor, and M. R. Bloase. (2000). Anthocyanic vacuolar inclusions—their nature and significance in flower colouration. *Phytochemistry* **55**:327–336.

Markovic, J., N. Petranovic, and J. Baranac. (2005). The copigmentation effect of sinapic acid on malvin: a spectroscopic investigation on colour enhancement. *Journal of Photochemistry and Photobiology B* **78**:223–228.

Marrs, K. A., M. R. Alfenito, A. M. Lloyd, and V. Walbot. (1995). A glutathione S-transferase involved in vacuolar transfer encoded by the maize gene *Bronze-2*. *Nature* **375**:397–400.

Martin, C., and T. Gerats. (1993). The control of flower coloration. In B. R. Jordan (ed), *The Molecular Biology of Flowering*, pp. 219–255. CAB International, Wallingford.

Martin, C., R. Carpenter, H. Sommer, H. Saedler, and E. Coen. (1985). Molecular analysis of instability in flower pigmentation of *Antirrhinum majus* following isolation of the *PALLIDA* locus by transposon tagging. *EMBO Journal* **4**:1625–1630.

Martin, C., A. Prescott, S. Mackay, J. Bartlett, and E. Vrijlandt. (1991). The control of anthocyanin biosynthesis in flowers of *Antirrhinum majus*. *Plant Journal* **1**:37–49.

Martin, W., A. Gierl, and H. Saedler. (1989). Molecular evidence for pre-Cretaceous angiosperm origins. *Nature* **339**:46–48.

Martin-Trillo, M., A. Lazaro, R. S. Poethig, C. Gomez-Mena, M. A. Pineiro, J. M. Martinez-Zapater, and J. A. Jarillo. (2006). *EARLY IN SHORT DAYS 1 (ESD1)* encodes ACTIN-RELATED PROTEIN 6 (AtARP6), a putative component of chromatin remodelling complexes that positively regulates *FLC* accumulation in *Arabidopsis*. *Development* **133**:1241–1252.

Martinez-Garcia, J., E. Huq, and P. Quail. (2000). Direct targeting of light signals to a promoter element-bound transcription factor. *Science* **288**:859–863.

Mas, P., F. Devlin, S. Panda, and S. A. Kay. (2000). Functional interaction of phytochrome B and cryptochrome 2. *Nature* **408**:207–211.

Mas, P., W. Kim, D. Somers, and S. Kay. (2003). Targeted degradation of TOC1 by ZTL modulates circadian function in *Arabidopsis thaliana*. *Nature* **426**:567–570.

Massinga, P., S. Johnson, and L. Harder. (2005). Heteromorphic incompatibility and efficiency of pollination in two distylous *Pentanisia* species (Rubiaceae). *Annals of Botany* **95**:389–399.

Mast, A., and E. Conti. (2006). The primrose path to heterostyly. *New Phytologist* **171**:439–442.

Mathieu, J., Warthmann, N., Kuttner, F. and Schmid, M. (2007). Export of FT protein from phloem companion cells is sufficient for floral induction in *Arabidopsis*. *Current Biology* **17**, 1055–1060.

Mayfield, M., N. Waser, and M. Price. (2001). Exploring the 'most effective pollinator principle' with complex flowers: bumblebees and *Ipomopsis aggregata*. *Annals of Botany* **88**:591–596.

Mazzurco, M., W. Sulaman, H. Elina, J. M. Cock, and D. Goring. (2001). Further analysis of the interactions between the *Brassica* S receptor kinase and three interacting proteins (ARC1, THL1 and THL2) in the yeast two-hybrid system. *Plant Molecular Biology* **45**:365–376.

Medrano, M., C. Herrera, and S. Barrett. (2005). Herkogamy and mating patterns in the self-compatible daffodil *Narcissus longispathus*. *Annals of Botany* **95**:1105–1111.

Meyerowitz, E. M. (1996). Plant development: local control, global patterning. *Current Opinion in Genetics and Development* **6**:475479.

Michaels, S. D., and R. M. Amasino. (1999). *FLOWERING LOCUS C* encodes a novel MADS domain protein that acts as a repressor of flowering. *Plant Cell* **11**:949–956.

Michaels, S. D., and R. M. Amasino. (2001). Loss of *FLOWERING LOCUS C* activity eliminates the late-flowering phenotype of *FRIGIDA* and autonomous pathway mutations but not responsiveness to vernalization. *Plant Cell* **13**:935–941.

Michaels, S. D., I. C. Bezerra, and R. M. Amasino. (2004). *FRIGIDA*-related genes are required for the winter-annual habit in *Arabidopsis*. *Proceedings of the National Academy of Sciences of the USA* **101**:3281–3285.

Millar, A. J., I. A. Carre, C. A. Strayer, N. Chua, and S. A. Kay. (1995). Circadian clock mutants in Arabidopsis identified by luciferase imaging. *Science* **267**:1161–1163.

Mitchell, R., and R. Shaw. (1993). Heritability of floral traits for the perennial wild flower *Penstemon centranthifolius* (Scrophulariaceae): clones and crosses. *Heredity* **71**:185–192.

Mizoguchi, T., K. Wheatley, Y. Hanzawa, L. Wright, M. Mizoguchi, H. Song, I. Carre, and G. Coupland. (2002). *LHY* and *CCA1* are partially redundant genes required to maintain circadian rhythms in *Arabidopsis*. *Developmental Cell* **2**:629–641.

Mizoguchi, T., L. Wright, S. Fujiwara, F. Cremer, K. Lee, H. Onouchi, A. Mouradov, S. Fowler, H. Kamada, J. Putterill, and G. Coupland. (2005). Distinct roles of *GIGANTEA* in promoting flowering and regulating circadian rhythms in Arabidopsis. *Plant Cell* **17**:2255–2270.

Mizukami, Y. (2001). A matter of size: Developmental control of organ size in plants. *Current Opinion in Plant Biology* **4**:533–539.

Mizukami, Y., and R. L. Fischer. (2000). Plant organ size control: *AINTEGUMENTA* regulates growth and cell numbers during organogenesis. *Proceedings of the National Academy of Sciences, USA* **97**:942–947.

Mockler, T., H. Yang, X. Yu, D. Parikh, Y. Cheng, S. Dolan, and C. Lin. (2003). Regulation of photoperiodic flowering by *Arabidopsis* photoreceptors. *Proceedings of the National Academy of Sciences, USA* **100**:2140–2145.

Moehs, C. P., L. Tian, K. W. Osteryoung, and D. DellaPenna. (2001). Analysis of carotenoid biosynthetic gene expression during marigold petal development. *Plant Molecular Biology* **45**:281–293.

Moise, A. R., J. von Lintig, and K. Palczewski. (2005). Related enzymes solve evolutionarily recurrent problems in the metabolism of carotenoids. *Trends in Plant Science* **10**:178–186.

Mol, J. N. M., M. P. Robbins, R. A. Dixon, and E. Veltkamp. (1985). Spontaneous and enzymic rearrangement of naringenin chalcone to flavanoe. *Phytochemistry* **24**:2267–2269.

Mol, J., E. Grotewold, and R. Koes. (1998). How genes paint flowers and seeds. *Trends in Plant Science* **3**:212–217.

Montalvo, A., and J. Ackermann. (1987). Limitations to fruit production in *Ionopsis utricularioides* (Orchidaceae). *Biotropica* **19**:24–31.

Moon, J., S. Suh, H. Lee, K. Choi, C. Hong, N. Paek, S. Kim, and I. Lee. (2003). The *SOC1* MADS-box gene integrates vernalization and gibberellin signals for flowering in Arabidopsis. *Plant Journal* **35**:613–623.

Moon, J., H. Lee, M. Kim, and I. Lee. (2005). Analysis of flowering pathway integrators in *Arabidopsis*. *Plant Cell Physiology* **46**:292–299.

Mooney, M., T. Desnos, K. Harrison, J. Jones, R. Carpenter, and E. Coen. (1995). Altered regulation of tomato and tobacco pigmentation genes caused by the *delila* gene of *Antirrhinum*. *Plant Journal* **7**:333–339.

Motte, P., H. Saedler, and Z. Schwarz-Sommer. (1998). *STYLOSA* and *FISTULATA*: regulatory components of the homeotic control of *Antirrhinum* floral organogenesis. *Development* **125**:71–84.

Mouradov, A., F. Cremer, and G. Coupland. (2002). Control of flowering time: interacting pathways as a basis for diversity. *Plant Cell* **14**:S111–S130.

Moyano, E., J. F. Martinez-Garcia, and C. Martin. (1996). Apparent redundancy in *myb* gene function provides gearing for the control of flavonoid biosynthesis in Antirrhinum flowers. *Plant Cell* **8**:1519–1532.

Murfett, J., T. Atherton, B. Mou, C. Gasser, and B. McClure. (1994). S-RNase expressed in transgenic *Nicotiana* causes S-allele-specific pollen rejection. *Nature* **367**:563–566.

Mylne, J. S., L. Barrett, F. Tessadori, S. Mesnage, L. Johnson, Y. V. Bernatavichute, S. E. Jacobsen, P. Fransz, and C. Dean. (2006). LHP1, the *Arabidopsis* homologue of HETEROCHROMATIN PROTEIN 1, is required for epigenetic silencing of *FLC*. *Proceedings of the National Academy of Sciences, USA* **103**:5012–5017.

Nagasawa, N., M. Miyoshi, Y. Sano, H. Satoh, H. Hirano, H. Sakai, and Y. Nagato. (2003). *SUPERWOMAN1* and *DROOPING LEAF* genes control floral organ identity in rice. *Development* **130**:705–718.

Nageli, C. (1884). *Mechanisch-physiologische Theorie der Abstammungslehre*.

Nagy, F., and E. Schafer. (2002). Phytochromes control photomorphogenesis by differentially regulated, interacting signalling pathways in higher plants. *Annual Review of Plant Biology* 53:329–355.

Nakajima, M., A. Shimada, Y. Takashi, Y. Kim, S. Park, M. Ueguchi-Tanaka, H. Suzuki, E. Katoh, S. Iuchi, M. Kobayashi, T. Maeda, M. Matsuoka, and I. Yamaguchi. (2006). Identification and characterisation of Arabidopsis gibberellin receptors. *Plant Journal* 46:880–889.

Nakayama, T. (2002). Enzymology of aurone biosynthesis. *Journal of Bioscience and Bioengineering* 94:487–491.

Nakayama, T., K. Yonekura-Sakakibara, T. Sato, S. Kikuchi, Y. Fukui, M. Fukuchi-Mizutani, T. Ueda, M. Nakao, Y. Tanaka,T. Kusumi, and T. Nishino. (2000). Aureusidin synthase: a polyphenol oxidase homolog responsible for flower colouration. *Science* 290:1163–1166.

Nath, U., B. C. W. Crawford, R. Carpenter, and E. Coen. (2003). Genetic control of surface curvature. *Science* 299:1404–1407.

Navarro, C., N. Efremova, J. Golz, R. Rubiera, M. Kuckenberg, R. Castillo, O. Tietz, H. Saedler, and Z. Schwarz-Sommer. (2004). Molecular and genetic interactions between *STYLOSA* and *GRAMINIFOLIA* in the control of *Antirrhinum* vegetative and reproductive development. *Development* 131:3649–3659.

Newman, D. A., and J. D. Thomson. (2005). Effects of nectar robbing on dynamics and bumblebee foraging strategies in *Linaria vulgaris* (Scrophulariaceae). *Oikos* 110:309–320.

Ni, M., J. Tepperman, and P. Quail. (1998). PIF3, a phytochrome-interacting factor necessary for normal photoinduced signal transduction, is a novel basic helix-loop-helix protein. *Cell* 95:657–667.

Noda, K., B. J. Glover, P. Linstead, and C. Martin. (1994). Flower colour intensity depends on specialised cell shape controll ed by a MYB-related transcription factor. *Nature* 369:661–664.

Noh, Y., and R. Amasino. (2003). *PIE1*, an ISWI family gene, is required for *FLC* activation and floral repression in Arabidopsis. *Plant Cell* 15:1671–1682.

Oh, S., H. Zhang, P. Ludwig, and S. van Nocker. (2004). A mechanism related to the yeast transcriptional regulator Paf1c is required for expression of the Arabidopsis FLC/MAF MADS box gene family. *Plant Cell* 16: 2940–2953.

Ohnishi, M., S. Fukada-Tanaka, A. Hoshino, J. Takada, Y. Inagaki, and S. Iida. (2005). Characterisation of a novel Na^+/H^+ antiporter gene *InNHX2* and comparison of *InNHX2* with *InNHX1*, which is responsible for blue flower colouration by increasing the vacuolar pH in the Japanese Morning Glory. *Plant Cell Physiology* 46:259–267.

Ohshima, S., M. Murata, W. Sakamoto, Y. Ogura, and F. Motoyoshi. (1997). Cloning and molecular analysis

of the *Arabidopsis* gene *Terminal Flower 1. Molecular and General Genetics* 254:186–194.

Okada, K., T. Saito, T. Kakagawa, M. Kawamukai, and Y. Kamiya. (2000). Five geranylgeranyl diphosphate synthases expressed in different organs are localised into three subcellular compartments in Arabidopsis. *Plant Physiology* 122:1045–1056.

Okamuro, J., B. Caster, R. Villarroel, M. van Montagu, and K. D. Jofuku. (1997). The AP2 domain of *APETALA2* defines a large new family of DNA binding proteins in *Arabidopsis. Proceedings of the National Academy of Sciences, USA* 94:7076–7081.

Ollerton, J. (1998). Sunbirds surprise for syndromes. *Nature* 394:726–727.

Ollerton, J., A. Stott, E. Allnutt, S. Shove, C. Taylor, and E. Lamborn. (2007a). Pollination niche overlap between a parasitic plant and its host. *Oecologia*, 151:473–485.

Ollerton, J., R. Alarcon, N. Waser, M. Price, S. Watts, L. Cranmer, A. Hingston, C. Peter, and J. Rotenberry. (2007b). A global test of the pollination syndrome hypothesis. Submitted.

Osterlund, M., C. Hardtke, N. Wei, and X. Deng. (2000). Targeted destabilization of HY5 during light-regulated development of *Arabidopsis. Nature* 405:462–466.

Oyama, T., Y. Shimura, and K. Okada. (1997). The *Arabidopsis HY5* gene encodes a bZIP protein that regulates stimulus-induced development of root and hypocotyl. *Genes and Development* 11:2983–2995.

Pagnussat, G., H. Yu, Q. Ngo, S. Rajani, S. Mayalagu, C. Johnson, A. Capron, L. Xie, D. Ye, and V. Sundaresan. (2005). Genetic and molecular identification of genes required for female gametophyte development and function in Arabidopsis. *Development* 132:603–614.

Pannell, J. (2002). The evolution and maintenance of androdioecy. *Annual Reviews of Ecology and Systematics* 33:397–425.

Parcy, F., O. Nilsson, O., M. Busch, I. Lee, and D. Weigel. (1998). A genetic framework for floral patterning. *Nature* 395:561–566.

Park, D., D. Somers, Y. Kim, Y. Choy, H. Lim, M. Soh, H. Kim, S. Kay, and H. Nam. (1999). Control of circadian rhythms and photoperiodic flowering by the *Arabidopsis GIGANTEA* gene. *Science* 285:1579–1582.

Park, S., R. Howden, and D. Twell. (1998). The *Arabidopsis thaliana* gametophytic mutation *gemini pollen 1* disrupts microspore polarity, division asymmetry and pollen cell fate. *Development* 125:3789–3799.

Pauw, A. (1998). Pollen transfer on birds' tongues. *Nature* 394:731–732.

Pauw, A. (2006). Floral syndromes accurately predict pollination by a specialised oil-collecting bee (*Rediviva peringueyi*, Melittidae) in a guild of South African orchids (Coryciinae). *American Journal of Botany* **93**:917–926.

Pedersen, M. W. (1967). Cross-pollination studies involving three purple-flowered alfalfas, one white-flowered line, and two pollinator species. *Crop Science* **7**:59–62.

Pelaz, S., G. Ditta, E. Baumann, E. Wisman, and M. Yanofsky. (2000). B and C floral organ identity functions require *SEPALLATA* MADS-box genes. *Nature* **405**:200–202.

Pena, L., M. Martin-Trillo, J. Juarez, J. A. Pina, L. Navarro, and J. M. Martinez-Zapater. (2001). Constitutive expression of Arabidopsis *LEAFY* or *APETALA1* genes in citrus reduces their generation time. *Nature Biotechnology* **19**:263–267.

Perez-Rodriguez, M., F. Jaffe, E. Butelli, B. J. Glover, and C. Martin. (2005). Development of three different cell types is associated with the activity of a specific MYB transcription factor in the ventral petal of *Antirrhinum majus* flowers. *Development* **132**:359–370.

Perl-Treves, R., A. Kahana, N. Rosenman, Y. Xiang, and L. Silberstein. (1998). Expression of multiple *AGAMOUS*-like genes in male and female flowers of cucumber (*Cucumis sativus* L.). *Plant Cell Physiology* **39**:701–710.

Ponomarenko, A. G. (1995). The geological history of beetles. In J. Pakaluk and S. A. Slipinski (eds), *Biology, Phylogeny and Classification of Coleoptera*. Muzeum I Instytut Zoologii PAN, Warsaw.

Portereiko, M., L. Sandaklie-Nikolova, A. Lloyd, C. Dever, D. Otsuga, and G. Drews. (2006). *NUCLEAR FUSION DEFECTIVE 1* encodes the Arabidopsis RPL21M protein and is required for karyogamy during female gametophyte development and fertilisation. *Plant Physiology* **141**:957–965.

Putterill, J. (2001). Flowering in time: genes controlling photoperiodic flowering in *Arabidopsis*. *Philosophical Transactions of the Royal Society of London Series B* **356**:1761–1767.

Putterill, J., F. Robson, K. Lee, R. Simon, and G. Coupland. (1995). The *CONSTANS* gene of Arabidopsis pomotes flowering and encodes a protein showing similarities to zinc finger transcription factors. *Cell* **80**:847–857.

Putterill, J., R. Laurie, and R. Macknight. (2004). It's time to flower: the genetic control of flowering time. *BioEssays* **26**:363–373.

Qiao, H., F. Weng, L. Zhao, J. Zhou, Z. Lai, Y. Zhang, T. Robbins, and Y. Xue. (2004). The F-box protein AhSLF-S2 controls the pollen function of S-RNase-based self incompatibility. *Plant Cell* **16**:2307–2322.

Qiu, Y., J. Lee, F. Bernasconi-Quadroni, D. E. Soltise, P. Soltis, M. Zanis, E. A. Zimmer, Z. Chen, V. Savolainen, and M. W. Chase. (1999). The earliest angiosperms. *Nature* **402**:404–407.

Quail, P. H. (2002). Phytochrome photosensory signalling networks. *Nature Reviews Molecular Cell Biology* **3**:85–93.

Quattrocchio, F., J. F. Wing, K. van der Woude, J. N. M. Mol, and R. Koes. (1998). Analysis of bHLH and MYB domain proteins: species-specific regulatory differences are caused by divergent evolution of target anthocyanin genes. *Plant Journal* **13**:475–488.

Quattrocchio, F., J. Wing, K. van der Woude, E. Souer, N. de Vetten, J. Mol, and R. Koes. (1999). Molecular analysis of the *anthocyanin2* gene of petunia and its role in the evolution of flower colour. *Plant Cell* **11**:1433–1444.

Quattrocchio, F., W. Verweij, A. Kroon, C. Spelt, J. Mol, and R. Koes. (2006). PH4 of Petunia is an R2R3 MYB protein that activates vacuolar acidification through interactions with basic-helix-loop-helix transcription factors of the anthocyaninin pathway. *Plant Cell* **18**:1274–1291.

Quesada, V., R. Macknight, C. Dean, and G. G. Simpson. (2003). Autoregulation of *FCA* pre-mRNA processing controls *Arabidopsis* flowering time. *EMBO Journal* **22**:3142–3152.

Ramsay, N. A., and B. J. Glover. (2005). MYB-bHLH-WD40 protein complex and the evolution of cellular diversity. *Trends in Plant Science* **10**:63–70.

Ramsay, N. A., A. R. Walker, M. Mooney, and J. C. Gray. (2003). Two basic-helix-loop-helix genes (*MYC-146* and *GL3*) from *Arabidopsis* can activate anthocyanin biosynthesis in a white flowered *Matthiola incana* mutant. *Plant Molecular Biology* **52**:679–688.

Ramsey, J., H. D. Bradshaw, and D. Schemske. (2003). Components of reproductive isolation between the monkeyflowers *Mimulus lewisii* and *M. cardinalis* (Phrymaceae). *Evolution* **57**:1520–1534.

Ratcliffe, O., D. Bradley, and E. Coen. (1999). Separation of shoot and floral identity in *Arabidopsis*. *Development* **126**:1109–1120.

Rathcke, B. J. (2000). Hurricane causes resource and pollination limitation of fruit set in a bird-pollinated shrub. *Ecology* **81**:1951–1958.

Raven, P. (1972). Why are bird-visited flowers predominantly red? *Evolution* **26**:674.

Razem, F. A., A. El-Kereamy, S. R. Abrams, and R. D. Hill. (2006). The RNA-binding protein FCA is an abscisic acid receptor. *Nature* **439**:290–294.

Reale, L., A. Porceddu, L. Lanfaloni, C. Moretti, S. Zenoni, M. Pezzotti, B. Romano, and F. Ferranti. (2002). Patterns of cell division and expansion in developing petals of *Petunia hybrida*. *Sexual Plant Reproduction* **15**:123–132.

Redei, G. P. (1973). *Arabidopsis* as a genetic tool. *Annual Reviews of Genetics* **9**:111–127.

Reichmann, J., B. Krizek, and E. Meyerowitz, (1996). Dimerization specificity of Arabidopsis MADS domain homeotic proteins APETALA1, APETALA3, PISTIL-LATA and AGAMOUS. *Proceedings of the National Academy of Sciences of the USA* **93**:4793–4798.

Richards, A. (1997). *Plant Breeding Systems*, 2nd edn. Chapman and Hall, London.

Rodriguez-Girones, M., and L. Santamaria. (2004). Why are so many bird flowers red? *PLoS Biology* **2**:e306.

Rolland-Lagan, A., A. J. Bangham, and E. Coen. (2003). Growth dynamics underlying petal shape and symmetry. *Nature* **422**:161–163.

Rouse, D. T., C. C. Sheldon, D. J. Bagnall, W. J. Peacock, and E. S. Dennis. (2002). FLC, a repressor of flowering, is regulated by genes in different inductive pathways. *Plant Journal* **29**:183–191.

Rubio, V., and X. Deng. (2005). Phy tunes: phosphorylation status and phytochrome-mediated signalling. *Cell* **120**:290–292.

Rudall, P. J. (2006). How many nuclei make an embryo sac in flowering plants? *BioEssays* **28**:1067–1071.

Rutishauser, R., and B. Isler. (2001). Developmental genetics and morphological evolution of flowering plants, especially bladderworts (*Utricularia*): Fuzzy Arberian morphology complements classical morphology. *Annals of Botany* **88**:1173–1202.

Rutledge, R., S. Regan, O. Nicolas, P. Fobert, C. Cote, W. Bosnich, C. Kauffeldt, G. Sunohara, A. Seguin, and D. Stewart. (1998). Characterization of an *AGAMOUS* homologue from the conifer black spruce (*Picea mariana*) that produces floral homeotic conversions when expressed in *Arabidopsis*. *Plant Journal* **15**:625–634.

Saddic, L. A., B. Huvermann, S. Bezhani, Y. Su, C. M. Winter, C. S. Kwon, R. P. Collum, and D. Wagner. (2006). The LEAFY target LMI1 is a meristem identity regulator and acts together with LEAFY to regulate expression of *CAULIFLOWER*. *Development* **133**:1673–1682.

Samach, A., H. Onouchi, S. Gold, G. Ditta, Z. Schwarz-Sommer, M. Yanofsky, and G. Coupland. (2000). Distinct roles of *CONSTANS* target genes in reproductive development in Arabidopsis. *Science* **288**:1613–1616.

Sanda, S., and R. M. Amasino. (1996). Ecotype-specific expression of a flowering mutant phenotype in *Arabidopsis thaliana*. *Plant Physiology* **111**:641–644.

Sargent, R., and S. Otto. (2004). A phylogenetic analysis of pollination mode and the evolution of dichogamy in angiosperms. *Evolutionary Ecology Research* **6**:1183–1199.

Sargent, R., M. Mandegar, and S. Otto. (2006). A model of the evolution of dichogamy incorporating sex-ratio selection, anther-stigma interference and inbreeding depression. *Evolution* **60**:934–944.

Sato, T., T. Nakayama, S. Kikuchi, Y. Fukui, K. Yonekura-Sakakibara, T. Ueda, T. Nishino, Y. Tanaka, and T. Kusumi. (2001). Enzymatic formation of aurones in the extracts of yellow snapdragon flowers. *Plant Science* **160**:229–236.

Schemske, D., and P. Bierzychudek. (2001). Evolution of flower colour in the desert annual *Linanthus parryae*: Wright revisited. *Evolution* **55**:1269–1282.

Schemske, D., and H. Bradshaw. (1999). Pollinator preference and the evolution of floral traits in monkeyflowers (*Mimulus*). *Proceedings of the National Academy of Sciences of the USA* **96**:11910–11915.

Schoenrock, N., R. Bouveret, O. Leroy, L. Borghi, C. Koehler, W. Gruissem, and Hennig. (2006). Polycomb-group proteins repress the floral activator *AGL19* in the *FLC*-independent vernalization pathway. *Genes and Development* **20**:1667–1678.

Schomburg, F. M., D. A. Patton, D. W. Meinke, and R. M. Amasino. (2001). *FPA*, a gene involved in floral induction in Arabidopsis, encodes a protein containing RNA-recognition motifs. *Plant Cell* **13**:1427–1436.

Schopfer, C., M. Nasralah, and J. Nasrallah. (1999). The male determinant of self-incompatibility in *Brassica*. *Science* **286**:1697–1700.

Schwarz-Sommer, Z., I. Hue, P. Huijser, P. Flor, H. Hansen, F. Tetens, W. Lonnig, H. Saedler, and H. Sommer. (1992). Characterisation of the Antirrhinum floral homeotic MADS-box gene deficiens: evidence for DNA binding and autoregulation of its persistent expression throughout flower development. *EMBO Journal* **11**:251–263.

Schwinn, K., J. Venail, Y. Shang, S. Mackay, V. Alm, E. Butelli, R. Oyama, P. Bailey, K. Davies, and C. Martin. (2006). A small family of *MYB*-regulatory genes controls floral pigmentation intensity and patterning in the genus *Antirrhinum*. *Plant Cell* **18**:831–851.

Scora, R. (1964). Dependency of pollination on patterns in *Monarda* (Labiatae). *Nature* **204**:1011–1012.

Searle, I., Y. He, F. Turck, C. Vincent, F. Fornara, S. Kroeber, R. A. Amasino, and G. Coupland. (2006). The transcription factor FLC confers a flowering response to vernalization by repressing meristem competence and systemic signalling in *Arabidopsis*. *Genes and Development* **20**:898–912.

Seitz, C., C. Eder, B. Deiml, S. Kellner, S. Martens, and G. Forkmann. (2006). Cloning, functional identification and sequence analysis of flavonoid 3'-hydroxylase and flavonoid 3',5'-hydroxylase cDNAs reveals independent evolution of flavonoid 3',5'-hydroxylase in the Asteraceae family. *Plant Molecular Biology* **61**:365–381.

Shalitin, D., H. Yang, T. Mockler, M. Maymon, H. Guo, G. Whitelam, and C. Lin. (2002). Regulation of *Arabidopsis* cryptochrome by blue-light-dependent phosphorylation. *Nature* **417**:763–767.

Shalitin, D., X. Yu, M. Maymon, T. Mockler, and C. Lin. (2003). Blue light-dependent in vivo and in vitro phosphorylation of Arabidopsis cryptochrome 1. *Plant Cell* **15**:2421–2429.

Sheldon, C., J. E. Burn, P. P. Perez, J. Metzger, J. A. Edwards, W. J. Peacock, and E. S. Dennis. (1999). The *FLF* MADS box gene: a repressor of flowering in *Arabidopsis* regulated by vernalization and methylation. *Plant Cell* **11**:445–458.

Sheldon, C., E. J. Finnegan, D. T. Rouse, M. Tadege, D. J. Bagnall, C. A. Helliwell, W. J. Peacock, and E. S. Dennis. (2000a). The control of flowering by vernalization. *Current Opinion in Plant Biology* **3**:418–422.

Sheldon, C., D. T. Rouse, E. J. Finnegan, W. J. Peacock, and E. S. Dennis. (2000b). The molecular basis of vernalization: the central role of *FLOWERING LOCUS C* (*FLC*). *Proceedings of the National Academy of Sciences of the USA* **97**:3753–3758.

Sheldon, C., A. Conn, E. S. Dennis, and W. J. Peacock. (2002). Different regulatory regions are required for the vernalization-induced repression of *FLOWERING LOCUS C* and for the epigenetic maintenance of repression. *Plant Cell* **14**:2527–2537.

Sheldon, C., E. J. Finnegan, E. S. Dennis, and W. J. Peacock. (2006). Quantitative effects of vernalization on *FLC1* and *SOC1* expression. *Plant Journal* **45**:871–883.

Shiba, H., M. Iwano, T. Entani, K. Ishimoto, H. Shimosato, F. Che, Y. Satta, A. Ito, Y. Takada, M. Watanabe, A. Isogai, and S. Takayama. (2002). The dominance of alleles controlling self-incompatibility in *Brassica* pollen is regulated at the RNA level. *Plant Cell* **14**:491–504.

Shinomura, T., K. Uchida, and M. Furuya. (2000). Elementary processes of photoperception by phytochrome A for high-irradiance response of hypocotyl elongation in Arabidopsis. *Plant Physiology* **122**:147–156.

Shiono, M., N. Matsugaki, and K. Takeda. (2005). Structure of the blue cornflower pigment. *Nature* **436**:791.

Shpak, E. D., C. T. Berthiaume, E. J. Hill, and K. U. Torii. (2003). Synergistic interaction of three ERECTA-family receptor-like kinases controls Arabidopsis organ growth and flower development by promoting cell proliferation. *Development* **131**:1491–1501.

Sieburth, L., and E. Meyerowitz. (1997). Molecular dissection of the *AGAMOUS* control region shows that *cis* elements for spatial regulation are located intragenically. *Plant Cell* **9**:355–365.

Sijcic, P., X. Wang, A. Skirpan, Y. Wang, P. Dowd, A. McCubbin, S. Huang, and T. Kao. (2004). Identification of the pollen determinant of S-RNase-mediated self-incompatibility. *Nature* **429**:302–305.

Simon, R., R. Carpenter, S. Doyle, and E. Coen. (1994). Fimbriata controls flower development by mediating between meristem and organ identity genes. *Cell* **78**:99–107.

Simon, R., M. Igeno, and G. Coupland. (1996). Activation of floral meristem identity genes in Arabidopsis. *Nature* **84**:59–62.

Simpson, G. G. (2003). Evolution of flowering in response to day length: flipping the *CONSTANS* switch. *BioEssays* **25**:829–832.

Simpson, G. G. (2004). The autonomous pathway: epigenetic and post-transcriptional gene regulation in the control of *Arabidopsis* flowering time. *Current Opinion in Plant Biology* **7**:570–574.

Simpson, G. G., and C. Dean. (2002). Arabidopsis, the rosetta stone of flowering time? *Science* **296**:285–289.

Simpson, G. G., P. P. Dijkwel, V. Quesada, I. Henderson, and C. Dean. (2003). FY is an RNA 3′ end-processing factor that interacts with FCA to control the *Arabidopsis* floral transition. *Cell* **113**:777–787.

Somers, D., A. Webb, M. Pearson, and S. Kay. (1998a). The short-period mutant, *toc1–1*, alters circadian clock regulation of multiple outputs throughout development in *Arabidopsis thaliana*. *Development* **125**:485–494.

Somers, D., P. Devlin, and S. A. Kay. (1998b). Phytochromes and cryptochromes in the entrainment of the *Arabidopsis* ciracidan clock. *Science* **282**:1488–1490.

Somers, D., T. Schultz, M. Milnamow, and S. A. Kay. (2000). *ZEITLUPE* encodes a novel clock-associated PAS protein from *Arabidopsis*. *Cell* **101**:319–329.

Sommer, H., P. Beltran, P. Huijser, H. Pape, W. Lonnig, H. Saedler, and Z. Schwarz-Sommer. (1990). Deficiens, a homeotic gene involved in the control of flower morphogenesis in *Antirrhinum majus*: the protein shows homology to transcription factors. *EMBO Journal* **9**:605–613.

Sonneveld, T., K. Tobutt, S. Vaughan, and T. Robbins. (2005). Loss of pollen-S function in two self-compatible selections of *Prunus avium* is associated with deletion/mutation of an S haplotype-specific F-box gene. *Plant Cell* **17**:37–51.

Spaethe, J., and A. Briscoe. (2005). Molecular characterisation and expression of the UV opsin in bumblebees: three omnatidial subtypes in the retina and a new photoreceptor organ in the lamina. *Journal of Experimental Biology* **208**:2347–2361.

Spaethe, J., J. Tautz, and L. Chittka. (2001). Visual constraints in foraging bumblebees: flower size and colour affects search time and flight behaviour. *Proceedings of the National Academy of Sciences of the USA* **98**, 3898–3903.

Spelt, C., F. Quattrocchio, J. N. M. Mol, and R. Koes. (2000). *anthocyanin1* of Petunia encodes a basic helix-loop-helix protein that directly activates transcription of structural anthocyanin genes. *Plant Cell* **12**:1619–1631.

Spelt, C., F. Quattrocchio, J. N. M. Mol, and R. Koes. (2002). ANTHOCYANIN1 of Petunia controls pigment synthesis, vacuolar pH, and seed coat development by genetically distinct mechanisms. *Plant Cell* **14**:2121–2135.

Stacey, M., S. Hicks, and A. von Arnim. (1999). Discrete domains mediate the light-responsive nuclear and cytoplasmic localization of Arabidopsis COP1. *Plant Cell* **11**:349–363.

Stanton, M., A. Snow, and S. Handel. (1986). Floral evolution: attractiveness to pollinators increases male fitness. *Science* **232**:1625–1627.

Stebbins, G. L. (1970). Adaptive radiation of reproductive characteristics in angiosperms, I: pollination mechanisms. *Annual Reviews in Ecology and Systematics* **1**:307–326.

Stein, J., B. Howlett, D. Boyes, M. Nasrallah, and J. Nasrallah. (1991). Molecular cloning of a putative receptor protein kinase gene encoded at the self-incompatibility locus of *Brassica oleracea*. *Proceedings of the National Academy of Sciences, USA* **88**:8816–8820.

Stein, J., R. Dixit, M. Nasrallah, and J. Nasrallah. (1996). SRK, the stigma-specific S locus receptor kinase of Brassica, is targeted to the plasma membrane in transgenic tobacco. *Plant Cell* **8**:429–445.

Stellari, G., M. A. Jaramillo, and E. M. Kramer. (2004). Evolution of the *APETALA3* and *PISTILLATA* lineages of MADS-box-containing genes in the basal angiosperms. *Molecular Biology and Evolution* **21**:506–519.

Stewart, R. N., K. H. Norris, and S. Asen. (1975). Microspectrophotometric measurement of pH and pH effect on color of petal epidermal cells. *Phytochemistry* **14**:937–942.

Stinchcombe, J., C. Weinig, M. Ungerer, K. Olsen, C. Mays, S. Halldorsdottir, M. Purugganan, and J. Schmitt. (2004). A latitudinal cline in flowering time in *Arabidopsis thaliana* modulated by the flowering time gene *FRIGIDA*. *Proceedings of the National Academy of Sciences of the USA* **101**:4712–4717.

Stone, G., P. Willmer, and J. A. Rowe. (1998). Partitioning of pollinators during flowering in an African *Acacia* community. *Ecology* **79**:2808–2827.

Stone, S., E. Anderson, R. Mullen, and D. Goring. (2003). ARC1 is an E3 ubiquitin ligase and promotes the ubiquitination of proteins during the rejection of self-incompatible *Brassica* pollen. *Plant Cell* **15**:885–898.

Stotz, G., P. de Vlaming, H. Wiering, A. W. Schram, and G. Forkman. (1985). Genetic and biochemical studies on flavonoid 3'-hydroxylation in flowers of *Petunia hybrida*. *Theoretical and Applied Genetics* **70**:300–305.

Strack, D., T. Vogt, and W. Schliemann. (2003). Recent advances in betalain research. *Phytochemistry* **62**:247–269.

Strayer, C., T. Oyama, T. Schultz, R. Raman, D. Somers, P. Mas, S. Panda, J. Kreps, and S. Kay. (2000). Cloning of the *Arabidopsis* clock gene *TOC1*, an autoregulatory response regulator homolog. *Science* **289**:768–771.

Stromberg, M., and P. Johnsen. (1990). Hummingbird sweetness preferences; taste or viscosity? *The Condor* **92**:606–612.

Suarez-Lopez, P., K. Wheatley, F. Robson, H. Onouchi, F. Valverde, and G. Coupland. (2001). *CONSTANS* mediates between the circadian clock and the control of flowering in *Arabidopsis*. *Nature* **410**:1116–1120.

Subramanian, C., B. Kim, N. Lyssenko, X. Xu, C. Johnson, and A. von Arnim. (2004). The *Arabidopsis* repressor of light signalling, COP1, is regulated by nuclear exclusion: Mutational analysis by bioluminescence resonance energy transfer. *Proceedings of the National Academy of Sciences of the USA* **101**:6798–6802.

Sugimoto-Shirasu, K., and K. Roberts. (2003). 'Big it up': endoreduplication and cell cycle control in plants. *Current Opinion in Plant Biology* **6**:544–553.

Sun, G., D. L. Dilcher, S. Zheng, and Z. Zhou. (1998). In search of the first flower: a Jurassic angiosperm, *Archaefructus*, from northeast China. *Science* **282**:1692–1695.

Sundstrom, J., and P. Engstrom. (2002). Conifer reproductive development involves B-type MADS-box genes with distinct and different activities in male organ primordia. *Plant Journal* **31**:161–169.

Sundstrom, J. F., N. Nakayama, K. Glimelius, and V. Irish. (2006). Direct regulation of the floral homeotic *APETALA1* gene by APETALA3 and PISTILLATA in Arabidopsis. *Plant Journal* **46**:593–600.

Sung, S., and R. M. Amasino. (2004a). Vernalization in *Arabidopsis thaliana* is mediated by the PHD finger protein VIN3. *Nature* **427**:159–163.

Sung, S., and R. M. Amasino. (2004b). Vernalization and epigenetics: how plants remember winter. *Current Opinion in Plant Biology* **7**:4–10.

Sung, S., Y. He, T. W. Eshoo, Y. Tamada, L. Johnson, K. Nakahigashi, K. Goto, S. E. Jacobsen, and R. M. Amasino. (2006). Epigenetic maintenance of the vernalized state in *Arabidopsis thaliana* requires LIKE HETEROCHROMATIN PROTEIN 1. *Nature Genetics* **38**:706–710.

Takasaki, T., K. Hatakeyama, G. Suzuki, M. Watanabe, A. Isogai, and K. Hinata. (2000). The S receptor kinase determines self-incompatibility in *Brassica* stigma. *Nature* **403**:913–916.

Takayama, S., H. Shiba, M. Iwano, H. Shimosato, F. Che, N. Kai, M. Watanabe, G. Suzuki, K. Hinata, and A. Isogai. (2000). The pollen determinant of self-incompatibility in *Brassica campestris*. *Proceedings of the National Academy of Sciences, USA* **97**:1920–1925.

Takayama, S., H. Shimosato, H. Shiba, M. Funato, F. Che, M. Watanabe, M. Iwano, and A. Isogai. (2001). Direct ligand-receptor complex interaction controls *Brassica* self-incompatibility. *Nature* **413**:534–538.

Takebayashi, N., D. Wolf, and L. Delph. (2006). Effect of variation in herkogamy on outcrossing within a population of *Gilia achilleifolia*. *Heredity* **96**:159–165.

Takeda, S., N. Matsumoto, and K. Okada. (2003). *RABBIT EARS*, encoding a SUPERMAN-like zinc-finger protein, regulates petal development in *Arabidopsis thaliana*. *Development* **131**:425–434.

Tanabe, Y., M. Hasebe, H. Sekimoto, T. Nishiyama, M. Kitani, K. Henschel, T. Muenster, G. Theissen, H. Nozaki, and M. Ito. (2005). Characterization of MADS-box genes in charophycean green algae and its implication for the evolution of MADS-box genes. *Proceedings of the National Academy of Sciences of the USA* **102**:2436–2441.

Tanahashi, T., N. Sumikawa, M. Ato, and M. Hasebe. (2005). Diversification of gene function: homologs of the floral regulator *FLO/LFY* control the first zygotic cell division in the moss *Physcomitrella patens*. *Development* **132**:1727–1736.

Tandre, K., M. Svenson, M. E. Svensson, and P. Engstrom. (1998). Conservation of gene structure and activity in the regulation of reproductive organ development of conifers and angiosperms. *Plant Journal* **15**:615–623.

Teeri, T., V. Albert, P. Elomaa, J. Hamalainen, M. Kotilainen, E. Pollanen, and A. Uimari. (2002). Involvement of non-ABC MADS-box genes in determining stamen and carpel identity in *Gerbera hybrida* (Asteraceae). In Q. C. B. Cronk, R. M. Bateman, and J. A. Hawkins (eds), *Developmental Genetics and Plant Evolution*, pp. 173–205. Taylor and Francis, London.

Tepperman, J., T. Zhu, H. Chang, X. Wang, and P. H. Quail. (2001). Multiple transcription factor genes are early targets of phytochrome A signalling. *Proceedings of the National Academy of Sciences, USA* **98**:9437–9442.

The Arabidopsis Genome Initiative. (2000). Analysis of the genome of the flowering plant *Arabidopsis thaliana*. *Nature* **408**:796–815.

Theissen, G., A. Becker, K. Winter, T. Munster, C. Kirchner, and H. Saedler. (2002). How the land plants learned their floral ABCs: the role of MADS-box genes in the evolutionary origin of flowers. In Q. C. B. Cronk, R. M. Bateman, and J. A. Hawkins (eds), *Developmental Genetics and Plant Evolution*, pp. 173–205. Taylor and Francis, London.

Thien, L. B., H. Azuma, and S. Kawano. (2000). New perspectives on the pollination biology of basal angiosperms. *International Journal of Plant Sciences* **161**:S225–S235.

Thornton, T., S. Swain, and N. Olszewski. (1999). Gibberellin signal transduction presents the SPY who O-GlcNAc'd me. *Trends in Plant Science* **4**:424–428.

Trevaskis, B., D. Bagnall, M. Ellis, W. J. Peacock, and E. Dennis. (2003). MADS box genes control vernalization-induced flowering in cereals. *Proceedings of the Natioanl Academy of Sciences, USA* **100**:13099–13104.

Twell, D., S. Park, T. Hawkins, D. Schubert, R. Schmidt, A. Smertenko, and P. Hussey. (2002). Mor1/Gem1 has an essential role in the plant-specific cytokinetic phragmoplast. *Nature Cell Biology* **4**:711–714.

Uehara, S., and H. Morita. (1972). The effect of temperature on the labellar chemoreceptors of the blowfly. *Journal of General Physiology* **59**:213–226.

Uimari, A., and J. Strommer. (1997). Myb26: a MYB-like protein of pea flowers with affinity for promoters of phenylpropanoid genes. *Plant Journal* **12**:1273–1284.

Uimari, A., M. Kotilainen, P. Elomaa, D. Yu, V. A. Albert, and T. H. Teeri. (2004). Integration of reproductive meristem fates by a *SEPALLATA*-like MADS-box gene. *Proceedings of the Natioanl Academy of Sciences of the USA* **101**:15817–15822.

Valverde, F., A. Mouradov, W. Soppe, D. Ravenscroft, A. Samach, and G. Coupland. (2004). Photoreceptor regulation of CONSTANS protein in photoperiodic flowering. *Science* **303**:1003–1006.

Vamosi, J., T. Knight, J. Steets, S. Mazer, M. Burd, and T. Ashman. (2006). Pollination decays in biodiversity hotspots. *Proceedings of the National Academy of Sciences of the USA* **103**:956–961.

Vanoosthuyse, V., G. Tichtinsky, C. Dumas, T. Gaude, and J. M. Cock. (2003). Interaction of calmodulin, a sorting nexin and kinase-associated protein phosphatase with the *Brassica oleracea* S locus receptor kinase. *Plant Physiology* **133**:919–929.

van Tunen, A. J., R. E. Koes, C. E. Spelt, A. R. van der Krol, A. R. Stuitje, and J. N. M. Mol. (1988). Cloning of two chalcone flavanone isomerase genes from *Petunia hybrida*: Coordinate light regulated and differential expression of flavonoid genes. *EMBO Journal* **7**:1257–1263.

Varela, F., A. Palacios, and T. Goldsmith. (1993). Colour vision of birds. In H. Zeigler (ed), *Vision, Brain and Behaviour in Birds*. MIT Press, Cambridge MA.

Vyskot, B., and R. Hobza. (2004). Gender in plants: sex chromosomes are emerging from the fog. *Trends in Genetics* **20**:432–438.

Wagner, D., R. Sablowski, and E. Meyerowitz. (1999). Transcriptional activation of APETALA1 by LEAFY. *Science* **285**:582–584.

Walker, A. R., R. A. Davison, A. C. Bolognesi-Winfield, C. M. James, N. Srinivasan, T. Blundell, J. J. Esch, M. D. Marks, and J. C. Gray. (1999). The TRANSPARENT TESTA GLABRA 1 locus which regulates trichome differentiation and anthocyanin biosynthesis in Arabidopsis encodes a WD40 repeat protein. *Plant Cell* **11**:1337–1345.

Wang, H., and X. Deng. (2003). Dissecting the phytochrome A-dependent signalling network in higher plants. *Trends in Plant Science* **8**:172–178.

Wang, H., L. Ma, J. Li, H. Zhao, and X. Deng. (2001). Direct interaction of *Arabidopsis* cryptochromes with COP1 in light control development. *Science* **294**:154–158.

Waser, N. (1978). Competition for hummingbird pollination and sequential flowering in two Colorado wildflowers. *Ecology* **59**:934–944.

Waser, N. (2006). Specialization and generalization in plant-pollinator interactions: a historical perspective. In N. Waser and J. Ollerton (eds), *Plant–Pollinator Interactions*, pp. 3–17. University of Chicago Press, London.

Waser, N., and J. Ollerton. (2006). *Plant–Pollinator Interactions*. University of Chicago Press, London.

Waser, N., and M. Price. (1981). Pollinator choice and stabilizing selection for flower colour in *Delphinium nelsonii*. *Evolution* **35**:376–390.

Waser, N., L. Chittka, M. Price, N. Williams, and J. Ollerton. (1996). Generalization in pollination systems, and why it matters. *Ecology* **77**:1043–1060.

Watanabe, M., G. Suzuki, S. Takayama, A. Isogai, and K. Hinata. (2000). Genomic organization of the SLG/SRK region of the S locus in *Brassica* species. *Annals of Botany* **85**:155–160.

Weigel, D., and O. Nilsson. (1995). A developmental switch sufficient for flower initiation in diverse plants. *Nature* **377**:495–500.

Weigel, D., J. Alvarez, D. Smyth, M. Yanofsky, and E. Meyerowitz. (1992). *LEAFY* controls floral meristem identity in Arabidopsis. *Cell* **69**:843–859.

Weiss, D., A. van der Luit, J. Kroon, J. Mol, and J. Kooter. (1993). The petunia homologue of the *Antirrhinum majus candi* and *Zea mays A2* flavonoid genes; homology to flavanone 3-hydroxylase and ethylene-forming enzyme. *Plant Molecular Biology* **22**:893–897.

Weiss, M. R. (1991). Floral colour changes as cues for pollinators. *Nature* **354**:227–229.

Wellensiek, S. J. (1962). Dividing cells as the locus for vernalization. *Nature* **195**:307–308.

Weller, J., J. Reid, S. Taylor, and I. Murfet. (1997). The genetic control of flowering in pea. *Trends in Plant Science* **2**:412–418.

Wessler, S. R. (2005). Homing into the origin of the AP2 DNA binding domain. *Trends in Plant Science* **10**:54–56.

Weterings, K., and S. Russell. (2004). Experimental analysis of the fertilisation process. *Plant Cell* **16**:S107–S118.

Whipple, C., P. Ciceri, C. Padilla, B. Ambrose, S. Bandong, and R. Schmidt. (2004). Conservation of B-class floral homeotic gene function between maize and *Arabidopsis*. *Development* **131**:6083–6091.

Whipple, C., M. Zanis, E. Kellogg, and R. Schmidt. (2007). Conservation of B class gene expression in the second whorl of a basal grass and outgroups links the origin of lodicules and petals. *Proceedings of the National Academy of Sciences of the USA* **104**:1081–1086.

Wigge, P. A., M. C. Kim, K. E. Jaeger, W. Busch, M. Schmid, J. U. Lohmann, and D. Weigel. (2005). Integration of spatial and temporal information during floral induction in *Arabidopsis*. *Science* **309**:1056–1059.

Wilkinson, M., and G. Haughn. (1995). *UNUSUAL FLORAL ORGANS* controls meristem identity and organ primordia fate. *Plant Cell* **7**:1485–1499.

Willis, K. J., and J. C. McElwain. (2002). *The Evolution of Plants*. Oxford University Press, Oxford.

Wilson, P., M. Castellanos, J. Hogue, J. Thomson, and W. S. Armbruster. (2004). A multivariate search for pollination syndromes among penstemons. *Oikos* **104**:345–361.

Winter, K., C. Weiser, K. Kaufmann, A. Bohne, C. Kirchner, A. Kanno, H. Saedler, and G. Theissen. (2002). Evolution of class B floral homeotic proteins: obligate heterodimerization originated from homodimerization. *Molecular Biology and Evolution* **19**:587–596.

Wolfe, K. H., M. Gouy, Y. Yang, P. Sharp, and W. Li. (1989). Date of the monocot-dicot divergence estimated from chloroplast DNA sequence data. *Proceedings of the National Academy of Sciences of the USA* **86**:6201–6205.

Wood, C. C., M. Robertson, G. Tanner, W. J. Peacock, E. S. Dennis, and C. A. Helliwell. (2006). The *Arabidopsis thaliana* vernalization response requires a polycomb-like protein complex that also includes VERNALIZATION INSENSITIVE 3. *Proceedings of the National Academy of Sciences of the USA* **103**:14631–14636.

Wuerschum, T., R. Gross-Hardt, and T. Laux. (2006). *APETALA2* regulates the stem cell niche in the *Arabidopsis* shoot meristem. *Plant Cell* **18**:295–307.

Yadegari, R., and G. Drews. (2004). Female gametophyte development. *Plant Cell* **16**:S133–S141.

Yamaguchi, T., S. Fukada-Tanaka, Y. Inagaki, N. Saito, K. Yonekura-Sakakibara, Y. Tanaka, T. Kusumi, and S. Iida. (2001). Genes encoding the vacuolar Na^+/H^+ exchanger and flower coloration. *Plant Cell Physiology* **42**:451–461.

Yamaguchi, T., D. Lee, A. Kiyao, H. Hirochika, G. An, and H. Hirano. (2006). Functional diversification of the two C-class MADS box genes *OsMADS3* and *OsMADS58* in *Oryza sativa*. *Plant Cell* **18**:15–28.

Yamamoto, Y., M. Matsui, L. Ang, and X. Deng. (1998). Role of a COP1 interactive protein in mediating light-regulated gene expression in Arabidopsis. *Plant Cell* **10**:1083–1094.

Yan, L., A. Loukoianov, G. Tranquilli, M. Helguera, T. Fahima, and J. Dubcovsky. (2003). Positional cloning of the wheat vernalization gene *VRN1*. *Proceedings of the National Academy of Sciences, USA* **100**:6263–6268.

Yan, L., A. Loukoianov, A. Blechl, G. Tranquilli, W. Ramakrishna, P. SanMiguel, J. Bennetzen, V. Echenique, and J. Dubcovsky. (2004). The wheat

VRN2 gene is a flowering repressor down-regulated by vernalization. *Science* **303**:1640–1644.

Yang, C., M. Spielman, J. Coles, Y. Li, S. Ghelani, V. Bourdon, R. Brown, B. Lemmon, R. Scott, and H. Dickinson. (2003). *TETRASPORE* encodes a kinesin required for male meiotic cytokinesis in Arabidopsis. *Plant Journal* **34**:220–240.

Yang, H., Y. Wu, R. Tang, D. Liu, Y. Liu, and A. Cashmore. (2000). The C termini of *Arabidopsis* cryptochromes mediate a constitutive light response. *Cell* **103**:815–827.

Yang, J., R. Lin, J. Sullivan, U. Hoecker, B. Liu, L. Xu, X. Deng, and H. Wang. (2005). Light regulates COP1-mediated degradation of HFR1, a transcription factor essential for light signalling in Arabidopsis. *Plant Cell* **17**:804–821.

Yang, S., L. Xie, H. Mao, C. Puah, W. Yang, L. Jiang, V. Sundaresan, and D. Ye. (2003). *TAPETUM DETERMINANT 1* is required for cell specialisation in the Arabidopsis anther. *Plant Cell* **15**:2792–2804.

Yano, M., Y. Katayose, M. Ashikari, U. Yamanouchi, L. Monna, T. Fuse, T. Baba, K. Yamamoto, Y. Umehara, Y. Nagamura, and T. Sasaki. (2000). *Hd1*, a major photoperiod sensitivity quantitative trait locus in rice, is closely related to the Arabidopsis flowering time gene *CONSTANS*. *Plant Cell* **12**:2473–2483.

Yanofsky, M., H. Ma, J. Bowman, G. Drews, K. Feldmann, and E. Meyerowitz. (1990). The protein encoded by the Arabidopsis homeotic gene *agamous* resembles transcription factors. *Nature* **346**:35–39.

Yanovsky, M. J., and S. A. Kay. (2003). Living by the calendar: how plants know when to flower. *Nature Reviews Molecular Cell Biology* **4**:265–275.

Yoshida, K., T. Kondo, Y. Okazaki, and K. Katou. (1995). Cause of blue petal colour. *Nature* **373**:291.

Yoshida, K., M. Kawachi, M. Mori, M. Maeshima, M. Kondo, M. Nishimura, and T. Kondo. (2005). The involvement of tonoplast proton pumps and $Na^+(K^+)/H^+$ exchangers in the change of petal colour during flower opening of morning glory, *Ipomoea tricolor* cv. Heavenly Blue. *Plant and Cell Physiology* **46**:407–415.

Yoshida, K., S. Kitahara, D. Ito, and T. Kondo. (2006). Ferric ions involved in the flower colour development of the Himalayan blue poppy *Meconopsis grandis*. *Phytochemistry* **67**:992–998.

Yu, D., M. Kotilainen, E. Pollanen, M. Mehto, P. Elomaa, Y. Helariutta, V. A. Albert, and T. H. Teeri. (1999). Organ identity genes and modified patterns of flower development in *Gerbera hybrida* (Asteraceae). *Plant Journal* **17**:51–62.

Yu, H., T. Ito, Y. Zhao, J. Peng, P. Kumar, and E. Meyerowitz. (2004). Floral homeotic genes are targets of gibberellin signaling in flower development. *Proceedings of the National Academy of Sciences, USA* **101**:7827–7832.

Zachgo, S., E. Andrade Silva, P. Motte, W. Trobner, H. Saedler, and Z. Schwarz-Sommer. (1995). Functional analysis of the *Antirrhinum* floral homeotic *DEFICIENS* gene in vivo and in vitro by using a temperature-sensitive mutant. *Development* **121**:2861–2875.

Zahn, L., J. Leebens-Mack, C. dePamphlis, H. Ma, and G. Theissen. (2005). To B or not to B a flower: the role of *DEFICIENS* and *GLOBOSA* orthologs in the evolution of the angiosperms. *Journal of Heredity* **9**:225–240.

Zhang, H., and S. van Nocker. (2002). The *VERNALIZATION INDEPENDENCE 4* gene encodes a novel regulator of *FLOWERING LOCUS C*. *Plant Journal* **31**:663–673.

Zhang, H., C. Ransom, P. Ludwig, and S. van Nocker. (2003). Genetic analysis of early flowering mutants in Arabidopsis defines a class of pleiotropic developmental regulator required for expression of the flowering-time switch *FLOWERING LOCUS C*. *Genetics* **164**:347–358.

Zhang, L, S. Barrett, J. Gao, J. Chen, W. Cole, Y. Liu, Z. Bai, and Q. Li. (2005). Predicting mating patterns from pollination syndromes: the case of 'sapromyiophily' in *Tacca chantrieri* (Taccaceae). *American Journal of Botany* **92**:517–524.

Zhang, P., H. Tan, K. Pwee, and P. Kumar. (2004). Conservation of class C function of floral organ development during 300 million years of evolution from gymnosperms to angiosperms. *Plant Journal* **37**:566–577.

Zhao, D., G. Wang, B. Speal, and H. Ma. (2002). The *EXCESS MICROSPOROCYTES 1* gene encodes a putative leucine-rich repeat receptor protein kinase that controls somatic and reproductive cell fates in the Arabidopsis anther. *Genes and Development* **16**:2021–2031.

Zhu, Y., J. Tepperman, C. Fairchild, and P. H. Quail. (2000). Phytochrome B binds with greater apparent affinity than phytochrome A to the basic helix-loop-helix factor PIF3 in a reaction requiring the PAS domain of PIF3. *Proceedings of the National Academy of Sciences, USA* **97**:13419–13424.

Zupan, J., T. R. Muth, O. Draper, and P. Zambryski. (2000). The transfer of DNA from *Agrobacterium tumefaciens* into plants: a feast of fundamental insights. *Plant Journal* **23**:11–28.

Index